Frontiers in Enzyme Inhibition

(Volume 1)

Enzyme Inhibition - Environmental and Biomedical Applications

Edited by

G. Baskar

Department of Biotechnology,
St. Joseph's College of Engineering,
Chennai,
India

K. Sathish Kumar

Department of Chemical Engineering,
SSN College of Engineering,
Chennai – 603110,
India

&

K. Tamilarasan

Department of Chemical Engineering,
SRM University,
Chennai – 603203,
India

Frontiers in Enzyme Inhibition

Volume # 1

Enzyme Inhibition - Environmental and Biomedical Applications

Editors: G. Baskar, K. Sathish Kumar and K. Tamilarasan

ISBN (Online): 978-981-14-6082-1

ISBN (Print): 978-981-14-6080-7

ISBN (Paperback): 978-981-14-6081-4

© 2020, Bentham Books imprint.

Published by Bentham Science Publishers Pte. Ltd. Singapore. All Rights Reserved.

need for a court order if at any point you breach any terms of this License Agreement. In no event will any delay or failure by Bentham Science Publishers in enforcing your compliance with this License Agreement constitute a waiver of any of its rights.

3. You acknowledge that you have read this License Agreement, and agree to be bound by its terms and conditions. To the extent that any other terms and conditions presented on any website of Bentham Science Publishers conflict with, or are inconsistent with, the terms and conditions set out in this License Agreement, you acknowledge that the terms and conditions set out in this License Agreement shall prevail.

Bentham Science Publishers Pte. Ltd.
80 Robinson Road #02-00
Singapore 068898
Singapore
Email: subscriptions@benthamscience.net

CONTENTS

PREFACE

Enzyme, a protein molecule exhibits specific activity and binding towards the substrate molecule for the completion of biocatalytic reaction. Enzyme inhibitors are molecules which prevents the functioning and interaction of enzyme by specific substrate or analogue. The enzyme inhibitors alters and slow down the catalytic action of enzyme in various modes. The activity is mainly inhibited by reversible, irreversible inhibition, covalent, non-covalent binding. The inhibitors also act on specific and non-specific site including bacteria, virus, plants and animals. The inhibitors plays a vital role in pharmaceutical and other bio-chemical industries. The enzyme inhibitors are widely used as herbicide and pesticide for the destruction of pathogens. The action of inhibitors has paved way for the development in drug discovery and pharmaceutical industries. The enzyme inhibitors acts as drug molecules (Eg.antimicrobial drug) in the treatment of various disease. These drug deactivates the enzyme needed for the survival of pathogens. Enzyme inhibitors also finds its application in metabolic process by inhibiting the action of enzyme against animal predators. The new insight of enzyme inhibitors in environmental monitoring has gained wide attention among researchers for the development of biosensor. The main objective of this book on "Frontiers in Enzyme Inhibition – Environmental and Biomedical Application" is to provide basic information on recent development in the field of enzymology. This book also explain about the applied inhibition method in drug discovery. This book will mainly focus on enzyme inhibition and its application in pharmaceutical and environment. The book will be highly valuable for students and researchers for enhancing their knowledge on basic concept of applied inhibition in the field drug discovery related projects. This book is compiled based on the basic concept of specific inhibitors in environmental and pharmaceutical applications. The main theme of this book includes

1] Enzyme inhibition in development and assessment of biosensors

2) Product inhibition during fermentation process

3) Inhibition in crop management

4) Inhibition in the development and formulation of drug

This book will give wide knowledge and understanding in the development of drug and biosensors. This book gives a collective information on various mechanism and alterations that are noted during the developmental stage. This book consists of 13 chapters and the summary of each chapters is given below.

Chapter 1 gives a general background and significance of enzyme inhibition. The basic mechanism and scope of inhibitions is listed out in this chapter. This chapter outlines the outstanding application of Serine protease inhibitors in anticancer treatment and regulating blood coagulation factor. The importance of other remedial inhibitors is also summarized in this chapter.

Chapter 2 gives information about the enzyme based biosensors and its typical component. The importance of coupling electrochemical sensors with metabolites and antibodies is summarized. This chapter gives an idea about the intendment parameters influencing during the design of biosensors. The role of biosensor in clinical and management of pesticides is explained in detailed fashion. This chapter also highlight the strategies of enzyme inhibition in bio-sensing frameworks. The purport of biosensors in light of catalyst and its resistant is explained in this chapter.

Chapter 3 discuss the role of enzyme inhibition in the design and assessment of biosensors. The recent advances and challenges faced during the fabrication of biosensors is explained in this chapter. This chapter explains about the different application of inhibition based biosensor focusing on clinical and environmental management.

Chapter 4 provides applications RNA Silencing in Enzyme Inhibition and its Role in Crop Improvement. Expression of *antisense* genes and the related gene silencing technique has been exploited as an applied *technique in plant* biotechnology for creating "metabolic engineered *plants*" in which the endogenous target gene, which is responsible for the unpleasant character is specifically suppressed. In the present chapter, the down regulations of the enzyme using the gene-silencing technique used in various crop varieties are discussed in detail.

Chapter 5 provides an overview about the history and development of biosensor based on different generation of biochemical signal including antigen –antibody interactions, nucleic acid and microbial cell. The role of enzyme inhibition and development of enzyme based biosensors is explained in this chapter. The importance of enzyme inhibition in the detection of pesticides and heavy metal is discussed in this chapter.

Chapter 6 deals with interesting insights of therapeutic accomplishments of matrix metalloproteinase inhibitors in arthritis, autoimmune disease, inflammations, cancer and cardiovascular disease. The functional role and inhibitors of metalloproteinase is explained in this chapter. This chapter also deals with the clinical implication of matrix metalloproteinase and its effect.

Chapter 7 reviews the potential biosensors based on enzyme inhibition and its application in various field. The kinetic parameter in the design of biosensor is explained in this chapter. The detection level of toxins, heavy metal and pesticides is detailed in this chapter. This chapter also highlights the significance of enzyme inhibition in assessing the drug and its development during various phases. The assessment of safety level of food by enzyme based biosensors is explained in this chapter.

Chapter 8 presents the detailed information on product inhibition during fermentation of ethanol. The cell growth and ethanol inhibitions and related kinetics is explained in this chapter. This chapter also explains about the validation of product inhibition.

Chapter 9 highlights the significance strategies to reduce the inhibition of bioethanol during fermentation process. The production of bioethanol from different strains and source is explained in this chapter. This chapter summarizes the different pre-treatment conditions and factors affecting the inhibition condition is explained in this chapter.

Chapter 10 explains on various Persistent Organic Pollutants (POPs) and its action with respect to the neurotoxic effects emphasising on dose-response and structure-activity relationships (SAR) are discussed in this chapter. The potential modes of actions and alteration in neurotransmitter systems and the mechanisms is discussed in detailed in this chapter.

Chapter 11 gives general introduction about the anatomy of breast carcinogenesis and the enzyme produced predominately produces in breast. The mechanism of different enzymes and the inhibitors used as drug in the treatment of breast cancer is given in this chapter.

Chapter 12 reviews about the various enzyme inhibitor molecule involved in the formulation and development of antiviral drug for HIV Chikungunya, Dengue, Ebola, Influenza, and Nipah viral diseases. The mode of action, inhibition entry and replication of host cells is discussed in this chapter. This chapter also explains about the emerging technologies like CRISPR used for the diagnosis, treatment and alleviation of viral disease progression.

Chapter 13 explains about the various neurological disorders and the enzyme inhibitors used as drug is explained in this chapter. This chapter summarizes the action and mechanism of various enzyme inhibition in the therapy of neurological disease.

The editors would like to express sincere thanks to all authors for their phenomenal contributions to this book. The editor thank all reviewers and well-wishers for the completion of this book.

Dr. G. Baskar
Department of Biotechnology
St. Joseph's College of Engineering
Chennai, India

Dr. K. Sathish Kumar
Department of Chemical Engineering
SSN College of Engineering
Chennai – 603110, India

K. Tamilarasan
Department of Chemical Engineering
SRM University
Chennai – 603203, India

List of Contributors

A. Saravanan   Department of Biotechnology, Rajalakshmi Engineering College, Chennai 600048, India

Ankur Khare   Environmental Impact and Sustainability Division, CSIR-NEERI, Nagpur, India
Academy of Scientific and Innovative Research (AcSIR), Ghaziabad, India

Aparna Madan   Department of Biotechnology, Sri Venkateswara College of Engineering, Sriperumbudur Tk. - 602 117, Kancheepuram District, Tamil Nadu, India

Arthi Udhayachandran   Department of Biotechnology, Sri Venkateswara College of Engineering (Autonomous), Sriperumbudur Tk, Tamilnadu, India

Balu Ranganathan   Palms Connect Sdn Bhd, Shah Alam 40460, Selangor Darul Ehsan, Malaysia
Palms Connect LLC, Showcase Lane, Sandy, UT, USA

Bethu Madhumitha   Department of Biotechnology, Sri Venkateswara College of Engineering (Autonomous), Sriperumbudur Tk - 602 117, Tamil Nadu, India

Carlos Ricardo Soccol   Biotechnology and Bioprocess Engineering Department, Federal University of Parana, Curitiba-PR, Brazil

Cristine Rodrigues   Biotechnology and Bioprocess Engineering Department, Federal University of Parana, Curitiba-PR, Brazil

D. Lokapriya   Department of Biotechnology, Sri Venkateswara College of Engineering (Autonomous), Sriperumbudur Tk, Tamilnadu, India

E. Raja Sathendra   Department of Biotechnology, Arunai Engineering College, Tiruvannamalai-606603, India

Ekambaram Nakkeeran   Department of Biotechnology, Sri Venkateswara College of Engineering (Autonomous), Sriperumbudur Tk - 602 117, Tamil Nadu, India

Elsa Cherian   Department of Food Technology, SAINTGITS College of Engineering, Kottayam, Pathamuttom P. O, Kerala-686532, India

G. Baskar   Department of Biotechnology, St. Joseph's College of Engineering, Chennai – 600119, India

Hariharan Jayaraman   Department of Biotechnology, Sri Venkateswara College of Engineering, Sriperumbudur Tk. - 602 117, Kancheepuram District, Tamil Nadu, India

Joyce Gueiros Wanderley Siqueira   Biotechnology and Bioprocess Engineering Department, Federal University of Parana, Curitiba-PR, Brazil

K. Sathish Kumar   Department of Chemical Engineering, SSN College of Engineering, Chennai – 603110, India

K. Tamilarasan   Department of Chemical Engineering, SRM Institute of Science and Technology, Chennai – 603203, India

Kanchan Kumari   Environmental Impact and Sustainability Division, CSIR-NEERI, Nagpur, India
Academy of Scientific and Innovative Research (AcSIR), Ghaziabad, India

Lakshmi Suresh	Department of Biotechnology, Sri Venkateswara College of Engineering, Sriperumbudur Tk. - 602 117, Kancheepuram District, Tamil Nadu, India
Luciana Porto de Souza Vandenberghe	Biotechnology and Bioprocess Engineering Department, Federal University of Parana, Curitiba-PR, Brazil
Mahalakshmi Varadan	Department of Biotechnology, Sri Venkateswara College of Engineering, Sriperumbudur Tk. - 602 117, Kancheepuram District, Tamil Nadu, India
Nelson Libardi Junior	Biotechnology and Bioprocess Engineering Department, Federal University of Parana, Curitiba-PR, Brazil
Nilavunesan Dhandapani	Department of Biotechnology, Sri Venkateswara College of Engineering (Autonomous), Sriperumbudur Tk - 602 117, Tamil Nadu, India
P.K. Praveen Kumar	Department of Biotechnology, Sri Venkateswara College of Engineering (Autonomous), Sriperumbudur Tk, Tamilnadu, India
P.R. Yaashikaa	Department of Chemical Engineering, SSN College of Engineering, Chennai 603110, India
P. Senthil Kumar	Department of Chemical Engineering, SSN College of Engineering, Chennai 603110, India
Pradip Jadhao	Environmental Impact and Sustainability Division, CSIR-NEERI, Nagpur, India Academy of Scientific and Innovative Research (AcSIR), Ghaziabad, India
Praveen Kumar Posa Krishnamoorthy	Department of Biotechnology, Sri Venkateswara College of Engineering, Sriperumbudur Tk. - 602 117, Kancheepuram District, Tamil Nadu, India
R. Aiswarya	Department of Biotechnology, St. Joseph's College of Engineering, Chennai – 600119, India
R. Jayasree	Department of Biotechnology, Rajalakshmi Engineering College, Chennai 600048, India
R. Praveen Kumar	Department of Biotechnology, Arunai Engineering College, Tiruvannamalai-606603, India
R.V. Hemavathy	Department of Biotechnology, Rajalakshmi Engineering College, Chennai 600048, India
Rajvikram Madurai Elavarasan	Department of Electrical and Electronics Engineering, Sri Venkateswara College of Engineering (Autonomous), Sriperumbudur Tk - 602 117, Tamil Nadu, India
Ravichandran Rathna	Department of Biotechnology, Sri Venkateswara College of Engineering (Autonomous), Sriperumbudur Tk - 602 117, Tamil Nadu, India
Ravichandran Viveka	Department of Biotechnology, Sri Venkateswara College of Engineering (Autonomous), Sriperumbudur Tk - 602 117, Tamil Nadu, India
S. Jeevanantham	Department of Biotechnology, Rajalakshmi Engineering College, Chennai 600048, India
S. Justin Packia Jacob	Department of Biotechnology, St. Joseph's College of Engineering, Chennai – 600119, India
Senthil Nagappan	Department of Biotechnology, Sri Venkateswara College of Engineering (Autonomous), Sriperumbudur Tk - 602 117, Tamil Nadu, India

Sonam Paliya   Environmental Impact and Sustainability Division, CSIR-NEERI, Nagpur, India
Academy of Scientific and Innovative Research (AcSIR), Ghaziabad, India

Subasree Sekar   Department of Biotechnology, Sri Venkateswara College of Engineering (Autonomous), Sriperumbudur Tk – 602117, Tamilnadu, India

T.R. Sundaraman   Department of Biotechnology, Rajalakshmi Engineering College, Chennai 600048, India

CHAPTER 1

Introduction to Enzyme Inhibition – Environmental and Biomedical Applications

G. Baskar[1,*], R. Aiswarya[1], K. Sathish Kumar[2] and K. Tamilarasan[3]

[1] *Department of Biotechnology, St. Joseph's College of Engineering, Chennai – 600119. India*

[2] *Department of Chemical Engineering, SSN College of Engineering, Chennai – 603110, India*

[3] *Department of Chemical Engineering, SRM Institute of Science and Technology, Chennai-603203, India*

Abstract: Enzyme inhibitors alter the activity of an enzyme molecule when an enzyme binds to it. The enzyme inhibition finds its applications in drug discovery and assessment of various environmental pollutants owing to its high specificity and potency. The study of enzyme inhibition mechanism is highly recommended as it mainly depends upon the structural requirement and site of enzyme action. The enzyme inhibition also plays a vital role in the design of biosensors for the detection and assessment of an analyte molecule.

Keywords: Biosensor, Drug discovery, Enzyme inhibitors, Environment assessments, Specificity.

The enzyme, a protein molecule acts as a catalyst in the various enzymatic reactions. The enzyme inhibitors inhibit the catalytic activity by modifying the amino acid. The design of the drug analogue is accomplished by a complete understanding of kinetics and structure-function relationship [1]. The combination of chemistry and high throughput screening technology against the catalytic site helps in the discovery of the drug [2]. The state of the art of enzyme inhibition plays a major quest in various fields. The development of enzyme-based inhibitions follows mainly 1) Slow-tight inhibition 2) Substrate and product inhibition. Antimetabolites, anti-enzyme, antibodies, and biosensors also follow enzyme inhibitions for the detection and assessment of a particular analyte molecule. The drug discovery is based on enzyme inhibitions represented by monolithic immobilized enzyme reactors (MIERs). The high throughput screening helps in the identification of target protein during drug development [3]. The

---

* **Corresponding author G. Baskar:** Department of Biotechnology, St. Joseph's College of Engineering, Chennai – 600119, India; E-mail: basg2004@gmail.com

major challenging factor in enzyme inhibition is the kinetic calculation using various software. The major impact on enzyme inhibition is based on the unpredictable synergistic mechanism. The enzyme inhibitor has created a new space for the therapeutics market and has potential therapeutic applications for various diseases like Chronic Obstructive Pulmonary Disease (COPD), gastrointestinal disorders, cardiovascular, and other inflammatory-related diseases. The conceptual models of the enzyme inhibition play a major role in the interaction of substrate-inhibitor. The use of immobilized enzyme technology has paved the way for the development of tools in drug discovery and the design of biosensors. The cytochrome (P450) (CYP) actively helps in drug interaction by enzyme induction [4]. The CYP1 enzyme inhibits the action of dimethylhydrazine and acts as chemo protectants [5]. The CYP1 and its regulation of aryl hydrocarbon receptors have been extensively studied for drug resistance of the carcinogenesis [6]. Matrix metalloproteinases (MMP) play a vital role in embryogenesis, wound healing, and stem cell mobilization. They cleave the intra and extracellular matrix molecules at the pericellular environment [7, 8]. MMP cleaves and regulates the enzyme involved during signal transduction at a particular site. The MMP actively helps in the regulation of tumor suppression and autoimmune disorders [9 - 12]. MMP as gelatinases helps in the progression of aneurysms by proteolytic activity of neutrophils [13]. The neutrophil extracellular traps HLE along with MMP-9 has therapeutic value, where oleoyl moiety is replaced by ß-Lactam [14]. The natural metabolites extracted from plant sources were reported to have the lead component of enzyme inhibitors known as acetylcholinesterase (AChE), glutathione S-transferase (GST), and α-glucosidase. The natural compounds exhibited their potential role in the treatment of Alzheimer's disease as evident from the structure-activity relationship [15].

The basic principle of biosensor lies in the interaction of specific chemical and biological agents in the form of inhibitors. The inhibition in biosensors helps in analysing the kinetic characterization of the process at a heterogeneous surface. The concentration of inhibitor helps in analysing the percentage of inhibition of the biocatalyst over immobilized biosensor. The inhibition biosensor plays a significant role in environmental assessment. Enzymes like alkaline phosphatase combined with electrochemical sensors help in the detection of heavy metals [16]. The enzymatic alterations of various Persistent Organic Pollutants (POPs) and the assessment of pollutant levels with respect to their cell signaling are formulated by inhibition mechanism [17]. Thus, the mechanism of enzyme inhibition plays a major in the discovery of drug and monitoring of environmental pollutants.

CONCLUSION

Enzyme inhibitors are acts as a significant tool to distinguish the enzyme reaction and its parameters in biological industries. The measurement and accuracy of enzyme detections have entered a new era by using pico technology. The immobilized enzyme technology was widely used in drug discovery and sensor development industry due to the availability of various new enzyme inhibitors. The use of a single MMP or combination of MMP with Quantum Dots (QD) helps in the treatment of cancer. Various *in-vitro* and *in-silico* models help in the analysis of the drug development process to minimize the uneventful interaction of drugs. The enzyme inhibition plays a major role in improving the target of drugs to a particular site during the chemoprevention process. The enzyme inhibitor at the heterogeneous surface helps in the analysis of persistent organic pollutants and also helps in other environmental assessments.

CONSENT FOR PUBLICATION

Not applicable.

CONFLICT OF INTEREST

The author(s) confirms that there is no conflict of interest.

ACKNOWLEDGEMENTS

Declared none.

REFERENCES

[1]   Sami AJ, Shakoori AR. Cellulase activity inhibition and growth retardation of associated bacterial strains of Aulacophora foviecollis by two glycosylated flavonoids isolated from Mangifera indica leaves. J Med Plants Res 2011; 5(2): 184-90.

[2]   El-Metwally TH, El-Senosi Y. Enzyme Inhibition Medical Enzymology: Simplified Approach. NY: Nova Publishers 2010; pp. 57-77.

[3]   Bartolini M, Greig NH, Yu QS, Andrisano V. Immobilized butyrylcholinesterase in the characterization of new inhibitors that could ease Alzheimer's disease. J Chromatogr A 2009; 1216(13): 2730-8.
      [http://dx.doi.org/10.1016/j.chroma.2008.09.100] [PMID: 18950780]

[4]   Delgoda R, Westlake AC. Herbal interactions involving cytochrome p450 enzymes: a mini review. Toxicol Rev 2004; 23(4): 239-49.
      [http://dx.doi.org/10.2165/00139709-200423040-00004] [PMID: 15898829]

[5]   Chang TK, Chen J, Benetton SA. *In vitro* effect of standardized ginseng extracts and individual ginsenosides on the catalytic activity of human CYP1A1, CYP1A2, and CYP1B1. Drug Metab Dispos 2002; 30(4): 378-84.
      [http://dx.doi.org/10.1124/dmd.30.4.378] [PMID: 11901090]

[6]   King HWS, Osbourne MR, Beland FA, Harvey RG, Brookes P. (+)-7α,8β- Dihydroxy-9β,10β-epoxy-7,8,9,10-tetrahydrobenzo[a]-pyrene is an intermediate in the metabolism and binding to DNA of benzo[a]pyrene. Proc Natl Acad Sci 1976; 73: 2679-81.

[7]   Cauwe B, Opdenakker G. Intracellular substrate cleavage: a novel dimension in the biochemistry, biology and pathology of matrix metalloproteinases. Crit Rev Biochem Mol Biol 2010; 45(5): 351-

423.
[http://dx.doi.org/10.3109/10409238.2010.501783] [PMID: 20812779]

[8]   Nagase H, Visse R, Murphy G. Structure and function of matrix metalloproteinases and TIMPs. Cardiovasc Res 2006; 69(3): 562-73.
[http://dx.doi.org/10.1016/j.cardiores.2005.12.002] [PMID: 16405877]

[9]   Hu J, Van den Steen PE, Sang QX, Opdenakker G. Matrix metalloproteinase inhibitors as therapy for inflammatory and vascular diseases. Nat Rev Drug Discov 2007; 6(6): 480-98.
[http://dx.doi.org/10.1038/nrd2308] [PMID: 17541420]

[10]   López-Otín C, Matrisian LM. Emerging roles of proteases in tumour suppression. Nat Rev Cancer 2007; 7(10): 800-8.
[http://dx.doi.org/10.1038/nrc2228] [PMID: 17851543]

[11]   Mandal M, Mandal A, Das S, Chakraborti T, Sajal C. Clinical implications of matrix metalloproteinases. Mol Cell Biochem 2003; 252(1-2): 305-29.
[http://dx.doi.org/10.1023/A:1025526424637] [PMID: 14577606]

[12]   Murphy G, Nagase H. Progress in matrix metalloproteinase research. Mol Aspects Med 2008; 29(5): 290-308.
[http://dx.doi.org/10.1016/j.mam.2008.05.002] [PMID: 18619669]

[13]   Brinkmann V, Reichard U, Goosmann C, *et al.* Neutrophil extracellular traps kill bacteria. Science 2004; 303(5663): 1532-5.
[http://dx.doi.org/10.1126/science.1092385] [PMID: 15001782]

[14]   Moroy G, JP, Alix A, Sapi J, Hornebeck W, Bourguet E. Neutrophil elastase as a target in lung cancer. Anti-Cancer Agents in Medicinal Chemistry (Formerly Current Medicinal Chemistry-Anti-Cancer Agents) 2012; 12(6): 565-79.
[http://dx.doi.org/10.2174/187152012800617696]

[15]   Ata A, Van Den Bosch SA, Harwanik DJ, Pidwinski GE. Glutathione S-transferase-and acetylcholinesterase-inhibiting natural products from medicinally important plants. Pure Appl Chem 2007; 79(12): 2269-76.
[http://dx.doi.org/10.1351/pac200779122269]

[16]   Renedo OD, Lomillo MA, Martinez MA. Optimisation procedure for the inhibitive determination of chromium (III) using an amperometric tyrosinase biosensor. Anal Chim Acta 2004; 521(2): 215-21.
[http://dx.doi.org/10.1016/j.aca.2004.06.026]

[17]   Freeman HC, Uthe J, Sangalong G. The use of steroid hormone metabolism studies in assessing the sublethal effects of marine pollution. Rapp P-v Reun Cons Int. Explor Mer 1980; 179: 16-22.

CHAPTER 2

Enzyme Inhibition in Therapeutic Applications

A. Saravanan[1,*], R. Jayasree[1], T.R. Sundaraman[1], R.V. Hemavathy[1], S. Jeevanantham[1], P. Senthil Kumar[2] and P. R. Yaashikaa[2]

[1] *Department of Biotechnology, Rajalakshmi Engineering College, Chennai 600048, India*

[2] *Department of Chemical Engineering, SSN College of Engineering, Chennai 603110, India*

Abstract: Enzyme is a protein fragment that catalyzes the biological reactions by reducing the activation energy required for the reactions to occur. Enzyme inhibition is a vital method for regulating movement in living cells. Enzyme inhibition occurs by the substrate called inhibitors that can bind to an enzyme and reduce its activity, endogenous mixes and xenobiotics are compound. There are three fundamental kinds of enzyme inhibition: competitive, non-competitive, and uncompetitive. Among the measuring time frame, enzyme repressing drugs are anticipated to be presented for new signs including asthma and interminable obstructive pneumonic ailment, aspiratory blood vessel hypertension, hepatitis C and discontinuous claudication. This chapter offers an expansive point by point diagram of compound inhibitors at present available and those in late-organize medical trials. The data and examination exhibited in this report are vital resources in basic leadership for chiefs engaged with business advancement, advertising, statistical surveying, item improvement, mergers, and acquisitions, authorizing, business administration, speculation managing an account and arrangement creation, and to specialists to the pharmacological and biotechnology industry. The investigation gives a complete examination of the present markets for compound hindering medications and, specifically, the market capability of promising medications and innovations a work in progress.

Keywords: Biosynthetic, Biochemical, Covalent, Chronic obstructive pulmonary disease, Enzyme, Hepatitis C, Inhibition, Inorganic metal, Medication, Non-competitive, Pharmaceutical, Therapeutic, Uncompetitive, Xenobiotics.

INTRODUCTION

Enzymes have great potency towards the application of the pharmaceutical industry particularly for the treatment of cardiovascular diseases, treatment of cancer, wound debridement, bleeding disorders and digestive aids [1]. The catalytic property of therapeutic enzymes plays an important role in converting various target molecules into other desired products [2]. Most of the therapeutic

* **Corresponding author Saravanan Anbalagan:** Department of Biotechnology, Rajalakshmi Engineering College, Sriperumbudur Tk - 602 105, Tamil Nadu, India; Tel: +91-9003838356; E-mail: saravanan.a@rajalakshmi.edu.in

G. Baskar, K. Sathish Kumar & K. Tamilarasan (Eds.)

enzymes are required in less amount when compared to industrially imperative enzymes.

Animal tissues, plants, and microorganisms are the major sources of the enzymes. Among the various sources, microbial enzymes are chosen as suitable due to their consistency, optimization and economic production. Few fungal, bacteria and yeast strains contribute to the production of enzymes with the therapeutic application. Enzymes for therapeutic application should have a high degree of purity and specificity. During the protein treatment, the transport of mixes inside the host considerable cells is basic as a result of the tremendous nuclear size. In addition, the immune response generated by the host cells after receiving foreign enzyme is the major concern contributing to life-threatening conditions with severe allergic reactions. The short half-life of enzymes is another problem associated with therapeutic application and microencapsulation and counterfeit liposomal capture are a portion of the strategies used to increase the strength and half-existence of chemical medications [3]. As an end, headways in medication improvement and conveyance in recent decades have upset another path for chemical treatment. Therapeutic enzymes are marketed as preparations of pure lyophilized substances with the addition of buffering salts [4]. Table **1** summarizes the list of enzymes with their therapeutic applications.

Table 1. List of some enzymes with therapeutic applications.

Enzyme	Application
Alteplase	Treatment of cardiovascular diseases
Urokinase	Treatment of cardiovascular diseases
Anistreplase	Treatment of cardiovascular diseases
Carboxypeptidase G2	Treatment of cancer
β–glucuronidase	Treatment of cancer
Alkaline Phosphatase	Treatment of cancer
β Galactosidase	Digestive Aids
Glutenases	Digestive Aids
Proteases	Wound Debridement

ENZYME INHIBITION

Enzyme inhibitors are organic or inorganic compounds that can control the substance reactant development either rescindable or permanent. The inhibitor can transform one amino destructive, or a couple of side-chain required in compound synergist activity. Safely, the engineered change should be conceivable to test inhibitor for any medicine regard.

Enzymes catalyse a reaction by reducing the activation energy needed for the reaction to occur. However, enzymes need to be tightly regulated to ensure that levels of the product do not rise to undesired levels. This is accomplished by enzyme inhibition. Some noticeable incredible delineations are medicine and toxic substance action and cure diagram for remedial utilizations *e.g.*, iodoacetamide deactivates Cys amino destructive in impetus side chain; methotrexate in threat chemotherapy through semi-explicitly control DNA mix of perilous cells; ibuprofen quells the amalgamation of the pro-inflammatory prostaglandins; drugs stifle the folic destructive amalgamation essential for the improvement of pathogenic tiny life forms in this way various distinctive prescriptions.

Protease inhibitors can work in many different ways to inhibit the action of proteases. These inhibitors can be classified by the type of proteases they inhibit and the mechanism by which they inhibit those enzymes. While commercial protease inhibitors are typically sold based on the class of protease they inhibit, understanding the various mechanisms by which inhibitors function is essential for a comprehensive understanding of inhibition and for developing protease inhibitors as a therapeutic strategy.

Regulation of enzyme activity can be attained through enzyme inhibition and usually resulted in reduced enzyme activity [5]. Enzyme inhibition can be reversible or irreversible. Reversible inhibition of enzyme can be distinct types of inhibition due to the possible non-covalent bond formed between the inhibitor molecule and the enzyme and thereby reduction in enzyme activity occurs completely inhibited or partially inhibited. Irreversible inhibition can cause chemical modification of the particular enzyme. There are diverse sorts of conceivable reversible restraints that may happen as focused kind, non-aggressive sort and uncompetitive, in spite of the fact that a blended kind now and then emerges. Degree of inhibition (i) of reversible inhibitor is calculated by using the following equation,

$$i = (V_o - V)/V_o \tag{1}$$

where V_o and V are the rates of uninhibited and inhibited reactions, respectively.

TYPES OF ENZYME INHIBITION

Reversible Enzyme Inhibition

Reversible inhibitors can bind to an enzyme simply by a noncovalent interaction is resulting in reversibly and can be reversed by dialysis or by dilution to lower the concentration of the inhibitor.

Generally, reversible enzyme inhibition is denoted by

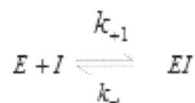

$$E + I \underset{k_{-1}}{\overset{k_{+1}}{\rightleftharpoons}} EI$$

The dissociation constant for the reversible enzyme inhibition (k_{-1} / k_{+1}), is defined as the inhibitor constant, k_i

Michaelis–Menten relationship can be expressed as,

$$v = \frac{V_{max}[S]}{K_m + [S]} \qquad (2)$$

Twofold complementary type of Michaelis– Menten is indicated by Lineweaver-Burk condition, which is utilized to foresee the sort of hindrance by showing the contrasts between the motor conduct of the distinctive inhibitor types [6],

$$\frac{1}{v} = \frac{K_m}{V_{max}[S]} + \frac{1}{V_{max}} \qquad (3)$$

The Lineweaver-Burk plot is a plot of 1/v vs 1/[S]. With the diminishment in the inhibitor focus, the chemical movement is recovered due to the non-covalent affiliation and the reversible harmony with the chemical.

Competitive Reversible Inhibition

In the forceful restriction type, the compound can tie either substrate or inhibitor yet not both. At particularly high substrate centers, all inhibitors will be removed from the impetus; along these lines, V_{max} is unaltered yet inside seeing inhibitors, more substrate will be relied upon to get to V_{max}. Along these lines, for focused inhibitors, V_{max} is unaltered yet K_m is expanded. Clearly, the level of hindrance is diminished by expanding the substrate focus if the inhibitor goes after the dynamic site with the substrate.

The inhibitor-enzyme bond is so strong that the inhibition cannot be reversed by the addition of excess substrate. The nerve gases, especially Diisopropyl fluorophosphate (DIFP), irreversibly inhibit biological systems by forming an enzyme-inhibitor complex with a specific OH group of serine situated at the active sites of certain enzymes. According to the law of mass action, a reasonably higher inhibitor obsession keeps the substrate official. Since the reaction rate is clearly with respect to [ES], diminishment in ES course of action for EI improvement cuts down the rate. Expanding substrate towards an immersing fixation mitigates focused restraint. In the time driving force, substrate complex discharges the free protein and a thing, the compound inhibitor complex releases neither free protein nor a thing.

Non-competitive Reversible Inhibition

In the non-aggressive restraint type, S and I can connect to the protein all the while since their structures are probably not going to be comparative so they don't vie for the dynamic site. Both I and S tie to the catalyst in various locales autonomously. The level of restraint is unaffected by the change in the substrate fixation.

The mixed kind inhibitor does not have essential comparability to the substrate yet rather it ties both of the free protein and the compound substrate complex. Thusly, its coupling way is not irrelevant with the substrate and the proximity of a substrate has no effect on the limit of a non-centered inhibitor to tie a compound and a different way. In any case, it is restricting - albeit far from the dynamic site - adjusts the adaptation of the catalyst and diminishes its reactant action because of changes in the idea of the synergist bunches at the dynamic site. EI and ESI buildings are ineffective and expanding substrate to an immersing focus does not turn around the hindrance prompting unaltered Km but rather lessened V_{max}. Inversion of the restraint requires an uncommon treatment, *e.g.*, dialysis or pH change. A few groupings separate between non-focused hindrance as char-acterized above and blended hindrance in that the EIS-complex has lingering enzymatic action in the blended restraint. In the blended restraint type, S and I can append to the catalyst at the same time in two unique destinations conditionally. The liking of restricting S or I is influenced by the authority of the other. The level of restraint is influenced by the change in the substrate focus relies upon the communication between two destinations of S and I.

Uncompetitive Reversible Inhibition

The uncompetitive inhibitor has no associate closeness to the substrate. It might tie the free driving force or protein substrate complex that uncovered the inhibitor constraining site (ESI). Its power, yet far from the dynamic site, causes right-hand bending of the dynamic and allosteric territories of the complexed to intensify that inactivates the catalysis. This prompts a diminishing in both Km and Vmax. Expanding substrate towards a sprinkling focus does not rotate this sort of obstruction and inversion requires uncommon treatment, *e.g.*, dialysis. This sort of obstacle is exorbitantly experienced in multi-substrate engineered substances, where the inhibitor battles with one substrate (S2) to which it has some crucial closeness and is uncompetitive for the other (S1). In the uncompetitive restraint type, I can append to the catalyst subsequent to the authoritative of S. It implies that an adjustment in the structure of protein ought to be happened by the official of S, which the inhibitor-restricting site winds up uncovered, so I can tie to its coupling site. The level of restraint is expanded by a change in the substrate fixation. The inhibitor does not attach to the free compound but instead just to an enzyme-substrate complex. Both the Km and the best speed regards are reduced by a comparable aggregate. Thusly, the obstruction will increase as the gathering of the ES complex augmentations.

Irreversible Inhibition

The irreversible apoenzyme inhibitors have no assistant relationship to the substrate and tie covalently. They moreover tie stable non-covalently with the dynamic site of the protein or destroy a key utilitarian get-together of a dynamic site. Thus, irreversible inhibitors are utilized to distinguish utilitarian gatherings of the protein dynamic destinations at which area they tie. In spite of the fact that inhibitors have constrained remedial applications since they are typical, go about as toxins. Notwithstanding, it does not discharge any item because of its irreversible official at the catalyst dynamic site. Inhibitors make utilize of the ordinary protein response system to be initiated and along these lines inactivate the catalyst. Subsequently, inhibitor abuses the change state settling impact of the chemical, bringing about a superior restricting fondness (bring down Ki) than substrate-based plans. In an irreversible restraint type, the inhibitor collaborates with various utilitarian gatherings on the compound surface by framing solid covalent securities that frequently endure notwithstanding amid complete protein breakdown.

The present craft of medication revelation and the plan of new medications depend on self-destructive irreversible inhibitors. Synthetic substances are orchestrated in view of learning of 3D compliance of substrate active site authoritative at particular restricting rates in the nearness of co-factors, co-protein (catalyst response instruments) to repress at particular catalyst dynamic site with negligible symptoms due to its non-particular restricting nature. Change state analogs are to a great degree intense and particular inhibitors of chemicals since they have higher partiality and more grounded official to the dynamic site of the objective compound than the characteristic substrates or items. Nevertheless, the correct outline of drugs that decisively mirror the progress state is a test due to the precarious structure of change state in the free-state.

ENZYME INHIBITION: MECHANISM AND SCOPE

Catalyst inhibitor is a particle that ties to the chemical and hindering a compound action in this way making metabolic irregularity. There is a demand for the discovery of enzyme inhibitors for therapeutic applications [5]. Enzyme activators bind to the enzymes in a similar way as such as enzyme inhibitor but they increase the enzyme activity. To support the idea, two examples were given such as the inhibition of asparaginase and glucosidase.

Inhibition of Asparaginase

Asparaginase is a compound acquired from contagious strain and utilized as cytotoxic chemotherapy medicate. It is an intense medication for the treatment of malignancy. The enzyme L-asparaginase (ASNase), which catalyzes the hydrolysis of L-asparagine, is a component of most therapeutic protocols for the treatment of acute lymphoblastic leukemia. In particular, cell-permeable compounds capable of specifically inhibiting the enzyme may be valuable tools in evaluating whether increased levels of intracellular asparagine biosynthesis might be the key change in cellular metabolism that underlies the appearance of drug resistance [6]. An improved delivery system of asparaginase for the treatment of lung cancer was reported by immobilization of asparaginase on to gold nanoparticles and resulted in evading leakage into other cells or tissues [7].

Inhibition of Glycosidases

Lysosomal α-glucosidase is a catalyst that has a place with exoglucosidase that assumes a vital job in sugar preparation by catalyzing the hydrolysis of lysosomal glycogen and brought about glycogen breakdown [8]. 1-Deoxynojirimycin (dnm) has a place with inhibitor for α-glucosidases with hostile to HIV action and utilize their organic impact by aggressively restraining explicit glycoside-preparing catalysts, strong [9].

SOME THERAPEUTIC APPLICATIONS

Serine Protease Inhibitors

Serine proteases alongside different go-betweens are engaged with formative phases of numerous sicknesses, for example, tumor, viral ailments, draining scatters, aggravation, joint inflammation, pancreatitis, strong dystrophy and emphysema [10, 11].

Protein protease inhibitors are found in all cells including living things. They can be amassed, in context on the proteases they control, as serine protease inhibitors, cysteine protease inhibitors, aspartate protease inhibitors and metalloprotease inhibitors. Not in the smallest degree like the greater part of the protein protease inhibitors, little atom protease inhibitors have wide unequivocal and can smother proteases of various classes [12]. They have an imperative part in these living animals by controlling irksome proteases that might harm their own cells [13]. Few protease inhibitors in the two creatures and plants were especially considered and their physiological part has been created. Protease inhibitors are extensively streamed over the families. In people, they even can control atomic pathways related to tissue homeostasis, progress and cell confirmation [14].

Serine Protease Inhibitors for Cancer Treatment

Proteases acknowledge an essential part of different frameworks identified with tumor interference and improvement. Among them, a few serine proteases are joined and consequently can be promising focuses on anticancer solutions [15].

Serine Protease Inhibitors as Antiviral Specialists

Viral proteases are required for dealing with the viral polyproteins that are required for viral replication. These proteases are besides associated with baffling the antiviral reaction of the cells. Serine protease inhibitors may have potential applications in antiviral treatment by focusing on these viral proteases [16].

Phosphodiesterases

Phosphodiesterases might be potential focuses for tumor cell development restraint and apoptosis enlistment, in light of following perceptions. To start with, the control of cyclic nucleotide flagging is believed to be one of a few segment pathways engaged with tumor cell dispersals and capacities. Second, different PDE isozymes are engaged with different tumor tissues. Third, nonselective PDE inhibitors, for example, theophylline or aminophylline, direct the development of in an assortment of malignancy cell lines [17]. The following is a concise survey of late advances in the articulation and control of each PDE isoform during the time spent tumorigenesis and they are hostile to tumor inhibitors, which may control the outline of novel remedial medications focusing on PDEs for anticancer specialist.

Matrix Metalloproteinases

Matrix metalloproteinases (MMPs), additionally called matrixins, are a group of fundamentally related zinc-containing compounds that intervene the breakdown of connective tissue and are subsequently focused for remedial inhibitors in numerous fiery, dangerous, furthermore, degenerative diseases.1-3 Long before the person proteins were secluded and portrayed, scientists had been keen on the action of tissue rebuilding, both in physiological and malady forms. These proteins can debase the fibrillar collagens, which are for the most part safe to proteolysis, making a trademark 3/4 length break in the R-chain. Its favored substrate remains a matter of level headed discussion, yet it has proceeded to be of enthusiasm because of its evident enlistment in harmful tissue. It doesn't give off an impression of being especially dynamic in corrupting known extracellular framework proteins, however, it is compelling in debasing the serine proteinase inhibitor (serpin) R-1 antitrypsin and in doing as such may potentiate the activity of serine proteinases, for example, urokinase-type plasminogen activator (uPA). This serpinase movement is moreover shown by different MMPs and backings the theory that the metallo-and serine proteinase families can work in covering falls of restraint and actuation.

Histone Deacetylases

Histone changes—incorporating diminishes in acetylation—have been accounted for to be modified in malignancy (counting CNS tumors), and are probably going to be imperative at a principal level in the pathogenesis of neoplasia. More or less, overexpression of Histone deacetylases (HDACs) connects with tumorigenesis [18]. Changes in histone acetylation can cause deviant interpretation of key qualities, which control capacities vital in cell expansion, cell cycle control, apoptosis, and numerous different procedures. Short-sightedly, Class I HDACs incite cell expansion and repress separation and apoptosis. Overexpression of these HDACs is available in numerous sorts of tumors.

Thinking about the numerous consequences for malignancy cells, it is valid, along these lines, that treating a growth understanding with an HDAC-I might be even more a "shotgun" approach, as opposed to a great degree exact focusing of a key abnormal pathway. However, this ought to be worthwhile in an ailment in which the cell trademarks are heterogeneity, versatility, and flexibility. A key perception (made by numerous agents) is that growth cells are more helpless to HDAC-Is than are ordinary cells, that is, non-neoplastic cells [19]. This, thusly, has critical ramifications for foundational poisonous quality, and may in certainty be connected to the central idea of neoplasia (such as the concealing of tumor silencer qualities inside heterochromatin).

The method of reasoning for utilizing HDAC-Is in disease treatment is in fact multifold. To begin with, HDAC-Is advance more open chromatin compliance which would be relied upon to allow better access of DNA-harming specialists (*e.g.*, chemotherapy furthermore, radiation treatment), subsequently expanding the affectability to executing by these operators. Second, HDAC-Is should turn around a portion of the tumor silencer quality hushing in tumor cells straightforwardly, prompting improved cell-cycle capture and apoptosis. Third, there are backhanded impacts on tumor cells, such as through restraint of angiogenesis and boosting the safe reaction [20].

CONCLUSION

Enzyme inhibitors are particles, which lessen the enzymatic movement, and these inhibitors may incorporate numerous medications. There are distinctive sorts of compound inhibitors, as per their structure and their site of activity in the protein. Moreover, unique hindrance components might be usable in one catalyst restraint. Concentrates with various chemicals and hindrance examinations are critical, to clear up the necessities in the skeleton of the inhibitor. Catalyst restraint is a huge organic procedure to describe the protein response, extraction of catalysis parameters in bio-industry and bioengineering. Applied models of hindrance characterize the communications of substrate-chemical or inhibitor-compound or both substrate protein inhibitor in the moiety of dynamic sites. As of late, utilization of chemicals and chemical hindrance science has gone in human services, pharmaceutical, bio-enterprises, condition, and biochemical chemical chip businesses with incredible effect on medicinal services and restorative business

CONSENT FOR PUBLICATION

Not applicable.

CONFLICT OF INTEREST

The author(s) confirms that there is no conflict of interest.

ACKNOWLEDGEMENTS

Authors thank chairperson - Dr. Mrs. Thangam Meganathan and Principal – Dr. S. N. Murugesan, Rajalakshmi Engineering College for their support and encouragement.

REFERENCES

[1]   Kumar SS, Abdulhameed S. Therapeutic enzymes. Bioresources and bioprocess in biotechnology. Singapore: Springer 2017; pp. 45-75.
[http://dx.doi.org/10.1007/978-981-10-4284-3_2]

[2]   Cooney DA, Rosenbluth RJ. Enzymes as therapeutic agents. Advances in Pharmacology 1975 Jan 1; 12: 185-289.
[http://dx.doi.org/10.1016/S1054-3589(08)60222-7]

[3]   Sabu A, Nampoothiri KM, Pandey A. L-Glutaminase as a therapeutic enzyme of microbial origin InMicrobial Enzymes and Biotransformations. Humana Press 2005; pp. 75-90.
[http://dx.doi.org/10.1385/1-59259-846-3:075]

[4]   Gurung N, Ray S, Bose S, Rai V. A broader view: microbial enzymes and their relevance in industries, medicine, and beyond. BioMed research international 2013; 2013
[http://dx.doi.org/10.1155/2013/329121]

[5]   Balbaa M, El Ashry ES. Enzyme inhibitors as therapeutic tools. Biochem Physiol 2012; 1(2)1000103
[http://dx.doi.org/10.4172/2168-9652.1000103]

[6]   Wriston JC Jr, Yellin TO. L-asparaginase: a review. Adv Enzymol Relat Areas Mol Biol 1973; 39: 185-248.
[PMID: 4583638]

[7]   Baskar G, Garrick BG, Lalitha K, Chamundeeswari M. Gold nanoparticle mediated delivery of fungal asparaginase against cancer cells. J Drug Deliv Sci Technol 2018; 44: 498-504.
[http://dx.doi.org/10.1016/j.jddst.2018.02.007]

[8]   Legler G. Glycoside hydrolases: mechanistic information from studies with reversible and irreversible inhibitors. Advances in carbohydrate chemistry and biochemistry 1990 Jan 1; 48: 319-84.
[http://dx.doi.org/10.1016/S0065-2318(08)60034-7]

[9]   Gruters RA, Neefjes JJ, Tersmette M, *et al.* Interference with HIV-induced syncytium formation and viral infectivity by inhibitors of trimming glucosidase. Nature 1987; 330(6143): 74-7.
[http://dx.doi.org/10.1038/330074a0] [PMID: 2959866]

[10]   Sabotič J, Kos J. Microbial and fungal protease inhibitors--current and potential applications. Appl Microbiol Biotechnol 2012; 93(4): 1351-75.
[http://dx.doi.org/10.1007/s00253-011-3834-x] [PMID: 22218770]

[11]   Imada C. Enzyme inhibitors of marine microbial origin with pharmaceutical importance. Mar Biotechnol (NY) 2004; 6(3): 193-8.
[http://dx.doi.org/10.1007/s10126-003-0027-3] [PMID: 15129325]

[12]   Overall CM, Blobel CP. In search of partners: linking extracellular proteases to substrates. Nat Rev Mol Cell Biol 2007; 8(3): 245-57.
[http://dx.doi.org/10.1038/nrm2120] [PMID: 17299501]

[13]   Kantyka T, Rawlings ND, Potempa J. Prokaryote-derived protein inhibitors of peptidases: A sketchy occurrence and mostly unknown function. Biochimie 2010; 92(11): 1644-56.
[http://dx.doi.org/10.1016/j.biochi.2010.06.004] [PMID: 20558234]

[14]   Gagaoua M, Hafid K, Boudida Y, *et al.* Caspases and thrombin activity regulation by specific serpin inhibitors in bovine skeletal muscle. Appl Biochem Biotechnol 2015; 177(2): 279-303.

[http://dx.doi.org/10.1007/s12010-015-1762-4] [PMID: 26208691]

[15]   Ilies MA, Supuran CT, Scozzafava A. Therapeutic applications of serine protease inhibitors. Expert Opin Ther Pat 2002; 12(8): 1181-214.
[http://dx.doi.org/10.1517/13543776.12.8.1181]

[16]   Richer MJ, Juliano L, Hashimoto C, Jean F. Serpin mechanism of hepatitis C virus nonstructural 3 (NS3) protease inhibition: induced fit as a mechanism for narrow specificity. J Biol Chem 2004; 279(11): 10222-7.
[http://dx.doi.org/10.1074/jbc.M313852200] [PMID: 14701815]

[17]   Hirsh L, Dantes A, Suh BS, *et al.* Phosphodiesterase inhibitors as anti-cancer drugs. Biochem Pharmacol 2004; 68(6): 981-8.
[http://dx.doi.org/10.1016/j.bcp.2004.05.026] [PMID: 15313391]

[18]   Eckschlager T, Plch J, Stiborova M, Hrabeta J. Histone deacetylase inhibitors as anticancer drugs. Int J Mol Sci 2017; 18(7): 1414.
[http://dx.doi.org/10.3390/ijms18071414] [PMID: 28671573]

[19]   Carrier F. Chromatin modulation by histone deacetylase inhibitors: impact on cellular sensitivity to ionizing radiation. Mol Cell Pharmacol 2013; 5(1): 51-9.
[PMID: 24648865]

[20]   Subramanian S, Bates SE, Wright JJ, Espinoza-Delgado I, Piekarz RL. Clinical toxicities of histone deacetylase inhibitors. Pharmaceuticals (Basel) 2010; 3(9): 2751-67.
[http://dx.doi.org/10.3390/ph3092751] [PMID: 27713375]

CHAPTER 3

Analytical Aspects of Biosensor Based on Enzyme Inhibition

A. Saravanan[1,*]**, S. Jeevanantham**[1]**, P. Senthil Kumar**[2]** and P. R. Yaashikaa**[2]

[1] *Department of Biotechnology, Rajalakshmi Engineering College, Chennai 602105, India*

[2] *Department of Chemical Engineering, SSN College of Engineering, Chennai 603110, India*

Abstract: An enzyme biosensor is an investigative device that joins an enzyme with a transducer to create a signal corresponding to target analyte fixation. An ideal enzyme activity is a basis for the maintenance of physiological homeostasis of biosensor. Both non-hereditary and hereditary disturbances can too much initiate or quiet characteristic enzyme activities. Due to its virtuous sensitivity, easy to operate, high precision and low instrumentation cost it can be used for recognition of various analytes in different fields than the traditional analytical methods. The enzyme based biosensor is most commonly used dominant tool for the determination of various biological importance such as antigens, antibodies, therapeutic drugs and metabolites. In regular enzyme-based biosensors, signal amplification is not satisfactory for the ultrasensitive identification of biomolecules and it was enhanced by consolidating enzymatic responses with redox cycling of multi-enzyme tags for every discovery test. This chapter explains the working principles, features, types, sensing methods and applications of enzyme based biosensor in various fields.

Keywords: Antigen, Antibody, Biomolecules, Chemical activators, Electrochemical, Enzyme biosensor, Homeostasis, Hereditary, Inhibitor, Metabolite, Nanomaterials, Transducer, Therapeutic drugs.

INTRODUCTION

The biosensor can be characterized as a minimal explanatory gadget fusing a natural or organically inferred detecting component either coordinated inside or personally connected with a physicochemical transducer. The biosensors have three essential components including transducer, signal processor and biologically recognition element (enzyme) for its effective performances in various fields. During the detection, enzyme in the biosensors can interact with the target molecule produces the biological signals. The produces signals reaches were converted into measurable form by the transducer [1, 2]. An enzyme biosensor is

* **Corresponding author A. Saravanan:** Department of Biotechnology, Rajalakshmi Engineering College, Chennai 602105, India; E-mail: saravanan.a@rajalakshmi.edu.in

a systematic device that combines a chemical with a transducer to create a standard corresponding to target analyte focus. The transducer changes over this flag into a quantifiable reaction, for example, present, potential, temperature change, or retention of light through electrochemical, warm, or optical means. This flag can be additionally increased, handled, or put away for later examination.

Most Sensors Comprise of Three Chief Segments

1. A receptor fit for perceiving the types of enthusiasm with a high level of selectivity. This is normally simultaneous with a coupling occasion between the receptor and an analyte;
2. A transducer, where the coupling occasion is converted into a quantifiable physical change. Cases could incorporate the age of electrons, protons, a change in conductivity or the age of an electrochemically dynamic concoction species, for example, hydrogen peroxide;
3. A strategy for estimating the change identified at the transducer and changing over this into valuable data.

Currently enzyme based biosensors gained many researchers' interest from various fields due to the utilization of enzyme as the recognition element in the sensor. Enzyme used in the biosensor shows remarkable selectivity towards specific analyte. For example, enzyme glucose oxidase will interact or bound only with glucose molecule among various sugars and produces the biological signals [3].

Numerous catalysts additionally show quick turnover rates and this is regularly fundamental to (a) maintain a strategic distance from immersion and (b) permit sufficient age of the dynamic species keeping in mind the end goal to be distinguishable. Sadly, there are likewise a few disservices. Though the catalyst gives the results within a short period of time, they may reduce the precision of the result and affects the enzyme activity. In some cases the amount spent for the catalyst will be higher. Immobilization of enzyme can increase or maintain the enzyme activity for longer periods [4, 5]. Inorganic examples, blood or salivation, there can likewise be solutes that are electrochemically dynamic and meddle with conclusions of the objective species. Then again, species might be available that predicament to the surface, causing fouling and loss of sensor reaction.

Enzyme - based sensors are more specific than cell-based sensors. Due to their shorter distribution behaviors, they react quickly. Glucose biosensor is the most commonly used enzyme based biosensor. Similarly chemical biosensor is a descriptive device which combines the chemical compound with a transducer and delivers the results like enzyme based biosensor. But they are too expensive than enzyme based biosensor [6]. Biosensors allow the investigation in complex natural media. The discovery of an enormous number of mixes is of colossal pertinence for logical research and for process control in the nourishment and compound industry [7].

Compound biosensors utilize the fondness and selectivity of chemically dynamic proteins, towards their objective atoms. Commonly, enzyme immobilization with the transducer gains more attention and shows good results for longer period which reduces the harmful effects caused by the various substrates during the detection process. Inhibitors, co-substrate and co-factors in the analyte mixture are the different substrates which affect or reduce the effectiveness of the biosensor. Contingent upon the measure write, two key classes of protein sensors can be recognized. To begin with, the protein identifies the nearness of a substrate or co-substrate/co-factor. This is at that point, by method for a transducer, used to screen the expansion of enzymatic action. A run of the mill illustration is a glucose biosensor [8]. The most widely recognized case of this approach is the recognition of organophosphate mixes utilized as pesticides or fighting nerve specialists. The method of flag transduction can be electrochemical, optical, full (acoustic), warm and so on.

There have been critical enhancements in the field of enzymatic biosensors; the utilization of new, hereditarily built compounds has taken into consideration enhanced execution attributes of current biosensors for the recognition of setting up analytes (glucose, pesticides and so forth). Headway has been the usage of hereditarily altered catalysts to distinguish novel markers. An extra gathering of enhancements is the use of "non-conventional" transducer materials, *e.g.* carbon nanotubes (CNT), or diverse conductive polymers. Wonderful basic and electrical property headways have empowered new alternatives for the most part in the territory of electrochemical detecting advancements.

The innovation of enzymatic biosensors offers a strong blend of execution and explanatory highlights not accessible in some other bioanalytical framework. The posting of only a couple of alternatives in this review can energize future advancement, which could yield new ages of enzymatic biosensors for an extensive variety of utilizations in clinical, natural or mechanical diagnostics [9].

WORKING OF BIOSENSORS

The favored organic material like chemical is ideal for ordinary techniques like physical or membrane based methods. The favored organic material is in contact with the transducer. When the analyte contact with the biological material or enzyme in the biosensor will produce biochemical signals can be estimated once they converted into measurable signals by the transducer. Working principle of the common biosensor was shown in Fig. (**1**).

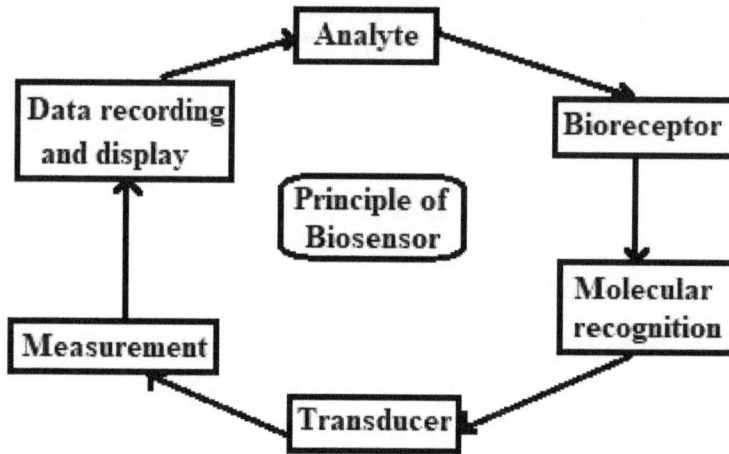

Fig. (1). Working principle of biosensor.

An effective biosensor must have at least a portion of the accompanying highlights:

a. Biosensor should be specific for the analyte.
b. The reaction utilized should be independent of different factors including pH, temperature, mixing and so on.
c. The response must be conventional or direct finished a helpful scope of analyte fixations.
d. The device should be small and biocompatible if it is to be exploited for investigations inside the body.
e. The device should be shabby, simple to utilize and equipped for rehashed utilize.

GENERAL FEATURES OF BIOSENSORS

A biosensor has two particular segments

1. Organic segment—protein, cell and so forth.
2. Physical part—transducer, speaker and so on.

The organic segment of the biosensor connects with the analyte to form bound analyte and produces biological signal that could be converted into measurable form by the transducer. Simultaneously, the biological material (enzyme) is properly immobilized on to the transducer and the immobilized enzyme based biosensors can be utilized over and again a few times (might be around 10,000 times) for a long period (numerous months). A diagrammatic representation of biosensor was shown in Fig. (2).

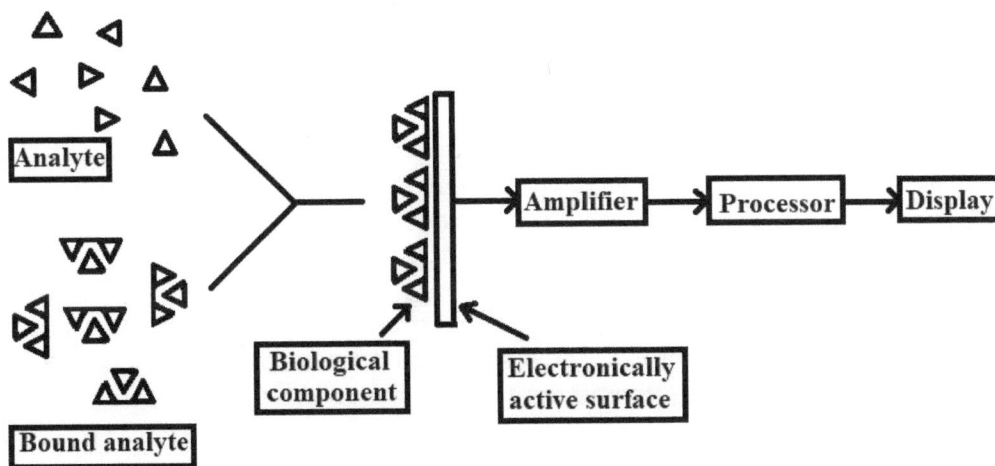

Fig. (2). A diagrammatic portrayal of a biosensor.

Enzyme in the biosensor can perform the following tasks;

i. It particularly perceives the analyte and
ii. It collaborates with the analyte in such a way and delivers the biological signals that could be recognised by the transducer.

The natural segment is reasonably immobilized on to the transducer. For the most part, the right immobilization of compounds upgrades their solidness. The organic segment communicates particularly to the analyte, which creates a physical change near the transducer surface. This physical change might be:

1. Warmth discharged or consumed by the response (calorimetric biosensors)
2. Generation of an electrical potential because of the changed circulation of electrons (potentiometric biosensors).
3. Production of electrons due to redox response (amperometric biosensors).
4. The light delivered or assimilated amid the response (optical biosensors).
5. Change in the mass of the organic part because of the response (acoustic wave biosensors).

BIOCOMPONENT

The particular acknowledgment of an analyte by an organic segment is basic to biosensor advancement. Most natural responses are portrayed by momentous specificity of the organic component towards an analyte or gathering of analytes. A few bio-receptors have been utilized as a part of biosensors including proteins, antibodies, receptors and DNA.

Biosensors might be characterized into:

i. Bio-partiality sensors that screen the official of an analyte with its profile receptor, for example, antibodies or DNA. For the detection process if antibodies are utilized as recognition element in the biosensors are named as immunosensors. If DNA is utilized for detection process that type of biosensors are called DNA sensors.

ii. Bio-reactant sensors are the sensors in which the biological materials like catalysts, microorganisms, or tissues are used as an organic segment. The analyte responds within the sight of the impetus to yield one or a few noticeable items. Because of the biosensor in light of enzymatic restraint, the association of the analyte with the catalyst is assessed.

Accordingly, biosensors change over into an electrical data stream from the synthetic data stream that includes the additional advances:

a. Diffusion of analyte at the surface of the biosensor for the reaction.
b. The analyte responds particularly and effectively with the natural segment of the biosensor.
c. This response alters the physico-concoction properties of the transducer surface.
d. This prompts an adjustment in the optical or electronic properties of the transducer surface.

PRINCIPLE OF A BIOSENSOR

The desired organic material (typically a particular catalyst) is immobilized by ordinary techniques. The immobilized biological material will attached with the transducer. Then the immobilized organic material will interact with the analyte to form a bound analyte, which thusly creates the electronic reaction that can be estimated. The basic principle of biosensor was presented in the Fig. (**3**).

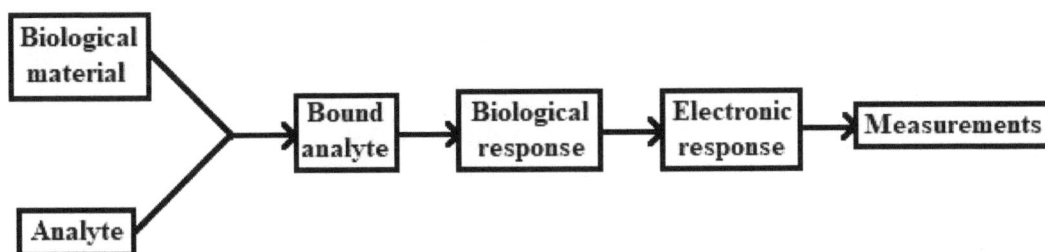

Fig. (3). Principle of Biosensor.

In a few cases, the analyte will be changed into different forms, which might be gas (oxygen), electron or hydrogen particles, *etc.* In that case the transducer and the recognition elements should be prepared respect to the analytes. Then only the analytes can be easily interacted by the organic materials and their signals were properly converted into a measurable form by the transducer.

Transducer Selection for Biosensors

The innovation utilized for transducer can be any of the four kinds recorded underneath and rely on the organic sensor utilized. In biosensors, reasonable transducers should be designed with the following features:

i. Specific wanted cooperation between the analyte and the organic components
ii. The expected utilization of the biosensors and the
iii. Manufacturing expense of the gadget.

BIO SENSING METHOD

The basic part of the biosensor is the coordination of the suitable organic and electronic segments to deliver an important flag amid the examination. Segregation of the natural part is particularly fundamental to guarantee that exclusive the particle of intrigue is bound or immobilized on the electronic segment or the transducer. The connection of the natural segment to the electronic segment is indispensable for the achievement of these gadgets [10]. The strength of the natural segment is likewise very basic since it is being utilized outside its typical organic condition.

Sorts of Biosensors

There are distinctive kinds of biosensors in light of the sensor gadgets and the organic materials and some of them are examined underneath.

Calorimetric Biosensors

Calorimetric biosensors measure the temperature change of the arrangement containing the analyte after the protein activity and decipher it regarding the analyte fixation in the arrangement. The analyte arrangement is passed through a bed section in which the immobilized proteins were arranged in a compressed manner; before this, the passage of analyte arrangement into the segment the temperature of the arrangement is estimated [11].

This is the most largely material sort of biosensor, and it can be utilized for turbid and firmly shaded arrangements. The best disservice is to keep up the temperature of the example stream, say $\pm 0.01°$ C, temperature. The affectability and the scope of such biosensors are very low for general applications. The affectability can be expanded by utilizing at least two catalysts of the pathway in the biosensor to connect a few responses to build the warmth yield.

Amperometric Biosensors

Amperometric biosensors are self-contained integrated devices based on the measurement of the current resulting from the oxidation or reduction of an electroactive biological element providing specific quantitative analytical information. The least difficult amperometric biosensors utilize the Clark oxygen cathode, which decides the diminishment of O_2 show in the example (analyte) arrangement [12]. These are the original biosensors. These biosensors are utilized to gauge redox responses, a run of the mill illustration being the assurance of glucose utilizing glucose oxidase.

A noteworthy issue of such biosensors is their reliance on the broke up O_2 focus in the analyte arrangement. This might be overwhelmed by utilizing arbiters; these particles exchange the electrons produced by the response straightforwardly to the anode instead of diminishing the O_2 broke up in analyte arrangement. These are additionally called second era biosensors. The present-day terminals, in any case, expel the electrons specifically from the lessened catalysts without the assistance of go-between and are covered with electrically directing natural salts.

Optical Biosensors

An optical sensor converts light rays into electronic signals. It measures the physical quantity of light and then translates it into a form that is readable by an instrument. An optical sensor is generally part of a larger system that integrates a source of light, a measuring device and the optical sensor. A most encouraging biosensor including iridescence utilizes firefly compound luciferase for the discovery of microscopic organisms in sustenance or clinical examples [13].

USES OF BIOSENSORS IN VARIOUS FIELDS

Biosensors are used in different fields for various aspects such as clinical and diagnostic applications, farming and food industries. From the Clinical and diagnostic perspective, they can be used to estimate the glucose concentration and lactic corrosiveness in the blood. In the food industries they could be used to measure the sugars, alcohols, and acid concentrations are economically estimated by using the biosensor. Among the distinctive biosensors utilized in ecological investigation, the main part is played by the hindrance-based biosensors. The rule of activity of these biosensors depends on the association that happens between particular compound and natural operators (inhibitors) show in the example, and the biocatalyst (a catalyst, a polyenzymatic succession and additionally even an entire tissue) immobilized on the biosensor itself [14].

The Biosensor in Light of Enzyme Inhibition

Various substances may cause a diminishment in the rate of a chemical catalyzed response (inhibitor). The level of hindrance is figured utilizing the condition.

$$\text{Percentage of inhibition} = \frac{A_o - A_i}{A_o}$$

A_o = the compound action before the presentation at the analyte.

A_i = the chemical movement after the introduction at the analyte.

In logical natural chemistry, scientists use *in vitro* tests to decide the impact of medications on catalysts to test their viability in single utilization and some cases in medicate blends. Enzymatic hindrance examines are utilized to take after pharmacokinetic changes that may influence medicate adequacy and security, particularly on the off chance that they have a low helpful file. Moreover, in silico strategy demonstrated an awesome intrigue for the advancement of new medications by anticipating the metabolic pathways of physiologically dynamic substances and their collaborations on an atomic level. The development of new scientific devices for observing medication focus and enzymatic hindrance rates is important to check their performance *in vitro* condition before the *in vivo* analyses, particularly those designed to test recently outlined medications. In like manner, quick and successful strategies for the assurance of medications are required in numerous territories, including clinical science, pharmaceutical, and nourishment.

Strategies in View of Enzyme Inhibition Biosensors

An element of intrigue, especially when performing routine examination (then again scarcely performed with the gadgets in this examined) is the cluster or persistent mode. In the meantime the cluster mode permits the test to dive into either the example or the framework under investigation, subsequently making estimations less demanding; in the training, the need for treatment previously bringing the test into the framework/test under contemplate makes ideal the utilization of a ceaseless approach. These frameworks empower in-line test pretreatment (specifically, weakening, pH modification, derivatizing responses, usage of persistent partition steps, and so forth) with a specific end goal to sufficient the example grid to biosensor prerequisites and simple usage of the recovery step.

A standout amongst the most unfortunate attributes of the compound hindrance based biosensors is the extraordinary restraint degree caused by various inhibitors. A case of this conduct is the amperometric corrosive phosphatase hindrance based biosensor, which indicates diverse restraint control towards organophosphorus and carbamates mixes. One of the courses for enhancing the selectivity of protein hindrance based biosensors is the utilization of a few proteins each covering a given terminal with ensuing chemometric treatment of the information gave by them. Artificially adjusted anodes have likewise been utilized with this point, or for upgrading affectability.

Strategies in Light of Enzyme Inhibition Biosensing Frameworks

Strategies in light of chemical hindrance can be produced either in persistent or spasmodic, bunch approaches. Every elective presents trademark focal points and hindrances, notwithstanding viewpoints straightforwardly related to the objective biochemical framework. More clump approaches are favoured when a couple of investigations required to be performed. In the intervening consistent methodologies bear the cost of for precise, schedule examinations, required more time for plan and development of the approach (if unique) or the higher expenses of the hardware.

Nonstop Strategies in View of Enzyme Inhibition

The nonstop biosensing frameworks in light of restraint make utilization of stream infusion (FI) most circumstances as this dynamic option enriches the strategy with awesome flexibility to actualize the fitting complex concurring to the highlights of the biochemical responses included.

At the point when strategies because of restraint have been connected to normal examples, similar to the case with the assurance of theophylline in blood serum because of its inhibitory impact on immobilized basic phosphatase computation of the focus of the analyte in the example depended on standard expansion, whose legitimacy relies upon the nearness of

Interferents, which have not been included in the obstruction, think about then this point is essential and frequently the creators did not consider the subjective nature of the hindrance reaction. The most reasonable utilization of inhibitory impacts is after a partition venture for legitimate expulsion of species, which can follow up on the action of the biocatalyst in pretty much expansion. Cases supporting this declaration are as per the following.

IMMOBILIZATION EFFECT ON BIOSENSOR RESPONSE

A few immobilization methods, for example, covalent official, physical entanglement, adsorption, cross-connecting, and embodiment have been accounted for in writing be that as it may, not a solitary strategy can be considered as a widespread technique for immobilization to accomplish the better biosensor reaction [15]. This is the most vital factor to be considered while planning a biosensor, to accomplish higher affectability, and useful solidness. The alleged immobilization impact adjusts the compound in the accompanying ways

  i. Substrate and catalyst bond will be weakens due to conformational change in chemical structure, caused by immobilization.
 ii. Immobilization may also bring changes in the separation balance of charged gatherings of the dynamic focus.
iii. Obvious dynamic constants tend to change by a non-uniform dispersion of substrate and or item between the chemical grid and the encompassing arrangement.
 iv. Immobilization in a multi-layer framework influences the response rate by making a diffusional obstruction amongst substrate and immobilized chemical.

PARAMETERS INFLUENCING THE EXECUTION OF ENZYMATIC BIOSENSORS

Impact of pH

The pH of the analyte mixture can influence the general enzymatic movement. Similar to every single regular protein, catalysts have a local tertiary structure that is delicate to pH. Therefore, denaturation of catalysts can happen at higher pH. Notably, the protein movement is profoundly pH subordinate also; the ideal pH for an enzymatic test must be decided exactly. It is best to pick a level area with the goal that the pH ought not to have any impact on protein action and won't meddle with the outcomes got relative to the hindrance of the protein by the inhibitor.

Impact of Substrate Concentration

The substrate focus can influence the level of hindrance. For the situation of an aggressive hindrance, at high substrate fixations, the hindrance impact is not seen since the substrate contends with the inhibitor.

ENZYME INHIBITION BASED BIOSENSOR

Glucose Oxidation Inhibition

Glucose oxidase is a discharged catalyst delivered prevalently by the organisms Aspergillus and Penicillium species. It catalyzes the oxidation of the sugar β-D-glucose to shape D-glucono-1,5-lactone and hydrogen peroxide:

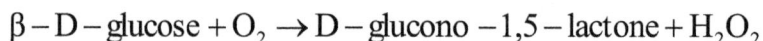

$$\beta - D - glucose + O_2 \rightarrow D - glucono - 1,5 - lactone + H_2O_2$$

By utilizing different substrates including atomic oxygen, quinines and an electron acceptor as analyte, glucose oxidase converts D-glucose into D-glucono-1, 5-lactone and H_2O_2 through an oxidation process. Then the produced D-glucono-1,5-lactone able to hydrolyze quickly to supply gluconic corrosive. Glucose oxidase is a protein fragment made up of two different monomers, where each monomer folds into two spaces. One space is interacting with the substrate and D-glucose and another space interacts non-covalently to a cofactor and Flavin adenine dinucleotide (FAD), which it utilizes as a ground-breaking oxidizing operator. The trend is a typical part of natural oxidation-diminishment (redox) responses, in which there is a pick-up or loss of electrons from a particle. In glucose oxidase, FAD acts as an electron acceptor. Once FAD accepts electron they can be reduced to $FADH_2$. There are three different amino acids such as His516, Glu412 and His559 associated along with catalysis at the dynamic site of glucose oxidase. Different substances, including sub-atomic oxygen, which instigates the interpretation of the catalyst, can incite the blend of glucose oxidase [16, 17].

Tyrosinase Inhibition

An amperometric biosensor for the assurance of hydrazine mixes, in view of their inhibitory impact on the movement of immobilized tyrosinase, has been produced.

The hydrazine-tyrosinase cooperation can be displayed as a reversible focused restraint. Motor parameters have been resolved for different hydrazine mixes. The tyrosinase-based carbon glue anode offers delicate estimations to the low-micromolar level and great accuracy [18]. Hydrazine mixes are broadly utilized as booster (in rocket drive frameworks), erosion inhibitors, impetuses and emulsifiers. Field estimations are favored, as they bear the cost of the choice of a fast cautioning capacity while maintaining a strategic distance from the mistakes, deferrals, and cost of research center based estimations [19]. Different artificially modified cathodes, in light of various electrocatalytic moieties, have consequently been produced for the discovery of various analogous biosensing plans for estimating hydrazines that have not been created, regardless of the considerable guarantee of biosensor innovation for on-location examinations.

The new method depends on the inhibitory activity that different hydrazines apply to the action of the compound tyrosinase. Inhibitor biosensors have been broadly utilized for observing lethal substances, for example, organophosphate pesticides, cyanides or overwhelming metals. The tweaked biocatalytic action accumulated from the inhibitor-compound connection has been checked from the diminished current reaction to the relating substrate. Likewise, we have found that tyrosinase is inactivated by hydrazine mixes and have misused this conduct for creating valuable hydrazine sensors.

Amperometric hydrazine recognition was depends on the biocatalytic action of tyrosinase within the sight of its substrate and inhibitor. Mushroom tyrosinase is a copper-containing monooxygenase that proselytes phenolic mixes by comparing quinones to the detriment of oxygen.

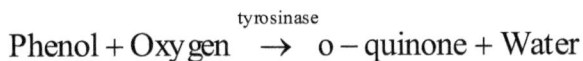

$$\text{Phenol} + \text{Oxygen} \xrightarrow{\text{tyrosinase}} \text{o} - \text{quinone} + \text{Water}$$

The freed quinone species can be advantageously identified by a low-potential diminishment. The expansion of hydrazine causes hindrance to the chemical, subsequently diminishing the measure of freed quinone. Tyrosinase offers a few preferences for such applications. It is very steady, can be utilized as a part of both watery and nonaqueous situations, and permits recognition at low possibilities where potential obstructions are insignificant. Overall, it has been outlined out of the blue that hydrazine biosensors can be composed in light of the bothered bioactivity of tyrosinase. Relevance to hydrazine estimations in nonaqueous conditions is additionally foreseen in view of the natural stage movement of the chemical. Tyrosinase movement in disgraceful situations may lead to the discovery of hydrazines in air. Such observing of hydrazine vapors ought to be greatly helpful for aviation applications. It is understood that tyrosinase is defenseless to an assortment of inhibitors. The little size of hydrazine mixes may consequently be favorable regarding distinctive size rejection films. On the other hand, it might be conceivable to utilize a multi-analyte chemometric approach in association with a chemical cluster biosensor.

A few chemicals, which are hindered to shifting degrees by diverse poisonous analytes, would thus be able to be utilized to give a particular unique mark example and blend investigation.

Urease Inhibition

Chelating operators (sequestrants) are critical sustenance added substances as they help to keep up the trustworthiness of numerous nourishment items. They work with cell reinforcements to anticipate oxidation in fats and oils.

It was indicated as of late that mixes of electrochemical sensors (*e.g.*, ISFET) with different natural operators frame specific and touchy biosensors, which can be utilized either for biomedical or nourishment investigation. A biosensor framework for estimating EDTA, where chelating properties of the sequestrant is utilized to re-establish reactant action of urease, beforehand restrained with copper particles

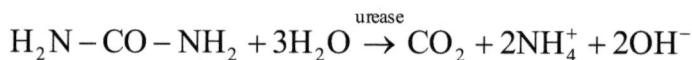

$$H_2N - CO - NH_2 + 3H_2O \xrightarrow{\text{urease}} CO_2 + 2NH_4^+ + 2OH^-$$

Since sulfhydryl bunches seem, by all accounts, to be associated with the dynamic site of a few compounds (urease, for instance), restraint of these catalysts by mixes which respond with sulfhydryl bunches has been revealed. Despite the fact that it is far-fetched that few sorts of sequestrant could be intentionally stored in similar nourishment or refreshment, there is a probability to discover a few chelating operators in the test. Therefore, we examined the impact of sequestrant blend on the rebuilding of compound movement. Particle delicate field impact transistor-based biosensor with urease immobilized into a decidedly charged polymeric film can be utilized for the assurance of EDTA particles in sustenance and refreshment tests.

Urease has been the most ordinarily utilized compound for the assurance of metal particles. It is steady furthermore, reasonable, and can be restrained by numerous metals, for example, mercury, copper, silver, cadmium, lead, nickel, cobalt, manganese, zinc and chromium. Notwithstanding their high affectability and selectivity towards substantial metals, urease restraint examines have barely been utilized for harmfulness evaluation and distinguishing proof because of issues with estimating urease hindrance in complex natural tests [20].

Though the immobilized enzyme based biosensor gives effective results, the immobilization techniques must be improved to boost protein soundness and affectability, alongside its fondness for both the substrate and inhibitor. Different factors, for example, the thickness of the compound layer, the chemical stacking and the conditions for ideal compound action should likewise be considered when immobilization is performed.

Cholinesterase Inhibition

Enzymatic strategies in light of the restraint of cholinesterase action have as of late been the protest of concentrated examination because of their affectability and specificity. The simplicity of execution of the measure and the ease of the gear make such strategies very appealing for research centers [21]. Protein restraint tests are of the incredible enthusiasm for natural contamination investigation because the poisonous quality of toxins, for example, pesticides is the aftereffect of *in vivo* catalyst restraint and organic tests including chemical hindrance utilized for the assurance of ecological poisons appear to be more solid than physical (*e.g.* chromatographic) strategies. Pesticide or any other substrates that can bind with cholinesterase and inhibits its action are commonly known as cholinesterase inhibitors. It also called as anticholinesterase agent. There are two types of pesticides such as organophosphates (OPs) and the carbamates (CMs) and chlorinated derivatives of nicotine are the major anticholinesterase agent or cholinesterase inhibitors.

Cholinesterase can be restrained pretty much particularly relying upon the wellspring of the compound. The estimation of the flag, corresponding to the focus of substrate or result of the enzymatic response, acquired after some hatching time when the chemical is in contact with pesticide relying upon the sort of pesticide and its focus, is completed after the clear time or the rate of the substrate or item changes amid the underlying time of enzymatic response is observed [22]. The estimation is frequently performed in non-stream conditions, be that as it may, as of late likewise stream infusion technique is increasingly regularly used.

The restraint of cholinesterase can be reversible (carbamate pesticides) or irreversible (organophosphorus pesticides). The proposed system of irreversible restraint incorporates the association of organophosphorus pesticide with serine hydroxyl gatherings of the chemical protein. In the field of biosensors in view of the compound hindrance, the highest quality level biocomponent is the cholinesterase catalyst.

The reason can be clarified considering diverse elements:

  i. Cholinesterase is portrayed by high turnover.
 ii. It is hindered by a few mixes, for example, organophosphorus pesticides and nerve operators, and a quick and in situ identification for these analytes is extremely valuable.
iii. Inhibitors of cholinesterase, for example, pesticides are generally disseminated in nature.
 iv. The substrate is dissolvable in the fluid arrangement and isn't so costly;

ASSURANCE OF PESTICIDES

The assurance of hints of toxins in natural materials and ecological tests (common water and air) has turned out to be progressively vital. One class of these toxins are organophosphorus and carbamate pesticides, generally utilized as a part of farming. They demonstrate an ecological steadiness lower than the organochlorine mixes yet have a higher intense poisonous quality which can be a major issue for the balance of oceanic biological systems. Another issue is nourishment sullying which could seriously affect human wellbeing. A high intense poisonous quality of these mixes makes the requirement for quick reacting recognition frameworks with a specific end goal to secure human wellbeing amid assembling and application. Pesticides hinder the activity of acetylcholinesterase by hydrolysing the neurotransmitter which creates acidic corrosive and choline, to restore the underlying condition in postsynaptic neuromuscular junction. At the point when this protein is repressed, nerve driving forces are upset since acetylcholine stays introduce in the synaptic locale.

In the present survey we might want to support the ongoing advances in biosensors because of chemical hindrance field, concentrating on:

• The examination of another hypothetical approach with a specific end goal to effortlessly comprehend the kind of hindrance and compute the motor parameters.
• The assessment of the exhibitions of the biosensor in light of catalyst hindrance for the situation of a reversible and irreversible hindrance, as far as time of examination, discovery constrain, framework impact.
• The utilization of nanomaterials to enhance the investigative exhibitions of the biosensors.
• The advancement of biosensors in light of catalyst restraint implanted in labs on a chip.

- The uses of biosensors in light of catalyst restraint in clinical, nourishment and natural examples.

CONCLUSION

The effectiveness of enzyme based biosensor is mainly depends on the compounds which interact with the transducer. The effective of the enzyme based biosensor is higher when enzyme immobilized with the transducer. A multitude of research endeavors was centered on the advancement of these biosensors for the location of their substrates, for example, glucose oxidase biosensor for glucose monitoring. The rule of this sort of biosensors depends on the measurement of the inhibitor, estimating the enzymatic movement in nonappearance and nearness of the inhibitor. Biosensors in light of compound hindrance are dependable instruments for the location of a considerable measure of lethal mixes. The recent development in enzyme based biosensors have been summarised in this chapter with the specific descriptive examples. The working principle and application of various enzyme based biosensors in different domains have been explicated in this chapter.

CONSENT FOR PUBLICATION

Not applicable.

CONFLICT OF INTEREST

The author(s) confirms that there is no conflict of interest.

ACKNOWLEDGEMENTS

Authors would like to thank the chairperson Dr. Mrs. Thangam Meganathan and Principal Dr. S. N. Murugesan, Rajalakshmi Engineering College for their support and encouragement.

REFERENCES

[1]   Upadhyay LSB, Verma N. Enzyme inhibition based biosensors: A Review. Anal Lett 2013; 46: 225-41.
[http://dx.doi.org/10.1080/00032719.2012.713069]

[2]   Luque de Castro MD, Herrera MC. Enzyme inhibition-based biosensors and biosensing systems: questionable analytical devices. Biosens Bioelectron 2003; 18(2-3): 279-94.
[http://dx.doi.org/10.1016/S0956-5663(02)00175-6] [PMID: 12485775]

[3]   Turner A, Karube I, Wilson GS. Biosensors: fundamentals and applications. Oxford University Press 1987.

[4]   Kobayashi T, Laidler KJ. Theory of the kinetics of reactions catalyzed by enzymes attached to membranes. Biotechnol Bioeng 1974; 16(1): 77-97.
[http://dx.doi.org/10.1002/bit.260160107] [PMID: 4813165]

[5]   Kuralay F, Özyörük H, Yıldız A. Inhibitive determination of Hg2+ ion by an amperometric urea biosensor using poly (vinylferrocenium) film. Enzyme Microb Technol 2007; 40(5): 1156-9.
        [http://dx.doi.org/10.1016/j.enzmictec.2006.08.025]

[6]   Wang Z, Liu S, Wu P, Cai C. Detection of glucose based on direct electron transfer reaction of glucose oxidase immobilized on highly ordered polyaniline nanotubes. Anal Chem 2009; 81(4): 1638-45.
        [http://dx.doi.org/10.1021/ac802421h] [PMID: 19170516]

[7]   Wang J. Glucose biosensors: 40 years of advances and challenges. Electroanalysis. An International Journal Devoted to Fundamental and Practical Aspects of Electroanalysis 2001; 13(12): 983-8.
        [http://dx.doi.org/10.1002/1521-4109(200108)13:12<983::AID-ELAN983>3.0.CO;2-#]

[8]   Newman JD, Warner PJ, Turner AP, Tigwell LJ. Biosensors: a clearer view. Biosensors 2004 – The 8th World Congress on Biosensors, Elsevier, New York.

[9]   Turner AP. Advances in Biosensors, I; II; Suppl I; III. : 1995; 1992: p. (1993)991.

[10]   Amine A, Arduini F, Moscone D, Palleschi G. Recent advances in biosensors based on enzyme inhibition. Biosens Bioelectron 2016; 76: 180-94.
        [http://dx.doi.org/10.1016/j.bios.2015.07.010] [PMID: 26227311]

[11]   Danielsson B. Calorimetric biosensors. J Biotechnol 1990; 15(3): 187-200.
        [http://dx.doi.org/10.1016/0168-1656(90)90026-8] [PMID: 1366673]

[12]   Arslan F, Ustabaş S, Arslan H. An amperometric biosensor for glucose determination prepared from glucose oxidase immobilized in polyaniline-polyvinylsulfonate film. Sensors (Basel) 2011; 11(8): 8152-63.
        [http://dx.doi.org/10.3390/s110808152] [PMID: 22164068]

[13]   Borisov SM, Wolfbeis OS. Optical biosensors. Chem Rev 2008; 108(2): 423-61.
        [http://dx.doi.org/10.1021/cr068105t] [PMID: 18229952]

[14]   Mehrotra P. Biosensors and their applications - A review. J Oral Biol Craniofac Res 2016; 6(2): 153-9.
        [http://dx.doi.org/10.1016/j.jobcr.2015.12.002] [PMID: 27195214]

[15]   Deng S, Jian G, Lei J, Hu Z, Ju H. A glucose biosensor based on direct electrochemistry of glucose oxidase immobilized on nitrogen-doped carbon nanotubes. Biosens Bioelectron 2009; 25(2): 373-7.
        [http://dx.doi.org/10.1016/j.bios.2009.07.016] [PMID: 19683424]

[16]   Guascito MR, Malitesta C, Mazzotta E, Turco A. Inhibitive determination of metal ions by an amperometric glucose oxidase biosensor: Study of the effect of hydrogen peroxide decomposition. Sens Actuators B Chem 2008; 131(2): 394-402.
        [http://dx.doi.org/10.1016/j.snb.2007.11.049]

[17]   Samphao A, Rerkchai H, Jitcharoen J, Nacapricha D, Kalcher K. Indirect determination of mercury by inhibition of glucose oxidase immobilized on a carbon paste electrode. Int J Electrochem Sci 2012; 7: 1001-10.

[18]   Wang X, Chen L, Xia S, Zhu Z, *et al.* Tyrosinase biosensor based on interdigitated electrodes for herbicides determination. Int J Electrochem Sci 2006; 1: 55-61.

[19]   Campanella L, Dragone R, Lelo D, Martini E, Tomassetti M. Tyrosinase inhibition organic phase biosensor for triazinic and benzotriazinic pesticide analysis (part two). Anal Bioanal Chem 2006; 384(4): 915-21.
        [http://dx.doi.org/10.1007/s00216-005-0175-6] [PMID: 16328240]

[20]   Domínguez-Renedo O, Alonso-Lomillo MA, Ferreira-Gonçalves L, Arcos-Martínez MJ. Development of urease based amperometric biosensors for the inhibitive determination of Hg (II). Talanta 2009; 79(5): 1306-10.
        [http://dx.doi.org/10.1016/j.talanta.2009.05.043] [PMID: 19635364]

[21]   Babkina SS, Medyantseva EP, Budnikov HC, Vinter VG. Enzyme amperometric sensor for the determination of cholinesterase inhibitors or activators. Anal Chim Acta 1993; 273(1-2): 419-24.

[http://dx.doi.org/10.1016/0003-2670(93)80185-N]

[22]   Botrè F, Lorenti G, Mazzei F, Simonetti G, *et al.* Cholinesterase based bioreactor for determination of pesticides. Sens Actuators B Chem 1994; 19(1-3): 689-93.
[http://dx.doi.org/10.1016/0925-4005(93)01131-M]

RNA Silencing in Enzyme Inhibition and its Role in Crop Improvement

S. Justin Packia Jacob[*]

Department of Biotechnology, St. Joseph's College of Engineering, Chennai, India

Abstract: Crop improvement represents the genetic alteration of plants to gratify human needs. Crop improvement, the art of engineering plants for the benefit of humankind, is as old as agriculture itself. Even though crop improvement programs focus on the development of novel crop varieties with enhanced quality and tolerance to ecological stresses (both biotic and abiotic) and making crops able to give more yield exhibiting good quality. Still, we cannot rely on the crops due to various reasons, like the irritating nature of onion, low lysine and threonine content in Maize, the presence of toxic Gossypol in cotton, *etc.* These irritating qualities of the crop must be reduced or removed by genetically modifying the crop plants, and then only it will be accepted for human consumption. Expression of *antisense* genes and the related gene-silencing technique has been exploited as an applied *technique in plant* biotechnology for creating "metabolic engineered *plants*" in which the endogenous target gene, which is responsible for the unpleasant character is specifically suppressed. The *antisense and its related technology* are used for various purposes such as silencing or ablating undesired genes. In the present chapter, the down-regulation of the enzyme using the gene-silencing technique used in various crop varieties are discussed in detail.

Keywords: Antisense, Crop, Down-regulation, Enzyme, Flavr-Savr, Gossypol, Improvement, Inhibition, Polygalacturonase, RNA-silencing, Tearless-onion.

INTRODUCTION

The global population was estimated to have reached 7.6 billion people as of May 2018 and it will be increased by more than 2 billion, till 2025. Hence to meet the worldwide demand for nourishment, the production of improved crops is required, particularly cereals, as they serve as the chief source of food for most of the human populace [1]. Antisense RNA is a recent technology and getting hold of popularity in the agricultural field. Genetic improvement of crops can be provided by RNAi (RNA interference) technology as it has established as a powerful appr-

---

[*] **Correspondence S. Justin Packia Jacob:** Professor, Department of Biotechnology, St. Joseph's College of Engineering, Chennai, Tamil Nadu, India; Email - drjpjacob@gmail.com

G. Baskar, K. Sathish Kumar & K. Tamilarasan (Eds.)

oach to improve the traits in crops by downregulating (silencing) genes. Antisense RNA, RNAi, and other related pathways retort to exogenous and endogenous genetic makeup (nucleic acids) along with basic cellular processes. Antisense RNA technology is an instrument for regulating the expression of the gene in the higher organisms and using this tool one can selectively inhibit the enzymes involved in a particular pathway. Thus, there is an enormous potential of Antisense RNA technology towards crop improvement and to encounter the agrarian-based demand of the growing population of our nation.

Antisense System in Nature

Naturally, arising antisense RNA was involved in the regulation of gene and this was confirmed during the study of E. coli ColE1 plasmid replication. E.coli based plasmid ColE1 replication includes the formation of an RNA primer, which is processed by RNase-H while bound to the DNA template. Inhibition of the processing of RNA primer and replication of the plasmid occurs by the binding of Antisense RNA with the primer; hence the plasmid copy number possibly will be regulated [2]. Replication of *Staphylococcus aureus* plasmid (pT 181) and copy number also appeared to be controlled by antisense RNA [3]. Antisense mRNA also inhibits E. coli Tn10 transposase mRNA translation.

Antisense control called an additional mechanism of translational control that occurs in bacterial cells. This form of regulation will be facilitated via antisense RNA since it comprises the sequence complementary to the sequence of a sense strand (mRNA). The initiation of translation is prevented by the pairing of the antisense RNA with the mRNA, which prevents the recognition of initiation codon and thereby 30S ribosomal subunit binding with the Shine Dalgarno sequence is prevented.

Transposase expression is encoded by the bacterial insertion sequence IS 10 is regulated by the antisense translation-control mechanism. The transposition of this mobile DNA element is catalyzed by Transposase. If the expression of transposase occurs considerably, as a result, abundant mutations may result from IS 10 transposition due to that the host cell may not be able to survive.

In general, this does not happen due to the occurrence of the antisense control, in which the IS 10 comprises two promoters: one called $P_{IN,}$ which directs transcription of the strand coding for transposase; the other called P_{OUT} lies within the transposase gene and directs transcription of the noncoding strand, producing an antisense RNA complementary to the 5' end of transposase mRNA. Since P_{OUT} is promoter a stronger than a P_{IN}, therefore that antisense mRNA is produced in greater abundance than transposase mRNA.Antisense RNA hybridization to most of the much rarer transposase mRNA prevents translation, thereby assuring that the rate of synthesis of transposase and, in turn, the frequency of transposition, is compatible with the survival of the host cells.

Antisense RNAs available naturally in plants are found to have a regulatory effect on gene expression. These include antisense RNA transcripts for the K-amylase mRNA of barley [4], antisense mRNA complementary to niv gene codes for the enzyme chalcone synthase (CHS) of the flavonoid pathway [5]. Two antisense transcripts were identified in barley [4] both were incorrectly complementary to the K-amylase gene whereas, in the case of niv gene, antisense transcripts occur due to an inverted duplication of un-translated leader sequences.

Therefore, a tentative mechanism has been proposed for the generation of antisense transcripts. Antisense RNA emerges when transcription of a gene proceeds in the sequence opposite to template in the absence of a strong termination site for transcription in the short intergenic region. In Brassica, self-incompatibility was controlled by antisense transcripts for the S locus receptor kinase gene [6]. There are different types of gene silencing mechanisms such as Ribozymes, RNAi technology, Antisense RNA technology, *etc.* let us discuss the role of these gene silencing mechanisms in the crop improvement program.

Ribozymes

A ribozyme is a catalytic RNA (RNA enzymes) which was first explained in *Tetrahymena thermophilic* in the early 1980s [7, 8]. The RNA processing competences of these ribozymes were subjugated as a possible antisense agent. Uhlenbeck [9] and Haseloff and Gerlach [10] first isolated this ribozyme from a viroid RNA. An excellent discussion about the ribozymes of hammerhead was given by Kurreck [11] and the mechanism of action of different ribozymes is elucidated by Doudna and Cech [12].

Antisense RNA

Antisense RNA (asRNA) [13] or oligonucleotide [14] is a single RNA strand that is complementary to the mRNA (codes for a protein) onto which the asRNA pairs and form the duplex structure, as a result, prevent its translation. Naturally occurring as RNAs were being identified in prokaryotic and eukaryotic organisms [13] antisense RNA can be classified into two types like shorter ones (less than 200 nucleotides) and longer ones (greater than 200 nucleotides) noncoding RNAs [15]. The prime function of asRNA is regulating the expression of a gene and found to have widespread usage as a research tool for gene down-regulation mechanism.

If a cloned gene is engineered in a cell, it will yield an antisense RNA strand that has a complementary sequence for the normal mRNA. When Antisense RNA is produced in enormous quantity, it will frequently be hybridized with the mRNA strand made by the normal genes and thus inhibit the expression of the corresponding protein. Synthesis of a short antisense nucleic acid strand by the chemical or enzymatic manner and then delivering this into the cells is a related technique, which will block the expression of the corresponding protein.

As RNA is a sequence having a single RNA strand, it is complementary to the RNA transcript (mRNA) transcribed within a cell. By gene silencing the activity of a targeted gene can be switched off, so it is imaginable to elucidate the exact function of that specific gene. To develop a particular novel transgenic crop variety, asRNA is introduced into the cell to prevent the translation event by base-pairing with the mRNA and activating the RNase H. By base pairing with the gene of interest, the respective enzyme (protein) can be inhibited and thereby the role of some unnecessary enzymes involved in the poor quality of food or yield can be eliminated to enhance the quality or yield of the crop.

Antisense Mechanism

The ultimate aim of presenting an antisense RNA into the cell by laboratory-based (*in-vitro*) or animal model (*in-vivo*) is to suppress or inhibit the expression of a gene into a protein/enzyme, which is responsible for the low quality or low yield in crop plants or diseased nature in animals. The antisense RNA effect is delivered through various mechanisms. This mechanism is accomplished in 3 stages in which the entire process occurs in a normal way and the message from the DNA strand gets transcribed into premature mRNA to mature mRNA which will be converted into a polypeptide sequence. In the first stage, the sense sequence of the DNA gets converted into the pre-mRNA. In the second stage the pre mRNA will be transformed into a mature RNA transcript (mRNA) through three important post-transcriptional events, these are the addition of a five-prime cap, removal of the intron (splicing), and polyadenylation. In the third stage, the mRNA is translated into an appropriate polypeptide.

Achieving antisense knockdown or knockout, the transcription stage is the first target, in that an antisense strand is targeted to the sense strand on the DNA and avoids the transcription of the target gene. Dagle and Weeks explained the antisense effect at this stage [16], this effect can be manifested in three different ways. This effect will be taken place by the polyamides (contain repeating amide) which binds on minor groove, displacing the strand by using peptide nucleic acids

(PNAs), and triplex-forming oligonucleotides which bind in the major groove. The pyrrole-imidazole polymers which are considered as a minor groove binding agent, the pairing of imidazole and pyrrole amino acid with the nucleotide sequence in the minor groove region of DNA are achieved in a sequence-specific way [17, 18]. Target specific pairing seems to limits short stretches of DNA, usually less than 7bp. PNAs are considerably longer ones and its mode of action is to pair with the complementary sequence of the DNA helical structure and relocate the real compliment. This procedure is supported by the fact that PNA: DNA hybrids are more stable and thermodynamically allowed than DNA: DNA hybrids. The third technique, which involves the pairing of triplex-forming oligonucleotides (TFO) with the major groove, which comprises a long stretch of DNA, these TFOs generate a stable triple helical structure of DNA. Till now, two triplex-forming motifs have been proved as efficacious. Which encompass pairing of the TFO with the purine base in a poly-purine: a poly-pyrimidine stretch of duplex DNA [16]. Although TFO was shown to have effectively arresting transcription both lab-based and animal models, the circumstances for forming stable triplexes are challenging. The target DNA strand is Watson-Crick bonded and the TFOs interact with the duplex DNA by Hoogsteen interaction; viz., triplets like T-A: T and C+-G: C. This approach demands that only pyrimidine-purine dsDNA must be targeted and the protonation of cytosine residue in the TFO is required. Protonation of cytosine is required for maintaining the acidic condition in the test. The involvement of locked nucleic acids (LNAs) may ease some of these difficulties encountered, however, it was [19] reported that triplex formation at physiologic pH will be stabilized by LNA-containing TFOs. It was observed that 7 LNAs with a size of 15-mer oligonucleotide induced the mechanism using animal models. Every antisense oligonucleotide strands are capable of forming a duplex (Fig. 1) with a mature form of mRNA resulted in the development of either RNA: RNA duplex or RNA: DNA duplex depends on the nature of oligonucleotide strand involved.

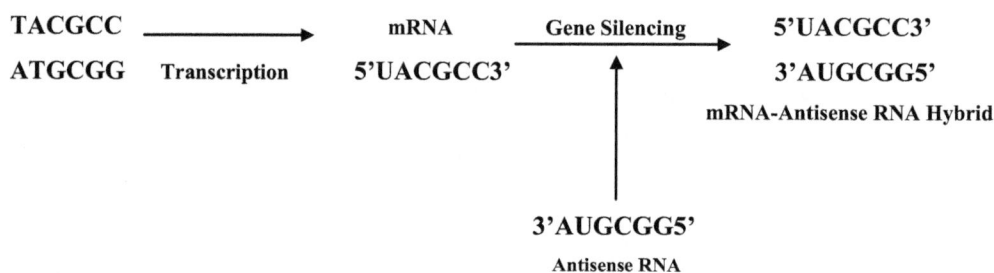

Fig. (1). Formation of mRNA/antisense RNA hybrid prevents the synthesis of the protein product from the mRNA.

In mRNAs, it seems to have an active translation mechanism in the cell that regularly involved in RNA: RNA duplex formation that naturally occurs [16]. The short stretch of RNA oligonucleotide may not be constant in the existence of helicases within the ribosomal complex and lengthy RNA oligonucleotides are involved in the activation of the RNA interference pathway. The sole exemption thus authenticated is the usage of morpholino oligonucleotide. These oligonucleotide modifications, which contain altered internucleoside linkages. Morpholino oligonucleotides are revealed to specifically lessening translation of the gene when it is situated closer to the 5'termini of the mRNA [20, 21].

Degradation of mRNA by RNAse H is the most validated and used antisense mechanism. It is an enzyme secreted by the organism (endogenous), which selectively cleaves the RNA strand of RNA: DNA duplexes [22 - 24]. The enzyme RNase-H is found within the nucleus and the cytoplasmic content of all cells and its usual function is displayed during the replication of DNA, it simply eliminates RNA primers from the Okazaki fragment. Due to the usual function of RNaseH, the oligonucleotides that will provoke an intentional and precise RNaseH response must be cautiously assembled. A chimeric oligonucleotide with a middle portion comprising DNA, either with or without phosphorothioate alterations, and nuclease resistant 5' and 3' flanking strands, usually 2'-O-methyl RNA but a broad range of 2' modification has been used [25].

Activating the antisense activity by RNase-H is a proven tool in evaluating gene functions, but is evolving as a method of choice for antisense therapeutics as well. There are fifteen different types of antisense oligonucleotides listed by Kurreck [11], which are either sanctioned or in the clinical trial phase for usage against diseases extending from asthma to cancer.

It was believed that naturally available as RNAs are known to regulate gene expression in plants also. These include antisense RNA transcripts against barley K-amylase mRNA [4], antisense mRNA complementary to niv gene encoding for the enzyme chalcone synthase (CHS) of the flavonoid pathway [5]. Rogers identified two antisense transcripts in barley, both were imperfectly complementary to the K-amylase gene whereas, in the case of NIV gene, antisense transcripts arose due to an inverted duplication of un-translated leader sequences.

Therefore, a tentative mechanism has been proposed for the generation of antisense transcripts. As RNA emerges when transcription of a gene sequence proceeds in the strand opposite to template in the nonappearance of a strong transcription termination site in the short intergenic sequence. In Brassica, the antisense transcripts are being identified for the S locus receptor kinase gene, which is responsible for the control of self-incompatibility [6].

In the present chapter, the role of Antisense RNA and its related mechanism in enzyme inhibition, which is being used in the crop improvement process of various economically valuable crop plants, is discussed in detail.

Delayed Fruit Ripening in Tomato by Silencing Polygalacturonase (PG) Enzyme using Antisense RNA

The tomato (*Solanum Lycopersicum*) is the palatable, red-colored, berry of the family Solanaceae, called as a tomato plant. Tomatoes are being modified using genetic engineering techniques and the first commercially available genetically modified food was a variety of tomato named the Flavr Savr; it was engineered to have a longer shelf life [26]. In tomato, the expression of Polygalacturonase (PG) enzyme occurs only at the ripening stage of the fruit. This enzyme is abundantly available during the later stage of fruit ripening and has a vital role in cell wall decay and softening of the fruit.

This cell wall degradation property reduces the shelf life of tomato, henceforth a full-length reverse oriented cDNA for PG enzyme was produced in transformed tomato plants to produce asRNA using a gene construct possessing the 35S promoter from CaMV (cauliflower mosaic virus). The *Agrobacterium*-Ti plasmid-mediated transformation was performed to integrate the gene of interest into the genome of tomato. The considerable level of decrease in the mRNA for PG gene and enzymatic activity in the fruit ripening was observed due to the constitutive production of PG genes asRNA in the transgenic plant. The levels of asRNA for the PG gene in green fruits of transgenic plants were lesser than the level of PG mRNA usually achieved during the ripening process. Though, transcriptional analysis of isolated nuclei established that the transcription of antisense RNA construct was at a higher degree than the normal gene for PGenzyme in tomato. Analysis of fruits from the transgenic plants also confirmed a great reduction in PG mRNA and around 70-90% enzymatic activity was also observed [27]. The reduction in the activity of the PG enzyme did not thwart the build-up of the red pigment lycopene, which is necessary for human health.

The suppression of Polygalacturonase in tomato fruit by fruit-specific RNAi-mediated method increases lycopene and β-carotene content in the fruit. The pathway of phytoene synthesis begins from two molecules of geranyl diphosphate (GGPP) in the isoprenoid pathway. Dietary carotenoids essentially require for humans where β-carotene is the dietetic precursor of Vit A, insufficiency of β-carotene leads to blindness, premature death, hence, an important step in the biosynthesis of ABA, for inhibiting 9-cis epoxycarotenoid dioxygenase (NCED), it was targeted through RNAi in the tomato fruits.

RNA Interference (RNAi)

The RNA interference (RNAi) effect was well explained in *Caenorhabditis elegans* [28]. Fire, *et al.* discovered a highly specific degradation of targeted RNAs in *C. elegans* cells by the introduction of long dsRNAs. This phenomenon was observed similar to what was termed as post-translational downregulation of gene in plants and suppressing in *Neurospora crassa* [29 - 31]. RNAi is a potentially powerful antisense tool and considered an olden eukaryotic cellular protection mechanism. Because of this curiosity, enormous progress has achieved in understanding RNA interference and exploiting this technique in antisense research for the improvement of various crop varieties.

Lysine and Threonine Increase in Maize by Silencing LKR/SDH Enzyme

The nutritional superiority of crop varieties is determined by screening the availability of essential amino acids existing in food for human beings and monogastric animal feed. Some important essential amino acids are deficient in plant proteins, which are essential in human and animal food. Normally, cereals grains, including sorghum, wheat, maize, *etc.* have an inadequate level of lysine and threonine. For livestock growth, lysine is regularly supplemented with the regular diet. Lysine and other few other essential amino acids are lacking in major crops like potato and oilseed rapeseed [32]. To increase the relative essential amino acid content chiefly lysine and threonine in crop plants, numerous approaches are being used to increase its yield.

Maize is considered as a chief source of food and feedstuff but inadequate in its nutritious value due to low lysine content. In plants, lysine, along with other amino acids like threonine, methionine, and isoleucine are produced from aspartate an amino acid. In plant lysine biosynthesis enzymes like Aspartate kinases (AK) and Dihydrodipicolinate synthases (DHDPS) are considered as the key regulators. Alternatively, high lysine-rich plants are engineered by reducing catabolism where the saccharopine pathway for lysine degradation. The deficiency of lysine in tobacco seed has attributed to improved catabolism of lysine observed by elevated activity of the major lysine catabolizing enzyme lysine-ketoglutarate reductase (LKR) [33]. It was observed that the generation of LKR in endosperm tissue is a preventive factor, which leads to a decrease in free lysine build-up [34] due to the increased lysine catabolism. The genes codes for the two first catabolic enzymes, LKR and saccharopine dehydrogenase (SDH) are cloned from Arabidopsis and maize [35, 36]. Hence, with an antisense or co-suppression method, it is likely to decrease the expression level of LKR and the buildup of lysine degradation products. In this background, it is important to mention that the production of LKR and SDH knock out mutant in Arabidopsis permitted us in noticing the greater accumulation of freely available lysine in seeds compared to the wild varieties [37]. By suppressing the LKR (lysine-ketoglutarate reductase)/SDH with recombinant RNAi, researchers were able to introduce genetically modified maize with high free lysine content.

High Amylose Sweetpotato using Silencing of 1,4-α-ᴅ-glucan-6-α- [1,4-glucan--transferase

In sweet potatoes, carbohydrates are the important constituents in terms of their industrial application and suitability for eating. The sweet potato, *Ipomoea batatas* (L.) Lam cultivation is recorded in around 8 million hectares and per annum, about 110 million tons of tubers are reaped all over the globe. It ranked tenth among the ordinary crops in the world in 2013 in terms of its production and quantity. It is chiefly cultivated in African and Asian countries, mainly in China, where about 70% of global manufacture happens. The storage root is used not only for the consumption of humans and for livestock, but also used in the production of starchy material, alcohol, and local household products. Starches are the essential constituent in the storage roots of the sweet potato and play a noteworthy role in diverse feet of its utilization. The edible values of sweet potato depend solely on its consistency and sweetness, and these merits are mostly based on the amount and superiority of carbohydrate present in it [38, 39].

Starch extracted from the storage root of the sweet potato is the utmost important carbohydrate constituent, while compared to other plant-based starches; the starch with lesser grain size, greater pasting temperature, and inferior pasting viscosity was observed in cereal starches like those found in corn. Further, tuber starches extracted from sweet potato tubers grown underground usually have a greater grainy size, inferior pasting temperature, and greater pasting viscosity. These properties of sweet potato starch emerge to acquire a middle position among the number of other starches [40]. Though, during the course of discovering the noticeable properties of sweet potato starches, a few fashionable qualities like low amylose content [41], a moderately low gelatinization temperature [42], a very low gelatinization temperature [43 - 45], and a rather high amylose content [46] were revealed in Japan.

High amylase content in starches displayed several useful effects for health profit, for instance, lesser postprandial glucose response [47] and lesser energy owing to its low calorific value [48]. Hence, foods comprising an elevated level of amylose starches are advantageous for human wellbeing. Antisense RNAs and RNAi mediated transgenic potatoes and wheat with great amylose levels for *SBE* genes respectively were introduced [49, 50].

The SBE enzyme plays a vital role in starch biosynthesis, it catalyzes the removal and relocation of glucan chains to either alike or different glucan residues by acting on glucose polymers [51, 52]. There are two classes of SBEs, like SBE-I, and SBE-II, with distinctive structure and enzymatic property [53]. The SBE-I class prefers a substrate - amylose and transfers relatively longer glucan chains [54, 55]. On the other hand, SBEII favors amylopectin as a substrate and transfer reasonably smaller chains (DP 6–14) [56]. To produce transgenic sweet potato plants with high amylose content, dsRNAs of *IbSBEII* was introduced in sweet potato [57]. In *IbSBE-II* dsRNA expressed transgenic plant, the amylose constituent in storage root was found to be enhanced to 242.4% while comparing the plants kept as control. The new sweet potato variety having starch with higher amylose content will be used in new dietary and industrial applications.

Low Gossypol Cotton Seed by Silencing

Cotton is the important fiber yielding plant, which has been cultivated in the past 7,000 years, despite the availability of artificial substitutes; textiles industries depend on cotton fiber as the best source of fiber. It is cultivated in 80 different countries and a major cash crop for 20 million farmers in the developing countries (Africa and Asia), where malnourishment and hunger are uncontrolled [58]. Global production of cottonseed can afford the protein necessities for half a billion people group for every year; however, it is desolately underutilized due to the existence of toxic gossypol in the seed glands. Hence, to remove gossypol is the longstanding goal of scientists working across the world. Hence, during 1950s efforts were being initiated to meet this purpose by evolving glandless cotton using the conventional plant breeding methods; but, the glandless cotton varieties are commercially unviable due to the augmented vulnerability of the plants to bestowing to the systemic lack of glands that consist of gossypol and other defensive terpenoids.

Apart from the oil content (21%), it is the reserve of comparatively high-class protein (23%). Though, the capability of usage of this nutrient-enriched reserve for foodstuff is held back by the existence of lethal gossypol that is exclusive to the tribe Gossypieae. This hepatic and cardiotoxic terpenoid, existing in the glands of cottonseed, render cottonseed unsafe for humans and other monogastric animal ingesting [59]. Suppose, if the seeds safe for humans consumption, around 44 million metric tons of cottonseed harvested every year could provide the total protein necessities of 500 million (half a billion) people in a year (50 g per day). Thus, in developing countries, gossypol less cottonseed may considerably add to human nourishment and wellbeing, chiefly in developing countries [60 - 62], and help to encounter the needs of the expected 50% upsurge in the world population in the next 50 years. Gossypol and related terpenoids are existing throughout the plant in the glands of foliar parts, floral structures, bolls and also in the underground parts. These terpenoids are induced in plants in response to the infection of microbes, where it gives resistance to the plant against insect and pathogen attack [63, 64]. After the introduction of glandless mutant varieties [65], several breeding programs were initiated in the U.S., Africa, and Asia countries to transfer the glandless feature into commercial varieties to produce gossypol free cotton [66 - 68]. Cottonseed equaled satisfactorily as a source of protein to other conventional food resources in many human nutrition-related studies [60, 62, 68]. Though the glandless cotton varieties were a commercial failure, under field conditions, glandless plants were extraordinarily vulnerable to attack by the insect pests, because they constitutively are deficient in protective terpenoids [69, 70] and were, hence, forbidden by the farmers. Thus, the perspective of cottonseed in contributing to human nourishment remains unsatisfied.

Gossypol and other sesquiterpenoids are derivative of δ-cadinene, the enzyme cadinene synthase catalyzes the initial stage concerning the cyclization of farnesyl diphosphate to δ-cadinene. Thus, inhibiting the terpenoid biosynthesis by the organ-specific exhibition of RNAi of cadinene synthase is the possible mechanism to get rid of gossypol a toxin, from the seed while holding a full set of this and associated terpenoids in the remaining parts of the plant for maintaining its defensive abilities against pests and diseases. This revealed the likelihood of a targeted RNAi based method to resolve a long-standing problem of cottonseed toxicity and provide an opportunity to exploit the substantial amounts of protein and oil existing in the cottonseed.

Low PhytateRice by Silencing Inositol Phosphate Kinases

In most of the cereals, the storage of phosphorus is mainly as *myo*-inositol 1,2,3,4,5,6-hexakisphosphate (InsP$_6$), called phytic acid [71]. Around 80% of the total phosphorus available in seeds is contributed by phytic acid and the remaining phosphorus is existing as soluble inorganic phosphates (Pi) and in several biomacromolecules (DNA, RNA, polypeptides, lipids, and sugars). Since phytase, the digestive enzyme, responsible for the stepwise discharge of phosphate residue from the phytic acid is lacking in monogastric animals, which cannot use the phosphorus in phytic acid proficiently [72]. Therefore, inorganic phosphates are, given as supplements to animal foods to avoid phosphorus scarcity, but this practice results in the water body pollution by the accumulation of high inorganic phosphorus concentration in waste material [73]. In addition to this, phytic acid strongly conjugated with several cationic minerals, like magnesium, potassium, iron, calcium, and zinc, to form the respective mixed salt called phytate [74], it renders these mineral nutrients unobtainable to animals. Feeding animals enriched with inorganic phosphate will hence not a solution to the problem of mineral nutrition insufficiency. Therefore, the degradation of phytate in the animal diet is important in overcoming both ecological and dietary issues.

To improve the bioavailability of phosphorus, the supplementary phytase from microbial source is normally used. The dietary phytases encourage the apparent discharge of available phosphorus from phytate [75]. To reduce environmental pollution from the manure, expression of the recombinant microbial phytase gene using a vector to degrade the phytic acid content accumulated in the seed is a promising substitute and economic strategy to improve the utilization of dietary phytate nursed to animals [76, 77]. Transgenic crop varieties like, maize [78], wheat [79], soybean [80] and rice [81 - 83] that accumulate microbial phytase in seeds also have been produced. However, to be more effective, these transgenic foods must be pulverized to a powder and incubated in the water at 37 °C before serving to animals [84]. Therefore, a cost-effective, easy and sustainable solution is essential to evade such meticulous processing stages.

Promising substitute scheme must be designed to use the metabolic pathway leads to the production of phytate in plants, and the enzymes concerned in this pathway has to be recognized to generate lpa genetically modified plants. D-glucose-6-phosphate is the source from which phytic acid is mostly synthesized [85], the biosynthesis of phytic acid and metabolism of inositol starts with the cyclization of D-glucose-6-phosphate to 1D-*myo*inositol-3-phosphate [Ins(3)P$_1$] which is catalyzed by Ins (3) P$_1$ synthase [86]. RNAi mediated organ (seed) specific downregulation of inositol 1, 3, 4, 5, 6- pentakisphosphate 2-kinase gene (IPK1) leads to the development of low phytate rice. Production of genetically modified rice by silencing the last stage of phytic acid synthesis in rice variety using the expression of the IPK1 gene using Ole18 a seed-specific promoter, in the RNAi mediated approach. RNAi mediated downregulation of myoinositol-3 phosphate synthase (MIPS) to produce low phytate rice. Here, seed-specific down-regulation of the MIPS gene using RNAi mediated method is the best way of producing low phytate in Rice.

Enhancing β-Carotene Content in Potato by RNAi-mediated Silencing of the β-Carotene Hydroxylase

In the various plant functions, the lipid-soluble pigments carotenoids play key roles. It also partakes an important role in the diet of human beings by serving as a precursor for vitamin-A production and by plummeting the incidence of certain diseases. The deficiency of this leads to various diseases. Thus, it is required for improving its presence in staple food. Potato is considered as the fourth staple food consumed in the world after wheat, rice, and maize. Plant-based carotenoid is the fat-soluble pigment, which plays a major role in various functions of a plant.

In plants, algae, in many bacteria and fungi, carotenoid is a diverse group of plant pigments, which serve up numerous functions, including light-harvesting, protection from oxidative damages, aid in attracting animals and pests to support pollination and seed dispersal mechanism in plants [87]. Carotenoids in plants are available in different colors like yellow, orange, red and a fat-soluble pigment, which is produced in chloroplasts of photosynthetic tissues and the chromoplasts of flowers, fruits, and underground parts (roots and tubers) of several species [88]. Most carotenoids are types of C40 terpenoids; it is made up of eight units of isoprene produced by the isoprenoid pathway. Significant progress has been made during the last decade, in the elucidation of its molecular biological mechanism and genetic makeup [89]. Apart from the various vital functions of carotenoids take part in plant structures; they also serve an indispensable role in the nourishment of individuals. Beta-carotene is the prime carotenoid available in lots of crops and takes part in an important role in the human diet. Beta-carotene is structurally made of essentially two units of vitamin-A (retinol) linked tail to tail and a major substrate for the production of vitamin A in the human system. Since it serves a vital role in human nutrition, it is defined as provitamin-A. In the current time, there is adequate reports on the health profit of other carotenoids, specifically lycopene and lutein, which are to be linked with reduced menace for various diseases such as coronary heart disease, cancer, and a few degenerative sicknesses [90 - 93]. The benefits of provitamin-A in human health are well studied since it is the chief source of vitamin-A in the diet of human beings [94]. It is an essential micronutrient and the deficiency of which in the diet may result in xerophthalmia, blindness, and premature death [95]. In reality, blindness and weakens in the immune system are the causes of the vitamin-A deficit is the chief global micronutrient deficiency associated diseases. WHO (World Health Organization) predicted that about 250 million preschool going children globally are scarce in vitamin A, and 250,000 - 500,000 of these children may become blind each year, about half of these affected children are facing death within 12 months after losing the eyesight.

Reports in the previous decade indicated that an increase in lycopene consumption can thwart the occurrence of prostate cancer [96, 97], and it was also observed that Zeaxanthin and lutein intake is shown to be related with a decline in the onset of age-related muscular degeneration. Thus, an increase in the content and composition of carotenoids in our daily diet may have a range of health benefits in both developing as well as developed nations.

Scientists across the world are involved in discovering an easy metabolic engineering approach to increase the content of β-carotene in potato. The RNA interference (RNAi) technique was used to down-regulate the gene for β-carotene hydroxylase (bch), which converts β-carotene to zeaxanthin. It is identified that β-carotene hydroxylase an only enzyme that is accountable for converting β-carotene to zeaxanthin, and nowadays the gene for this enzyme (bch) has been successfully cloned into plants. Gene silencing technologies offer an approach to get rid of the activity of the bch gene. This single modification should end metabolic flux through the carotenoid pathway at β-carotene, resulting in the conversion of zeaxanthin accumulating potato lines into β-carotene accumulators. Thereby, RNAi mediated gene downregulation of the enzyme β-Carotene Hydroxylase can be efficiently used for increasing β-Carotene Content in Potato.

Lachrymatory Synthase Enzyme Suppression in Tearless Onion

Generally, onion (*Allium cepa* L.,) is called the bulb onion or common onion, a vegetable that is the most extensively cultivated crop species from the genus *Allium*. Around the world, onion is cultivated and used as a food item; which is used in the preparation of various dishes, as a vegetable or it can also be eaten raw or used in pickle making or preparation of chutney. When chopped it is pungent due to a chemical substance, which irritates the eyes. Peculiar secondary sulfur-containing compounds are synthesized from *Allium* species, the notable ones are the *S*-alk(en)yl-L-Cyssulfoxides, including *S*-2-propenyl-L-cysteine sulfoxide (2-PRENCSO) and trans-*S*-1-propenyl-L-cysteine sulfoxide (1-PRENCSO) [98]. Alliinase an enzyme cleaves the amino acid derivatives to subsequent sulfenic acids, and volatile sulfur compounds when the tissue of onion is disrupted, which is responsible for the distinctive flavor and bioactivity.

In garlic (*Allium sativum*), 2-PRENCSO was found to be the major sulfoxide [99], and which generate di-2-propenyl thiosulfinate (allicin) upon tissue disturbance. The di-2-propenyl disulfide a prevailing volatile component generated is the decomposition product of allicin [100, 101, 98]. The major sulfoxide produced from onion is 1-PRENCSO [98]; it was forecasted to yield di-1-propenyl thiosulfinate and di-1-propenyl disulfide. Though di-1-propenyl thiosulfinate has never been reported in onion and di-1- propenyl disulfide has tentatively been reported in the meager amount [102]. Instead, the dominance of propanethial *S*-oxide (LF), 1-propenyl methanethiosulfonate, and dipropyl disulfide are observed [98, 100, 101]. LF is the substance accountable for inducing tearing events in onion, it is an undesirable irritant and hypothesized that its production causes the absence of sulfur-containing volatile compounds [103], similar to which present in garlic is known for their health aspect [104].

Lachrymatory factor synthase (LFS) an enzyme [105] mediates the synthesis of LF from 1-propenyl sulfenic acid. It (LFS) specifically acts on 1-propenyl sulfenic acid, followed by the action of alliinase on 1-PRENCSO to produce LF. The selectivity of this reaction suggested that the production of LF might be minimized by the genetic manipulation of the LFS transcript using RNAi mediated gene silencing. In the nonexistence of LFS, the unsteady 1-propenyl sulfenic acid would undergo spontaneous self-condensation to di-1-propenyl thiosulfinate [105]. This elevated thiosulfinate level may be available for conversion, through the nonenzymatic pathway, into a predicted secondary compound in scarce amounts, which have not been detected in onion previously. The above compounds are accountable for the distinctive flavor, sensory and odor [106] as well as for the health-promoting attribute of the *Allium* species [107 - 110]. By the falling of LFS and preventing the alteration of 1-propenyl sulfenic acid to the unattractive LF, the 1-propenyl sulfenic acid to be available for the spontaneous conversion into thiosulfinate and thiosulfonate derived sulfur compounds, analogs of which are well renowned for its desirable sensory and health-promoting quality. The onion thus produced by inhibiting the LFS enzyme using RNAi technology is not irritating and tearless.

CONCLUSION

Among the different recent biotechnological tools used, gene silencing has been playing a significant role in crop improvement. Gene silencing techniques include RNA interference, antisense RNA and usage of ribozymes to inhibit the selected gene, thereby silencing the activity of an enzyme, which is accountable for the unwanted character of the crop plant. RNA Interference (RNAi) is a technique, in which exogenous or endogenous double-stranded RNA is capable of suppressing the gene expression, which is corresponding to the sequence in the double-stranded RNA. Antisense RNA also called as antisense transcript, is a single RNA strand, which is complementary to a protein-encoding messenger RNA (mRNA) onto which it hybridizes, and by this means blocks its translation into a protein. RNA interference and antisense RNA, the highly valuable and influential tools of functional genomics for silencing the gene expression for crop improvement are discussed with a neat example in this chapter. The appropriate usage of these technologies may go a long way to narrow the gap through the manufacture of insect, virus and disease resistant, nutritionally enriched and toxin-free crop varieties for safe human consumption.

CONSENT FOR PUBLICATION

Not applicable.

CONFLICT OF INTEREST

The author(s) confirms that there is no conflict of interest.

ACKNOWLEDGEMENTS

My sincere thanks are to Dr. L. F. A. Anandaraj, Associate Professor, Department of Biotechnology, St.Joseph's College of Engineering for his valuable help in preparing the figure on time.

REFERENCES

[1]   Saurabh S, Vidyarthi AS, Prasad D. RNA interference: concept to reality in crop improvement. Planta 2014; 239(3): 543-64.
[http://dx.doi.org/10.1007/s00425-013-2019-5] [PMID: 24402564]

[2]   Tomizawa J, Itoh T, Selzer G, Som T. Inhibition of ColE1 RNA primer formation by a plasmid-specified small RNA. Proc Natl Acad Sci USA 1981; 78(3): 1421-5.
[http://dx.doi.org/10.1073/pnas.78.3.1421] [PMID: 6165011]

[3]   Kumar CC, Novick RP. Plasmid pT181 replication is regulated by two countertranscripts. Proc Natl Acad Sci USA 1985; 82(3): 638-42.
[http://dx.doi.org/10.1073/pnas.82.3.638] [PMID: 2579377]

[4]   Rogers JC. RNA complementary to α-amylase mRNA in barley. Plant Mol Biol 1988; 11(2): 125-38.
[http://dx.doi.org/10.1007/BF00015665] [PMID: 24272255]

[5]   Coen ES, Carpenter R. A semi-dominant allele, niv-525, acts in trans to inhibit expression of its wild-type homologue in Antirrhinum majus. EMBO J 1988; 7(4): 877-83.
[http://dx.doi.org/10.1002/j.1460-2075.1988.tb02891.x] [PMID: 3402437]

[6]   Cock JM, Swarup R, Dumas C. Natural antisense transcripts of the S locus receptor kinase gene and related sequences in Brassica oleracea. Mol Gen Genet 1997; 255(5): 514-24.
[http://dx.doi.org/10.1007/s004380050524] [PMID: 9294036]

[7]   Cech TR, Zaug AJ, Grabowski PJ. *In vitro* splicing of the ribosomal RNA precursor of Tetrahymena: involvement of a guanosine nucleotide in the excision of the intervening sequence. Cell 1981; 27(3 Pt 2): 487-96.
[http://dx.doi.org/10.1016/0092-8674(81)90390-1] [PMID: 6101203]

[8]   Kruger K, Grabowski PJ, Zaug AJ, Sands J, Gottschling DE, Cech TR. Self-splicing RNA: autoexcision and autocyclization of the ribosomal RNA intervening sequence of Tetrahymena. Cell 1982; 31(1): 147-57.
[http://dx.doi.org/10.1016/0092-8674(82)90414-7] [PMID: 6297745]

[9]   Uhlenbeck OC. A small catalytic oligoribonucleotide. Nature 1987; 328(6131): 596-600.
[http://dx.doi.org/10.1038/328596a0] [PMID: 2441261]

[10]   Haseloff J, Gerlach WL. Simple RNA enzymes with new and highly specific endoribonuclease activities. Nature 1988; 334(6183): 585-91.
[http://dx.doi.org/10.1038/334585a0] [PMID: 2457170]

[11]   Kurreck J. Antisense technologies. Improvement through novel chemical modifications. Eur J Biochem 2003; 270(8): 1628-44.
[http://dx.doi.org/10.1046/j.1432-1033.2003.03555.x] [PMID: 12694176]

[12]   Doudna JA, Cech TR. The chemical repertoire of natural ribozymes. Nature 2002; 418(6894): 222-8.
[http://dx.doi.org/10.1038/418222a] [PMID: 12110898]

[13] Pelechano V, Steinmetz LM. Gene regulation by antisense transcription. Nat Rev Genet 2013; 14(12): 880-93.
[http://dx.doi.org/10.1038/nrg3594] [PMID: 24217315]

[14] Kole R, Krainer AR, Altman S. RNA therapeutics: beyond RNA interference and antisense oligonucleotides. Nat Rev Drug Discov 2012; 11(2): 125-40.
[http://dx.doi.org/10.1038/nrd3625] [PMID: 22262036]

[15] Wahlestedt C. Targeting long non-coding RNA to therapeutically upregulate gene expression. Nat Rev Drug Discov 2013; 12(6): 433-46.
[http://dx.doi.org/10.1038/nrd4018] [PMID: 23722346]

[16] Dagle JM, Weeks DL. Oligonucleotide-based strategies to reduce gene expression. Differentiation 2001; 69(2-3): 75-82.
[http://dx.doi.org/10.1046/j.1432-0436.2001.690201.x] [PMID: 11798068]

[17] White S, Baird EE, Dervan PB. Orientation preferences of pyrrole-imidazole polyamides in the minor groove of DNA. J Am Chem Soc 1997; 119: 8756-65.
[http://dx.doi.org/10.1021/ja971569b]

[18] White S, Szewczyk JW, Turner JM, Baird EE, Dervan PB. Recognition of the four Watson-Crick base pairs in the DNA minor groove by synthetic ligands. Nature 1998; 391(6666): 468-71.
[http://dx.doi.org/10.1038/35106] [PMID: 9461213]

[19] Sørensen JJ, Nielsen JT, Petersen M. Solution structure of a dsDNA:LNA triplex. Nucleic Acids Res 2004; 32(20): 6078-85.
[http://dx.doi.org/10.1093/nar/gkh942] [PMID: 15550567]

[20] Ekker SC, Larson JD. Morphant technology in model developmental systems. Genesis 2001; 30(3): 89-93.
[http://dx.doi.org/10.1002/gene.1038] [PMID: 11477681]

[21] Xanthos JB, Kofron M, Wylie C, Heasman J. Maternal VegT is the initiator of a molecular network specifying endoderm in Xenopus laevis. Development 2001; 128(2): 167-80.
[PMID: 11124113]

[22] Walder RY, Walder JA. Role of RNase H in hybrid-arrested translation by antisense oligonucleotides. Proc Natl Acad Sci USA 1988; 85(14): 5011-5.
[http://dx.doi.org/10.1073/pnas.85.14.5011] [PMID: 2839827]

[23] Eder PS, Walder JA. Ribonuclease H from K562 human erythroleukemia cells. Purification, characterization, and substrate specificity. J Biol Chem 1991; 266(10): 6472-9.
[PMID: 1706718]

[24] Eder PS, Walder RY, Walder JA. Substrate specificity of human RNase H1 and its role in excision repair of ribose residues misincorporated in DNA. Biochimie 1993; 75(1-2): 123-6.
[http://dx.doi.org/10.1016/0300-9084(93)90033-O] [PMID: 8389211]

[25] Crooke ST. Progress in antisense technology. Annu Rev Med 2004; 55: 61-95.
[http://dx.doi.org/10.1146/annurev.med.55.091902.104408] [PMID: 14746510]

[26] Redenbaugh K, Hiatt B, Martineau B, Kramer M, *et al*. Safety Assessment of Genetically Engineered Fruits and Vegetables: A Case Study of the Flavr Savr Tomato. CRC Press 1992; p. 288.

[27] Sheehy RE, Kramer M, Hiatt WR. Reduction of polygalacturonase activity in tomato fruit by antisense RNA. Proc Natl Acad Sci USA 1988; 85(23): 8805-9.
[http://dx.doi.org/10.1073/pnas.85.23.8805] [PMID: 16593997]

[28] Fire A, Xu S, Montgomery MK, Kostas SA, Driver SE, Mello CC. Potent and specific genetic interference by double-stranded RNA in *Caenorhabditis elegans*. Nature 1998; 391(6669): 806-11.
[http://dx.doi.org/10.1038/35888] [PMID: 9486653]

[29] Jorgensen R. Altered gene expression in plants due to trans interactions between homologous genes.

Trends Biotechnol 1990; 8(12): 340-4.
[http://dx.doi.org/10.1016/0167-7799(90)90220-R] [PMID: 1366894]

[30] Fagard M, Boutet S, Morel JB, Bellini C, Vaucheret H. AGO1, QDE-2, and RDE-1 are related
      proteins required for post-transcriptional gene silencing in plants, quelling in fungi, and RNA
      interference in animals. Proc Natl Acad Sci USA 2000; 97(21): 11650-4.
      [http://dx.doi.org/10.1073/pnas.200217597] [PMID: 11016954]

[31] Waterhouse PM, Wang MB, Finnegan EJ. Role of short RNAs in gene silencing. Trends Plant Sci
      2001; 6(7): 297-301.
      [http://dx.doi.org/10.1016/S1360-1385(01)01989-6] [PMID: 11435167]

[32] Jaynes JM, Yang MS, Espinoza N, Scorza R. Plant protein improvement by genetic engineering: use
      of synthetic genes. Trends Biotechnol 1986; 4: 314-20.
      [http://dx.doi.org/10.1016/0167-7799(86)90183-6]

[33] Karchi H, Shaul O, Galili G. Lysine synthesis and catabolism are coordinately regulated during
      tobacco seed development. Proc Natl Acad Sci USA 1994; 91(7): 2577-81.
      [http://dx.doi.org/10.1073/pnas.91.7.2577] [PMID: 8146157]

[34] Arruda P, Kemper EL, Papes F, Leite A. Regulation of lysine catabolism in higher plants. Trends Plant
      Sci 2000; 5(8): 324-30.
      [http://dx.doi.org/10.1016/S1360-1385(00)01688-5] [PMID: 10908876]

[35] Epelbaum S, McDevitt R, Falco SC. Lysine-ketoglutarate reductase and saccharopine dehydrogenase
      from Arabidopsis thaliana: nucleotide sequence and characterization. Plant Mol Biol 1997; 35(6): 735-
      48.
      [http://dx.doi.org/10.1023/A:1005808923191] [PMID: 9426595]

[36] Tang G, Miron D, Zhu-Shimoni JX, Galili G. Regulation of lysine catabolism through lysine-
      ketoglutarate reductase and saccharopine dehydrogenase in Arabidopsis. Plant Cell 1997; 9(8): 1305-
      16.
      [PMID: 9286108]

[37] Zhu X, Tang G, Granier F, Bouchez D, Galili G. A T-DNA insertion knockout of the bifunctional
      lysine-ketoglutarate reductase/saccharopine dehydrogenase gene elevates lysine levels in Arabidopsis
      seeds. Plant Physiol 2001; 126(4): 1539-45.
      [http://dx.doi.org/10.1104/pp.126.4.1539] [PMID: 11500552]

[38] Walter WM Jr, Purcell A, Nelson AM. Effects of amylolytic enzymes on "moisture" and carbohydrate
      changes of baked sweet potato cultivars. J Food Sci 1975; 40: 793-6.
      [http://dx.doi.org/10.1111/j.1365-2621.1975.tb00558.x]

[39] Walter WM Jr. Effect of curing on sensory properties and carbohydrate composition of baked sweet
      potatoes. J Food Sci 1987; 52: 1026-9.
      [http://dx.doi.org/10.1111/j.1365-2621.1987.tb14267.x]

[40] Suganuma T, Fujimoto S, Kitahara K, Nagahama T. Classification of starches from wild plants in
      Japan by cluster analysis and their visualization by radar chart. J Appl Glycosci 1996; 43: 525-33.

[41] Kitahara K, Mizukami S, Suganuma T, Nagahama T, et al. A new line of sweet potato with low
      amylose content. J Appl Glycosci 1996; 43: 551-4. b

[42] Kitahara K, Ueno J, Suganuma T, Ishiguro K, Yamakawa O. Physicochemical properties of root
      starches from new types of sweet potato. J Appl Glycosci 1999; 46: 391-7.
      [http://dx.doi.org/10.5458/jag.46.391]

[43] Katayama K, Komae K, Kohyama K, Kato T, et al. New sweet potato line having low gelatinization
      temperature and altered starch structure. Starke 2002; 54: 51-7.
      [http://dx.doi.org/10.1002/1521-379X(200202)54:2<51::AID-STAR51>3.0.CO;2-6]

[44] Katayama K, Tamiya S, Kuranouchi T, Komaki K, Nakatanni M. New sweet potato cultivar "Quick
      Sweet". Bull NARO Inst Crop Sci 2003; 35-52.

[45] Katayama K, Tamiya S, Ishiguro K. Starch properties of new sweet potato lines having low pasting temperature. Starch/Särke 2004; 56: 563-9.
[http://dx.doi.org/10.1002/star.200400304]

[46] Katayama K, Kitahara K, Sakai T, Kai Y, Yoshinaga M. Resistant and digestible starch contents in sweet potato cultivars and lines. J Appl Glycosci 2011; 58: 53-9.
[http://dx.doi.org/10.5458/jag.jag.JAG-2010_016]

[47] Nestel PJ, Noakes M, Clifton P, McIntosh G. High amylose starch on bowel function, insulin sensitivity, and lipids. FASEB J 1996; 10: 3168.

[48] Brown IL. Applications and uses of resistant starch. J AOAC Int 2004; 87(3): 727-32.
[http://dx.doi.org/10.1093/jaoac/87.3.727] [PMID: 15287672]

[49] Regina A, Bird A, Topping D, et al. High-amylose wheat generated by RNA interference improves indices of large-bowel health in rats. Proc Natl Acad Sci USA 2006; 103(10): 3546-51.
[http://dx.doi.org/10.1073/pnas.0510737103] [PMID: 16537443]

[50] Schwall GP, Safford R, Westcott RJ, et al. Production of very-high-amylose potato starch by inhibition of SBE A and B. Nat Biotechnol 2000; 18(5): 551-4.
[http://dx.doi.org/10.1038/75427] [PMID: 10802625]

[51] Hussain H, Mant A, Seale R, et al. Three isoforms of isoamylase contribute different catalytic properties for the debranching of potato glucans. Plant Cell 2003; 15(1): 133-49.
[http://dx.doi.org/10.1105/tpc.006635] [PMID: 12509527]

[52] Pan D, Nelson OE. A debranching enzyme deficiency in endosperms of the sugary-1 mutants of maize. Plant Physiol 1984; 74(2): 324-8.
[http://dx.doi.org/10.1104/pp.74.2.324] [PMID: 16663417]

[53] Martin C, Smith AM. Starch biosynthesis. Plant Cell 1995; 7(7): 971-85.
[PMID: 7640529]

[54] Kuriki T, Stewart DC, Preiss J. Construction of chimeric enzymes out of maize endosperm branching enzymes I and II: activity and properties. J Biol Chem 1997; 272(46): 28999-9004.
[http://dx.doi.org/10.1074/jbc.272.46.28999] [PMID: 9360973]

[55] Takeda Y, Guan HP, Preiss J. Branching of amylose by the branching isoenzymes of maize endosperm. Carbohydr Res 1993; 240: 253-63.
[http://dx.doi.org/10.1016/0008-6215(93)84188-C]

[56] Guan HP, Preiss J. Differentiation of the properties of the branching isozymes from maize (Zea mays). Plant Physiol 1993; 102(4): 1269-73.
[http://dx.doi.org/10.1104/pp.102.4.1269] [PMID: 12231902]

[57] Shimada T, Otani M, Hamada T, Kim SH. Increase of amylose content of sweet potato starch by RNA interference of the starch branching enzyme II gene (IbSBEII). Plant Biotechnol 2006; 23: 85-90.
[http://dx.doi.org/10.5511/plantbiotechnology.23.85]

[58] de Onís M, Monteiro C, Akré J, Glugston G. The worldwide magnitude of protein-energy malnutrition: an overview from the WHO Global Database on Child Growth. Bull World Health Organ 1993; 71(6): 703-12.
[PMID: 8313488]

[59] Risco CA, Chase CC Jr. Handbook of Plant and Fungal Toxicants, edD'Mello JPF. Boca Raton, FL: CRC Press 1997; pp. 87-98.

[60] Bressani R. The use of cottonseed protein in human foods. Food Technol 1965; 19: 1655-62.

[61] Lambo MG, Shaw RL, Decossas KM, Vix HLE. Cottonseed's role in a hungry world. Econ Bot 1966; 20: 256-67.
[http://dx.doi.org/10.1007/BF02904276]

[62]   Alford BB, Liepa GU, Vanbeber AD. Cottonseed protein: what does the future hold? Plant Foods Hum Nutr 1996; 49(1): 1-11.
[http://dx.doi.org/10.1007/BF01092517] [PMID: 9139299]

[63]   Hedin PA, Parrott WL, Jenkins JN. Relationships of glands, cotton square terpenoid aldehydes, and other allelochemicals to larval growth of Heliothis virescens (Lepidoptera: Noctuidae). J Econ Entomol 1992; 85: 359-64.
[http://dx.doi.org/10.1093/jee/85.2.359]

[64]   Stipanovic RD, Bell AA, Benedict CR. Biologically Active Natural Products: Agrochemicals. Boca Raton, FL: CRC Press 1999; pp. 211-20.

[65]   McMichael SC. Glandless boll in Upland Cotton and its use in the study of the natural crossing. Agron J 1954; 46: 527-8.
[http://dx.doi.org/10.2134/agronj1954.00021962004600110016x]

[66]   McMichael SC. Combined effects of the glandless genes gl2 and gl3 on pigment glands in the Cotton plant. Agron J 1960; 52: 385-6.
[http://dx.doi.org/10.2134/agronj1960.00021962005200070005x]

[67]   Miravalle RJ, Hyer AH. Identification of the Gl2 gl2 Gl3 gl3 genotype in breeding for glandless cottonseed. Crop Sci 1962; 2: 395-7.
[http://dx.doi.org/10.2135/cropsci1962.0011183X000200050009x]

[68]   Lusas EW, Jividen GM. Glandless cottonseed: A review of the first 25 years of processing and utilization research. J Am Oil Chem Soc 1987; 64: 839-54.
[http://dx.doi.org/10.1007/BF02641491]

[69]   Bottger GT, Sheehan ET, Lukefahr MJ. Relation of gossypol of cotton plants to insect resistance. J Econ Entomol 1964; 57: 283-5.
[http://dx.doi.org/10.1093/jee/57.2.283]

[70]   Jenkins JN, Maxwell FG, Lafever HN. The comparative preference of insects for glanded and glandless cottons. J Econ Entomol 1966; 59: 352-6.
[http://dx.doi.org/10.1093/jee/59.2.352]

[71]   Raboy V. The ABCs of low-phytate crops. Nat Biotechnol 2007; 25(8): 874-5. b
[http://dx.doi.org/10.1038/nbt0807-874] [PMID: 17687363]

[72]   Mullaney EJ, Ullah AHJ. Phytases: attributes, catalytic mechanisms, and applications. Inositol Phosphates: Linking Agriculture and the Environment. Wallingford, Oxfordshire: CAB International 2007; pp. 97-110.
[http://dx.doi.org/10.1079/9781845931520.0097]

[73]   Sims JT, Edwards AC, Schoumans OF, Simard RR. Integrating soil phosphorus testing into environmentally based agricultural management practices. J Environ Qual 2000; 29: 60-71.
[http://dx.doi.org/10.2134/jeq2000.00472425002900010008x]

[74]   Lott JN, Greenwood JS, Batten GD. Mechanisms and regulation of mineral nutrient storage during seed development.Seed Development and Germination. New York: Marcel Dekker 1995; pp. 215-35.

[75]   Lei XG, Porres JM. Phytase and inositol phosphates in animal nutrition: dietary manipulation and phosphorus excretion by animals. Inositol Phosphates: Linking Agriculture and the Environment. Wallingford, Oxfordshire: CAB International 2007; pp. 133-49.
[http://dx.doi.org/10.1079/9781845931520.0133]

[76]   Lei XG, Stahl CH. Biotechnological development of effective phytases for mineral nutrition and environmental protection. Appl Microbiol Biotechnol 2001; 57(4): 474-81.
[http://dx.doi.org/10.1007/s002530100795] [PMID: 11762591]

[77]   Brinch-Pedersen H, Sørensen LD, Holm PB. Engineering crop plants: getting a handle on phosphate. Trends Plant Sci 2002; 7(3): 118-25.

[http://dx.doi.org/10.1016/S1360-1385(01)02222-1] [PMID: 11906835]

[78] Drakakaki G, Marcel S, Glahn RP, *et al.* Endosperm-specific co-expression of recombinant soybean ferritin and Aspergillus phytase in maize results in significant increases in the levels of bioavailable iron. Plant Mol Biol 2005; 59(6): 869-80.
[http://dx.doi.org/10.1007/s11103-005-1537-3] [PMID: 16307363]

[79] Brinch-Pedersen H, Hatzack F, Sørensen LD, Holm PB. Concerted action of endogenous and heterologous phytase on phytic acid degradation in seed of transgenic wheat (*Triticum aestivum* L.). Transgenic Res 2003; 12(6): 649-59.
[http://dx.doi.org/10.1023/B:TRAG.0000005113.38002.e1] [PMID: 14713194]

[80] Chiera JM, Finer JJ, Grabau EA. Ectopic expression of a soybean phytase in developing seeds of Glycine max to improve phosphorus availability. Plant Mol Biol 2004; 56(6): 895-904.
[http://dx.doi.org/10.1007/s11103-004-5293-6] [PMID: 15821988]

[81] Lucca P, Hurrell R, Potrykus I. Genetic engineering approach to improve the bioavailability and the level of iron in rice grains. Theor Appl Genet 2001; 102: 392-7.
[http://dx.doi.org/10.1007/s001220051659]

[82] Hong CY, Cheng KJ, Tseng TH, Wang CS, Liu LF, Yu SM. Production of two highly active bacterial phytases with broad pH optima in germinated transgenic rice seeds. Transgenic Res 2004; 13(1): 29-39.
[http://dx.doi.org/10.1023/B:TRAG.0000017158.96765.67] [PMID: 15070073]

[83] Hamada A, Yamaguchi K, Ohnishi N, Harada M, Nikumaru S, Honda H. High-level production of yeast (Schwanniomyces occidentalis) phytase in transgenic rice plants by a combination of signal sequence and codon modification of the phytase gene. Plant Biotechnol J 2005; 3(1): 43-55.
[http://dx.doi.org/10.1111/j.1467-7652.2004.00098.x] [PMID: 17168898]

[84] Lucca P, Hurrell R, Potrykus I. Genetic engineering approaches to improve the bioavailability and the level of iron in rice grains. Theor Appl Genet 2001; 102: 392-7.
[http://dx.doi.org/10.1007/s001220051659]

[85] Loewus FA, Loewus MW. Myo-inositol; its biosynthesis and metabolism. Annu Rev Plant Physiol 1983; 34: 137-61.
[http://dx.doi.org/10.1146/annurev.pp.34.060183.001033]

[86] Loewus FA, Murthy PPN. Myo-Inositol metabolism in plants. Plant Sci 2000; 150: 1-19.
[http://dx.doi.org/10.1016/S0168-9452(99)00150-8]

[87] Goodwin TW. Carotenoids. Bell E, Charlwood BV. Encyclopedia of Plant Physiology. New York. P1: Springer- Verlag 1980; pp. 257-87.

[88] Marano MR, Serra EC, Oreilano EG, Carrillo N. The path of chloroplast development in fruits and flowers. Plant Sci 1993; 94: 1-17.
[http://dx.doi.org/10.1016/0168-9452(93)90002-H]

[89] Cunningham FX Jr, Gantt E. Genes and enzymes of carotenoid biosynthesis in plants. Annu Rev Plant Physiol Plant Mol Biol 1998; 49: 557-83.
[http://dx.doi.org/10.1146/annurev.arplant.49.1.557] [PMID: 15012246]

[90] Krinsky NI, Landrum JT, Bone RA. Biologic mechanisms of the protective role of lutein and zeaxanthin in the eye. Annu Rev Nutr 2003; 23: 171-201.
[http://dx.doi.org/10.1146/annurev.nutr.23.011702.073307] [PMID: 12626691]

[91] Mayne ST. Beta-carotene, carotenoids, and disease prevention in humans. FASEB J 1996; 10(7): 690-701.
[http://dx.doi.org/10.1096/fasebj.10.7.8635686] [PMID: 8635686]

[92] Osganian SK, Stampfer MJ, Rimm E, Spiegelman D, Manson JE, Willett WC. Dietary carotenoids and risk of coronary artery disease in women. Am J Clin Nutr 2003; 77(6): 1390-9.
[http://dx.doi.org/10.1093/ajcn/77.6.1390] [PMID: 12791615]

[93]   Rock CL, Jacob RA, Bowen PE. Update on the biological characteristics of the antioxidant micronutrients: vitamin C, vitamin E, and the carotenoids. J Am Diet Assoc 1996; 96(7): 693-702.
[http://dx.doi.org/10.1016/S0002-8223(96)00190-3] [PMID: 8675913]

[94]   West CE, Eilander A, van Lieshout M. Consequences of revised estimates of carotenoid bioefficacy for dietary control of vitamin A deficiency in developing countries. J Nutr 2002; 132(9) (Suppl.): 2920S-6S.
[http://dx.doi.org/10.1093/jn/132.9.2920S] [PMID: 12221270]

[95]   West KP Jr. Extent of vitamin A deficiency among preschool children and women of reproductive age. J Nutr 2002; 132(9) (Suppl.): 2857S-66S.
[http://dx.doi.org/10.1093/jn/132.9.2857S] [PMID: 12221262]

[96]   Barber N. The tomato: an important part of the urologist's diet? BJU Int 2003; 91(4): 307-9.
[http://dx.doi.org/10.1046/j.1464-410X.2003.04097.x] [PMID: 12603399]

[97]   Giovannucci E, Ascherio A, Rimm EB, Stampfer MJ, Colditz GA, Willett WC. Intake of carotenoids and retinol in relation to risk of prostate cancer. J Natl Cancer Inst 1995; 87(23): 1767-76.
[http://dx.doi.org/10.1093/jnci/87.23.1767] [PMID: 7473833]

[98]   Rose P, Whiteman M, Moore PK, Zhu YZ. Bioactive S-alk(en)yl cysteine sulfoxide metabolites in the genus Allium: the chemistry of potential therapeutic agents. Nat Prod Rep 2005; 22(3): 351-68.
[http://dx.doi.org/10.1039/b417639c] [PMID: 16010345]

[99]   Fritsch RM, Keusgen M. Occurrence and taxonomic significance of cysteine sulphoxides in the genus Allium L. (Alliaceae). Phytochemistry 2006; 67(11): 1127-35.
[http://dx.doi.org/10.1016/j.phytochem.2006.03.006] [PMID: 16626766]

[100]   Block E, Naganathan S, Putman D, Zhao SH. Allium chemistry: HPLC analysis of thiosulfinates from onion, garlic, wild garlic (ramosoms), leek, scallion, shallot, elephant (great-headed) garlic, chive, and Chinese chive. Uniquely high allyl to methyl rations in some garlic samples. J Agric Food Chem 1992; 40: 2418-30. a
[http://dx.doi.org/10.1021/jf00024a017]

[101]   Block E, Putman D, Zhao SH. Allium chemistry: GC-MS analysis of thiosulfinates and related compounds from onion, leek, scallion, shallot, chive, and Chinese chive. J Agric Food Chem 1992; 40(12): 2431-8.
[http://dx.doi.org/10.1021/jf00024a018]

[102]   Boelens M, de Valois PJ, Wobben HJ, van der Gen A. Volatile flavor compounds from the onion. J Agric Food Chem 1971; 19: 984-91.
[http://dx.doi.org/10.1021/jf60177a031]

[103]   Randle WM, Lancaster JE. Sulphur compounds in alliums. Allium Crop Science: Recent Advances. Wallingford, UK: CAB International 2002; pp. 329-56.
[http://dx.doi.org/10.1079/9780851995106.0329]

[104]   Griffiths G, Trueman L, Crowther T, Thomas B, Smith B. Onions--a global benefit to health. Phytother Res 2002; 16(7): 603-15.
[http://dx.doi.org/10.1002/ptr.1222] [PMID: 12410539]

[105]   Imai S, Tsuge N, Tomotake M, et al. Plant biochemistry: an onion enzyme that makes the eyes water. Nature 2002; 419(6908): 685.
[http://dx.doi.org/10.1038/419685a] [PMID: 12384686]

[106]   Block E. The organosulfur chemistry of the genus Allium: implications for organic chemistry. Angew Chem Int Ed Engl 1992; 31: 1135-78.
[http://dx.doi.org/10.1002/anie.199211351]

[107]   Morimitsu Y, Morioka Y, Kawakishi S. Inhibitors of platelet aggregation generated from mixtures of Allium species and/or S-alk(en) nylL-cysteine sulfoxides. J Agric Food Chem 1992; 40: 368-72.
[http://dx.doi.org/10.1021/jf00015a002]

[108]   Block E, Thiruvazhi M, Toscano PJ, Bayer T, *et al.* Allium chemistry: structure, synthesis, a natural occurrence in onion (*Allium cepa*), and reactions of 2,3-dimethyl-5,6-dithiabicyclo [2.1.1] hexane S-oxides. J Agric Food Chem 1996; 118: 2790-8. b

[109]   Randle WM, Lancaster JE. Sulphur compounds in alliums.Allium Crop Science: Recent Advances. Wallingford, UK: CAB International 2002; pp. 329-56.
[http://dx.doi.org/10.1079/9780851995106.0329]

[110]   Lanzotti V. The analysis of onion and garlic. J Chromatogr A 2006; 1112(1-2): 3-22.
[http://dx.doi.org/10.1016/j.chroma.2005.12.016] [PMID: 16388813]

Frontiers in Enzyme Inhibition, 2020, *Vol. 1*, 59-72

Enzyme Biosensors Based on Enzyme Inhibition for Pesticide and Heavy Metal Detection

Elsa Cherian[1,*] and **G Baskar**[2]

[1] *Department of Food Technology, SAINTGITS College of Engineering, Kottayam, Pathamuttom P. O, Kerala-686532, India*

[2] *Department of Biotechnology, St. Joseph's College of Engineering, Chennai – 600119, India*

Abstract: A biosensor is an analytical apparatus, which connects a biologically sensitive and selective component and a physiochemical transducer. Biologically sensitive elements include organisms, tissues, cells, *etc.* and the transducer includes electrochemical, optical, thermal or mechanical signals which are received and converted into a measurable signal. There are different categories of biosensors based on the principle of working. Some of them are electrochemical, amperometric, thermometric, optical, microbial and immunosensors. Biosensors are broadly used in different areas like the food industry, fermentation industry, pharmaceutical industry, *etc.* The present chapter explains the use of different types of biosensors which are based upon enzyme inhibition, as an investigative and diagnostic tool. Some of the enzyme inhibitors such as pollutants are strongly associated with human as well as environmental health, so these have to be monitored with strong significance. Thus enzyme inhibition based enzyme biosensors will be a precious tool for very fast performing and accurate for the above applications.

Keywords: Biosensor, Enzyme inhibition, Environmental applications, Heavy metals, Inhibitors, Pollutants.

INTRODUCTION

A biosensor is an investigative and diagnostic device with an immobilized biological material. Some of the examples for immobilized biological elements include enzymes, hormones, organelle or whole cell. This biological element particularly reacts with an analyte and produces physical, chemical or electrical signals as output that can be measured. The unknown compound for which concentration has to be measured is the analyte. Also a third element present in the biosensor acts as the reference signal or the control element. So the difference

---

* **Corresponding author Elsa Cherian:** Department of Food Technology, SAINTGITS College of Engineering, Kottayam, Pathamuttom P. O, Kerala-686532. India; E-mail: elsa.c@saintgits.org

between the control signal and the analyte will be proportional to the concentration of the material that has to be measured. Thus a characteristic biosensor connects two elements that are, the biological sensing element and a transducer which can detect analyte concentration. Thus biosensor is advantageous. Some of the advantages of biosensors include fast and continuous measurement, high specificity, easy calibration, fast response time, *etc.* This chapter explains the importance of biosensor with its history, types, and applications and also focuses on the influence of enzyme inhibition on the working of the biosensor.

HISTORY OF BIOSENSOR

The biosensor was first invented by Professor Leland C Clark Jnr, who is recognized as the father of the biosensor theory. Leland C Clark Jnr put forward a biosensor with an oxygen electrode in the year 1956. In the biosensor, Glucose oxidase was immobilized on the Clark oxygen electrode using the dialysis membrane [1]. The glucose concentration was calculated by the reduction in the dissolved oxygen concentration. The first potentiometric biosensor was made into reality by immobilizing urease enzyme on the ammonia electrode to confirm urea, in the year 1969 by Guilbault and Montalvo. A brief note on the history of the biosensor is given in Table **1**.

Table 1. History of biosensor.

Sl.No	Event	Year	Reference
1.	The first description on the immobilisation of proteins (invertase on activated charcoal)	1916	[2]
2.	Glass electrode	1922	[3]
3.	First glass electrode for analysis of blood samples	1925	[4]
4.	Carbon dioxide electrode	1954	[5]
5.	The invention of the oxygen electrode	1956	[6]
6.	The invention of a glucose electrode	1962	[7]
7.	The initial usage of the potentiometric biosensor using ammonia electrode immobilized with urease which was intended to confirm the presence of urea	1969	[8]
8.	The innovation of the ion-selective field-effect transistor (ISFET)	1970	[9]
9.	Microbe-based immunosensor	1975	[10]
10.	First biosensor for glucose-based upon fiber optics	1982	[11]
11.	Surface plasmon resonance (SPR) immunosensor	1983	[12]
12.	First mediated amperometric biosensor where ferrocene was used with glucose oxidase for the finding glucose	1984	[13]

(Table 1) cont.....

Sl.No	Event	Year	Reference
13.	Starting off the Pharmacia BIACore SPR-based biosensor system	1990	[14]

TYPES OF BIOSENSOR

Biosensors could be classified based on the mechanism used for biological signaling or the type of signal transduction they employ. Different biological signals that are generated for measurement include antigen-antibody interaction, enzymes, nucleic acid and microbial cells.

Antigen-antibody Interaction-based Biosensors

Biosensors working on the principle of antigen-antibody interaction are known as immunosensors where antigen-antibody complexes are confirmed and transformed using a transducer into an electrical signal, which is then processed, recorded and displayed. There are many types of transducers based on the signal that is generated for the complex formation. Immunosensors are mainly based on specific antigen-antibody interactions and detected, either directly or indirectly by the immunochemical reactions. For example, speedy, non-expensive and numerous assays can be carried out with immunosensors and this could play a major role during epidemics to make a correct judgment and follow the epidemic spreading.

Biosensors Based on Nucleic Acid

Another breakthrough in the biosensor field is the nucleic acid-based biosensor. Here DNA, RNA or nucleic acid analogue is used as the probe, interaction of which results in the generation of the signal [15]. The major trouble in this type of biosensor is the immobilization of the probe on the biosensor which affects analysis performance. Nucleic acid biosensors are mainly used in various fields, such as genotyping and gene expression studies [16], disease diagnosis [17], drug discovery [18], *etc.* The increased development of nucleic acid analogues results in its widespread application.

Biosensors Based on Microbial Cells

The microbial biosensors are based on the interaction between microbial cells and transducers. The microbial cells are highly biosensitive to the external environment, so they are considered to be efficient bioreceptors and also tend to attach to some surfaces. The major transducers that can be used include amperometric, potentiometric, calorimetric, conductometric, colorimetric, luminescence and fluorescence [19 - 21]. In this technique, choice and strategy of microbial immobilization play a major role since it greatly influences the response, stability and long term usage. Many microbial cells have been efficiently in biosensors. One example is Pseudomonas sp. which was procured from the corroded material surface and attached on acetyl cellulose membrane [22].

Enzyme Based Biosensors

Enzyme based biosensors use immobilized enzymes for specific reactions that are measured using specific transducers. Different types of enzymes belonging to the classes like oxido-reductases, hydrolases, ligases, isomerases and lyases have been connected with definite transducers for the construction of biosensor for the variety of applications in clinical, medical and pharmaceutical areas, food and fermentation processes, environmental monitoring, *etc.* [23]. Commonly used biosensor with immobilized enzyme is for measuring blood glucose. It uses immobilized glucose oxidase or glucose dehydrogenase [24]. Apart from this many enzymes like invertase for sucrose determination [25], Amino acid oxidase for amino acids, Urease for urea have been used extensively. Some of the factors that influence the act of enzyme-based biosensors include enzyme loading, pH, temperature, type of immobilization method used to hold on to the enzyme as well as the thickness of the enzyme on the sensor, *etc.*

Although the biosensors are simple, portable, and a continuously operational configuration, these biosensors have some limitations too, when applied to environmental monitoring. Some of the major drawbacks include the inadequate amount of environmental pollutants that can act as the substrate for the enzyme and the high detection limits. To overcome this disadvantage substance or inhibitors that specifically interact with immobilized enzymes and inhibit its biocatalytic properties can be used. Such inhibitors will bind either to the enzyme or enzyme-substrate complex and further interfere with the enzymatic reactions. The important advantage of this method is that most of the enzymes are liable to a very low concentration of inhibitors, thus augment the sensitivity of the biosensor. The inhibition of enzyme can either be reversible, in which the binding of an inhibitor can be reversed by decreasing inhibitor concentration, or irreversible, in which binding of the inhibitor results in perpetual inhibition of enzyme activity [26].

INFLUENCE OF ENZYME INHIBITION ON THE WORKING OF BIOSENSOR

Biosensor based on enzyme inhibition plays an important role in the analysis of recent researches. The finding of an analyte is done by estimating the variation in enzyme activity according to the equation:

$$I = \frac{Ao - Ai}{Ao}$$

where Ao is the activity in the lack of inhibitor and Ai in the occurrence of inhibitor. This procedure is valid for reversible inhibition and, sometimes, in the case of irreversible inhibition. Also, the activity of inhibition based biosensor is influenced by many parameters like Enzyme loading, incubation time, substrate concentration, *etc.* This type of biosensor can be used in many areas.

APPLICATION OF ENZYME INHIBITION BASED BIOSENSOR IN ENVIRONMENTAL FIELD

Pesticide Detection

Pesticides are mandatory to the current agriculture and are used widely in agricultural fields to manage the attack of insects, microbes and rodents on crops; lessen the augmentation of weeds; enhance agricultural productivity, thus improving crop yields, and diminish post-harvest losses thereby crop production enhancement [27, 28]. A huge amount of pesticides are deposited in the environment every year. Due to the nature of achieving maximum effectiveness, pesticides continue in nature for an extensive duration. Apart from their refractory structure and agricultural payback, pesticides also enforce sharp toxicological impacts onto the various life forms. This may be harmful to the various life forms when it affects the food chains [29, 30]. Many of the pesticides are highly toxic [31] carcinogenic, and tumorogenic [32, 33].

Thus pesticides that are posing many problems, has to be analysed with much accuracy. Some of the commonly used techniques like chromatographic techniques (HPLC, GC, *etc.*) are associated with various restrictions like time consumption, difficult nature, the requirement of high-priced equipments and highly trained technicians, *etc.* As a result a novel method has to be used to detect these compounds in a delicately, selectively and quick way. Due to the outstanding performance index, easiness in operation, many biosensors have been extensively used for bio-monitoring various environmental samples for pesticide.

Some of the important biosensors working on the principle of enzyme inhibition for pesticide detection are discussed below.

Cholinesterase-Based Biosensors

Cholinesterases (ChEs) are a group of enzymes which help in the hydrolysis of choline-based esters, most of them act as neurotransmitters. It is chiefly responsive to the inhibition by organophosphates and carbamate pesticides. In the absence of an analyte in the solution, the substrate acetylthiocholine is changed into thiocholine and acetic acid. Applied voltage further oxidizes thiocholine. In the presence of the inhibitor, the conversion is worthless.

The serine hydrolases fit into the esterase family which comes in the α/β-hydrolase superfamily [34] and play an important task in several significant areas, like neurobiology, pharmacology, and toxicology [35]. ChEs help in the hydrolysis of choline esters at a superior pace than other esters.

Cholinesterase based biosensors can work as monoenzymatic or bienzymatic biosensors. Main types of cholinesterase enzyme in nature are acetylcholinesterase (AChE) and butyrylcholine-esterase (BChE) which act on its own substrate. Acetylcholinesterase act upon acetylcholine releasing choline and acetic acid, similarly butyrylcholine-esterase gives choline and butyric acid. AChE is established in conducting tissues and in membranes of red blood cells and BChE is mainly present in the liver [36].

The arrangement of ChEs is common. Both the enzymes are characterized by the presence of typical hydrophobic gorge extending into the surface of the protein molecule. A catalytic triad with three amino acids: Ser, His, Glu are present in ChEs. The variation in the substrate specificity of ChEs is most likely due to the difference in the architecture of the base of active site gorge [37].

An enzymatic finding of pesticides is mostly based on the inhibition of cholinesterase (ChE). Carbamate and organophosphate insecticides are the chief cholinesterase inhibitors. The first biosensor working on the principle of inhibition of cholinesterase was put forward in 1962 by Guilbault *et al.* for checking organophosphorus compounds [38, 39]. Even other compounds, like heavy metals, fluoride or nicotine, also will inhibit ChE enzyme.

This biosensor can also be used for the detection of artificial substrates like acetyl thiocholine and butyrlythiocholine. The major advantages for this are simpler design and lower detection limits when compared to bienzymatic biosensor. In bienzymatic biosensor cholinesterase is attached to choline oxidase (ChOD) [40].

A biosensor with immobilized AChE and ChOD on Au-Pt bimetallic NPs was put forward for the finding pesticides and nerve agents [41]. The combined effect of these nanoparticles increased the surface area and can result in the electron transfer process, dropping the applied potential for the finding H_2O_2. Besides detecting H_2O_2 oxidation, ChE inhibition can be estimated using a Clark electrode which can calculate the oxygen utilized by the ChOD catalyzed reaction [42].

Alkaline Phosphatase Based Biosensor

Alkaline phosphatase is inhibited by different compounds. So it can be effectively used for finding such pesticides. A biosensor was put forward for the detection of paraxon by Ayyagari which works on the principle of chemiluminescent ALP [43]. Paraxon is an organophosphate oxon which is an effective metabolite of the insecticide parathion. The biosensor work on the principle of measurement of the intensity of the light produced by ALP-catalyzed dephosphorylation of a chemiluminescent substrate, chloro 3-(4-methoxy spiro [1,2-dioxetane-3-2'-tricyclo-[3.3.1.1]-decan]-4- yl) phenyl phosphate.

A fluorescent ALP-based biosensor that can be used for finding the presence of organochlorine, pesticides (carbamate and fenitrothion), heavy metals and CN– was put forward by Garcia Sanchez *et al.* [44]. A novel biosensor was designed using carbon nanopowder paste for fabricating alkaline phosphatase inhibition-based amperometric biosensor for the detection of carbofuran (2,3-dihydro-2-2-dimethyl-7-benzofuranyl methylcarbamate). Carbofuran is one of the most toxic carbamate pesticides because it inhibits cholinesterase effectively. It can result in noteworthy toxicity to humans and wildlife animals. It works on the principle of interaction between alkaline phosphatase cross-linked with bovine serum albumin and glutaraldehyde immobilized on the surface of a carbon nanopowder paste electrode [45].

Vanadium is one of the important transition elements in biological systems, which is present as an ultra-trace metal in some marine organisms, in the prosthetic group of bromoperoxidases in certain marine algae and as part of the nitrogenase system of some bacteria and plants. Many industries working with fossil fuels discharge vanadium into the environment. So, mining areas and areas related to crude oil will release more vanadium into the atmosphere. A screen-printed based amperometric biosensor which can be simply used in an analytical laboratory for the discovery of vanadium was developed in the year 2014. Here alkaline phosphatase was cross-linked with the working electrode of screen-printed carbon electrodes which was formerly tailored by gold nanoparticles. To improvise the existing biosensor with improved conductivity and performance for the finding of vanadium, Gold nanoparticles were deposited onto the working electrode before the enzyme immobilization [46].

Peroxidase-Based Biosensors

In this type of biosensors H_2O_2 first, oxidize peroxidise which is then reduced by phenolic compounds. Several compounds are found to inhibit the enzyme activity of peroxidase. A biosensor based on the inhibition of peroxidase was developed for the finding of thiodicarb, a carbamate pesticide [47]

Several electrochemical biosensors working on the principle of inhibiting tyrosinase are currently in use. For the determination of glyphosate, a biosensor based on inhibition of peroxidise, immobilised on modified nanoclay was developed. Square-wave voltammetry was used for the optimisation and application of the biosensor, and several parameters were studied to conclude the optimum experimental conditions [48].

Tyrosinase-Based Biosensors

Another type of biosensor for pesticide detection is based on enzyme tyrosinase. The oxidation reaction of tyrosinase is usually carried out in two steps. The first o-hydroxylation of monophenol to o-diphenol is carried out by the enzyme and further it is oxidized to its corresponding o-quinone.

A tyrosinase-based biosensor was made into reality by immobilizing the enzyme on diazonium-functionalized screen-printed gold electrodes. The biosensor showed a rapid response to the changes in the concentration of all the tested phenolic compounds under optimized conditions. Sensitivity and limit of detection (LOD) were determined, and catechol was found to display the highest sensitivity (36.3 mA M−1) and the lowest LOD (0.1 μmol L−1). This biosensor was successfully used for the determination of polyphenols in tea samples [49].

A tyrosinase-based amperometric biosensor was developed for finding bisphenol in a flow-batch monosegmented sequential injection system. The enzyme was immobilized in a sol-gel TiO_2 matrix which is tailored with multi-walled carbon nanotubes, polycationic polymer poly(diallyl dimethylammonium chloride), (PDDA) and Nafion. The electrochemical behavior of the biosensor for bisphenol A was studied and analytical features were confirmed with respect to linear range, biosensor sensitivity, the limit of detection, long term stability, repeatability and reproducibility [49, 50].

Tyrosinase is usually inhibited by several compounds, like carbamate pesticides and atrazine. Some of the biosensors based on the inhibition of tyrosinase activity are stated in Table 2.

Table 2. Examples of biosensors based on the inhibition of tyrosinase activity.

Sample	Linear Range, IU	Detection Limit, IU	References
Dichlorvos	Up to 8×10^{-6}	**6×10^{-8}**	[51]
Pirimicarb	2×10^{-5} to 5×10^{-3}	10^{-5}	[52]
Carbofuran	$10^{-5} - 10^{-2}$	5×10^{-6}	[52]
Thiodicarb	3.75×10^{-7} to 2.23×10^{-6}	1.58×10^{-7}	[53]

Although Tyrosinase biosensors can stand high temperatures, it suffers from poor specificity because many substrates and inhibitors can obstruct its action.

Heavy Metal Detection

In the current environmental situation the amount of heavy metal deposition is increasing. Many of the heavy metals are important but some of them are highly toxic to humans. So it is very important to detect these harmful agents and reduce their activity. There are many methods to detect heavy metals. Some of them include spectroscopic methods like AAS, AES, *etc.* [54, 55]. These types of sensors monitor the presence of heavy metals based on the inhibitory properties on biocatalytic properties. But one of the main obstructions is the number of enzymes that are sensitive to heavy metals.

Major hindrances for the usage of these biosensors are either its expensive nature or not useful for detecting metal ions with low concentrations. In this situation, biosensors play a key role in the determination of heavy metals. Several enzymes like alkaline phosphatase [56], glucose oxidase (GOD) [57], peroxidase, *etc.* have been combined with the electrochemical and optical transducer to determine the number of metal ions like arsenic, bismuth, beryllium, zinc, mercury, cadmium, lead, and copper [58]. Some of the examples of biosensor used for heavy metal detection are listed in Table **3**.

Table 3. Examples of biosensors for heavy metal detection.

Heavy Metal	Inhibited Enzyme	Detection	References
Mercury, Copper	Urease	Optical	[59]
Mercury	Acetylcholine	Amperometric	[60]
Cadmium	Urease	Optical	[61]
Mercury	Urease	Potentiometric	[62]
Copper	Acetylcholine	Amperometric	[63]

(Table 3) cont.....

Heavy Metal	Inhibited Enzyme	Detection	References
Chromium	Glucose Oxidase	Amperometric	[64]
Silver	Alkaline Phosphatase ase	Fluorescence	[65]
Chromium	Tyrosinase	Amperometric	[66]

CONCLUSION

Biosensors are basic analytical tool that are found to have many applications in detecting the presence of specific analytes. The basic principle of action of biosensor is based on different mechanisms. Application of biosensor based on enzyme inhibition benefits a wide range of application in variety of areas for the qualitative improvement of human lives.

CONSENT FOR PUBLICATION

Not applicable.

CONFLICT OF INTEREST

The author(s) confirms that there is no conflict of interest.

ACKNOWLEDGEMENTS

Declared none.

REFERENCES

[1]   Turner APF, Karube I, Wilson GS, Eds. Biosensors: Fundamentals and Applications. Oxford: Oxford University Press 1987; p. 770.

[2]   Hughes WS. The potential difference between glass and electrolytes in contact with water. J Am Chem Soc 1922; 44: 2860-6.
        [http://dx.doi.org/10.1021/ja01433a021]

[3]   Griffin EG, Nelson JM. The influence of certain substances on the activity of invertase. J Am Chem Soc 1916; 38: 722.
        [http://dx.doi.org/10.1021/ja02260a027]

[4]   Kerridge PT, Tookey M. The use of the glass electrode in biochemistry. Biochem J 1925; 19(4): 611-7.
        [http://dx.doi.org/10.1042/bj0190611] [PMID: 16743549]

[5]   Stow RW, Randall BF. Electrical measurement of the pCO_2 of blood. Am J Physiol 1954; 179: 678.

[6]   Bradley AF, Severinghaus JW, Stupfel M. Accuracy of blood pH and pCO_2 determinations. J Appl Physiol 1956; 9(2): 189-96.
        [http://dx.doi.org/10.1152/jappl.1956.9.2.189] [PMID: 13376426]

[7]   Clark LC Jr, Lyons C. Electrode systems for continuous monitoring in cardiovascular surgery. Ann N Y Acad Sci 1962; 102: 29-45.

[http://dx.doi.org/10.1111/j.1749-6632.1962.tb13623.x] [PMID: 14021529]

[8] Guilbault GG, Montalvo JG Jr. A urea-specific enzyme electrode. J Am Chem Soc 1969; 91(8): 2164-5.
      [http://dx.doi.org/10.1021/ja01036a083] [PMID: 5784180]

[9] Bergveld P. Development of an ion-sensitive solid-state device for neurophysiological measurements. IEEE Trans Biomed Eng 1970; 17(1): 70-1.
      [http://dx.doi.org/10.1109/TBME.1970.4502688] [PMID: 5441220]

[10] Suzuki S, Takahashi F, Satoh I, Sonobe N. Ethanol and lactic acid sensors using electrodes coated with dehydrogenase–collagen membranes. Bull Chem Soc Jpn 1975; 48: 3246-9.
      [http://dx.doi.org/10.1246/bcsj.48.3246]

[11] Schultz JS. Optical sensor of plasma constituents. US Pat 4,344,438 A 1982.

[12] Liedberg B, Nylander C, Lunström I. Surface plasmon resonance for gas detection and biosensing. Sens Actuators 1983; 4: 299-304.
      [http://dx.doi.org/10.1016/0250-6874(83)85036-7]

[13] Cass AE, Davis G, Francis GD, *et al.* Ferrocene-mediated enzyme electrode for amperometric determination of glucose. Anal Chem 1984; 56(4): 667-71.
      [http://dx.doi.org/10.1021/ac00268a018] [PMID: 6721151]

[14] Vestergaard MC, Kerman K, Hsing IM, Tamiya E, Eds. Nanobiosensors and Nanobioanalyses. Tokyo: Springer 2015.
      [http://dx.doi.org/10.1007/978-4-431-55190-4]

[15] Sassolas A, Leca-Bouvier BD, Blum LJ. DNA biosensors and microarrays. Chem Rev 2008; 108(1): 109-39.
      [http://dx.doi.org/10.1021/cr0684467] [PMID: 18095717]

[16] Xie H, Yu YH, Xie F, Lao YZ, Gao Z. A nucleic acid biosensor for gene expression analysis in nanograms of mRNA. Anal Chem 2004; 76(14): 4023-9.
      [http://dx.doi.org/10.1021/ac049839d] [PMID: 15253638]

[17] Callewaert N, Van Vlierberghe H, Van Hecke A, Laroy W, Delanghe J, Contreras R. Noninvasive diagnosis of liver cirrhosis using DNA sequencer-based total serum protein glycomics. Nat Med 2004; 10(4): 429-34.
      [http://dx.doi.org/10.1038/nm1006] [PMID: 15152612]

[18] Debouck C, Goodfellow PN. DNA microarrays in drug discovery and development. Nat Genet 1999; 21(1) (Suppl.): 48-50.
      [http://dx.doi.org/10.1038/4475] [PMID: 9915501]

[19] Tran MC. Biosensors. Paris: Chapman and Hall and Masson 1993.

[20] Mikkelsen SR, Corton E. Bioanalytical Chemistry. New Jersey: John Wiley and Sons 2004.
      [http://dx.doi.org/10.1002/0471623628]

[21] Blum LJ, Coulet PR, Eds. Biosensor Principles and Applications. New York: Marcel Dekker 1991.

[22] Dubey RS, Upadhyay SN. Microbial corrosion monitoring by an amperometric microbial biosensor developed using whole cell of Pseudomonas sp. Biosens Bioelectron 2001; 16(9-12): 995-1000.
      [http://dx.doi.org/10.1016/S0956-5663(01)00203-2] [PMID: 11679280]

[23] Guilbault GG. Analytical Uses of Immobilized Enzymes. NY: Marcel Dekker 1984; pp. 53-5.

[24] Clark LC Jr, Lyons C. Electrode systems for continuous monitoring in cardiovascular surgery. Ann N Y Acad Sci 1962; 102: 29-45.
      [http://dx.doi.org/10.1111/j.1749-6632.1962.tb13623.x] [PMID: 14021529]

[25] Xu DP, Sung SJ, Loboda T, Kormanik PP, Black CC. Characterization of sucrolysis *via* the uridine diphosphate and pyrophosphate-dependent sucrose synthase pathway. Plant Physiol 1989; 90(2): 635-

42.
[http://dx.doi.org/10.1104/pp.90.2.635] [PMID: 16666820]

[26]   Sheo L, Upadhyay B, Verma N. Enzyme inhibition based biosensors: a review. Anal Lett 2013; 46: 225-41.
[http://dx.doi.org/10.1080/00032719.2012.713069]

[27]   Hua F, Yunlong Y, Xiaoqiang C, Xiuguo W, Xiaoe Y, Jingquan Y. Degradation of chlorpyrifos in laboratory soil and its impact on soil microbial functional diversity. J Environ Sci (China) 2009; 21(3): 380-6.
[http://dx.doi.org/10.1016/S1001-0742(08)62280-9] [PMID: 19634452]

[28]   Odukkathil G, Vasudevan N. Toxicity and bioremediation of pesticides in agricultural soil. Rev Environ Sci Biotechnol 2013; 12: 421-44.
[http://dx.doi.org/10.1007/s11157-013-9320-4]

[29]   McEwen FL, Stephenson GR. The use and significance of pesticides in the environment. New York: John Wiley and Sons 1979.

[30]   Viswanathan PN. Environmental toxicology in India. Membrane Biology 1985; 11: 88-97.

[31]   De Flora S, Viganò L, D'Agostini F, *et al.* Multiple genotoxicity biomarkers in fish exposed in situ to polluted river water. Mutat Res 1993; 319(3): 167-77.
[http://dx.doi.org/10.1016/0165-1218(93)90076-P] [PMID: 7694138]

[32]   Kuroda K, Yamaguchi Y, Endo G. Mitotic toxicity, sister chromatid exchange, and rec assay of pesticides. Arch Environ Contam Toxicol 1992; 23(1): 13-8.
[http://dx.doi.org/10.1007/BF00225990] [PMID: 1637192]

[33]   Rehana Z, Malik A, Ahmad M. Mutagenic activity of the Ganges water with special reference to the pesticide pollution in the river between Kachla to Kannauj (U.P.), India. Mutat Res 1995; 343(2-3): 137-44.
[http://dx.doi.org/10.1016/0165-1218(95)90079-9] [PMID: 7791807]

[34]   Lenfant N, Hotelier T, Bourne Y, Marchot P, Chatonnet A. Proteins with an alpha/beta hydrolase fold: Relationships between subfamilies in an ever-growing superfamily. Chem Biol Interact 2013; 203(1): 266-8.
[http://dx.doi.org/10.1016/j.cbi.2012.09.003] [PMID: 23010363]

[35]   Miao Y, He N, Zhu JJ. History and new developments of assays for cholinesterase activity and inhibition. Chem Rev 2010; 110(9): 5216-34.
[http://dx.doi.org/10.1021/cr900214c] [PMID: 20593857]

[36]   Colović MB, Krstić DZ, Lazarević-Pašti TD, Bondžić AM, Vasić VM. Acetylcholinesterase inhibitors: pharmacology and toxicology. Curr Neuropharmacol 2013; 11(3): 315-35.
[http://dx.doi.org/10.2174/1570159X11311030006] [PMID: 24179466]

[37]   Vellom DC, Radić Z, Li Y, Pickering NA, Camp S, Taylor P. Amino acid residues controlling acetylcholinesterase and butyrylcholinesterase specificity. Biochemistry 1993; 32(1): 12-7.
[http://dx.doi.org/10.1021/bi00052a003] [PMID: 8418833]

[38]   Kramer DN, Cannon PL, Guilbault GG. Electrochemical determination of cholinesterase and thiocholine esters. Anal Chem 1962; 34: 842-5.
[http://dx.doi.org/10.1021/ac60187a038]

[39]   Audrey S, Beatriz PS, Jean-Louis M. Biosensors for pesticide detection: new trends. Am J Anal Chem 2012; 3(3): 210-32.
[http://dx.doi.org/10.4236/ajac.2012.33030]

[40]   [Thévenot DR., Toth K, Durst RA, Wilson GS. Electrochemical Biosensors: Recommended Definitions and Classification. Pure Appl Chem 1999; 71(12): 2333-48.
[http://dx.doi.org/10.1351/pac199971122333]

[41]   Upadhyay S, Rao GR, Sharma MK, Bhattacharya BK, Rao VK, Vijayaraghavan R. Immobilization of acetylcholineesterase-choline oxidase on a gold-platinum bimetallic nanoparticles modified glassy carbon electrode for the sensitive detection of organophosphate pesticides, carbamates and nerve agents. Biosens Bioelectron 2009; 25(4): 832-8.
[http://dx.doi.org/10.1016/j.bios.2009.08.036] [PMID: 19762223]

[42]   Campanella L, Achilli M, Sammartino MP, Tomassetti M. Butyrylcholine enzyme sensor for determining organophosphorus inhibitors. J Electroanal Chem 1991; 321(2): 237-49.
[http://dx.doi.org/10.1016/0022-0728(91)85599-K]

[43]   Ayyagari MS, Kamtekar S, Pande R, Marx KA, et al. Biosensors for pesticide detection based on alkaline phosphatase-catalyzed chemiluminescence. Mater Sci Eng C 1995; 2(4): 191-6.
[http://dx.doi.org/10.1016/0928-4931(95)00077-1]

[44]   Sánchez FG, Diaz AN, Peinado MR, Belledone C. Free and sol-gel immobilized alkaline phosphatase-based biosensor for the determination of pesticides and inorganic compounds. Anal Chim Acta 2003; 484(1): 45-51.
[http://dx.doi.org/10.1016/S0003-2670(03)00310-6]

[45]   Samphao A, Suebsanoh P, Wongsa Y, et al. Alkaline phosphatase inhibition-based amperometric biosensor for the detection of carbofuran. Int J Electrochem Sci 2013; 8: 3254-64.

[46]   Alvarado-Gámez AL, Alonso-Lomillo MA, Domínguez-Renedo O, Arcos-Martínez MJ. A disposable alkaline phosphatase-based biosensor for vanadium chronoamperometric determination. Sensors (Basel) 2014; 14(2): 3756-67.
[http://dx.doi.org/10.3390/s140203756] [PMID: 24569772]

[47]   Moccelini SK, Vieira IC, de Lima F, et al. Determination of thiodicarb using a biosensor based on alfalfa sprout peroxidase immobilized in self-assembled monolayers. Talanta 2010; 82(1): 164-70.
[http://dx.doi.org/10.1016/j.talanta.2010.04.015] [PMID: 20685452]

[48]   Grasielli CO, Sally KM, Marilza C, Ailton JT, et al. Biosensor based on atemoya peroxidase immobilized on modified nanoclay for glyphosate biomonitoring. Talanata 2012; 98: 130-6.
[http://dx.doi.org/10.1016/j.talanta.2012.06.059]

[49]   Cortina-Puig M, Muñoz-Berbel X, Calas-Blanchard C, Marty JL. Diazonium-functionalized tyrosinase-based biosensor for the detection of tea polyphenols. Mikrochim Acta 2010; 171(1–2): 187-93.
[http://dx.doi.org/10.1007/s00604-010-0425-y]

[50]   Kochana J, Wapiennik K, Kozak J, et al. Tyrosinase-based biosensor for determination of bisphenol A in a flow-batch system. Talanta 2015; 144: 163-70.
[http://dx.doi.org/10.1016/j.talanta.2015.05.078] [PMID: 26452806]

[51]   Vidal JC, Esteban S, Gil J, Castillo JR. A comparative study of immobilization methods of a tyrosinase enzyme on electrodes and their application to the detection of dichlorvos organophosphorus insecticide. Talanta 2006; 68(3): 791-9.
[http://dx.doi.org/10.1016/j.talanta.2005.06.038] [PMID: 18970392]

[52]   Campanella L, Lelo D, Martini E, Tomassetti M. Organophosphorus and carbamate pesticide analysis using an inhibition tyrosinase organic phase enzyme sensor; comparison by butyrylcholinesterase+choline oxidase opee and application to natural waters. Anal Chim Acta 2007; 587(1): 22-32.
[http://dx.doi.org/10.1016/j.aca.2007.01.023] [PMID: 17386749]

[53]   De Lima F, Lucca B, Barbosa AMJ, Ferreira VS, et al. Biosensor based on pequi polyphenol oxidase immobilized on chitosan crosslinked with cyanuric chloride for thiodicarb determination. Enzyme Microb Technol 2010; 47(4): 153-8.
[http://dx.doi.org/10.1016/j.enzmictec.2010.05.006]

[54]   de Albuquerque YDT, Ferreira LF. Amperometric biosensing of carbamate and organophosphate

pesticides utilizing screen-printed tyrosinase-modified electrodes. Anal Chim Acta 2007; 596(2): 210-21.
[http://dx.doi.org/10.1016/j.aca.2007.06.013] [PMID: 17631099]

[55]   Wang HT, Kang BS. Selective detection of Hg (II) ions from Cu (II) and Pb (II) using AlGaN / GaN high electron mobility transistors. Electrochem Solid-State Lett 2007; 10(11): J150-3.
[http://dx.doi.org/10.1149/1.2778997]

[56]   Lee S, Lee W. Determination of heavy metal ions using conductometric biosensor based on sol-gel immobilized urease. Bull Korean Chem Soc 2002; 23(8): 1169-72.
[http://dx.doi.org/10.5012/bkcs.2002.23.8.1169]

[57]   Koncki R, Rudnicka K, Tymecki Ł. Flow injection system for potentiometric determination of alkaline phosphatase inhibitors. Anal Chim Acta 2006; 577(1): 134-9.
[http://dx.doi.org/10.1016/j.aca.2006.05.100] [PMID: 17723664]

[58]   Kukla AL, Kanjuk NI, Starodub NF, Shrishov YM. Multienzyme electrochemical sensor array for determination of heavy metal ions. Sens Actuators B Chem 1999; 57: 213-8.
[http://dx.doi.org/10.1016/S0925-4005(99)00153-7]

[59]   Tsai HC, Doong RA. Simultaneous determination of pH, urea, acetylcholine and heavy metals using array-based enzymatic optical biosensor. Biosens Bioelectron 2005; 20(9): 1796-804.
[http://dx.doi.org/10.1016/j.bios.2004.07.008] [PMID: 15681196]

[60]   Stoytcheva M, Sharkova V. Kinetics of the inhibition of immobilized acetylcholinesterase with Hg (II). Electroanalysis 2002; 14(14): 1007-10.
[http://dx.doi.org/10.1002/1521-4109(200208)14:14<1007::AID-ELAN1007>3.0.CO;2-1]

[61]   Lee MM, Russel DA. Novel determination of cadmium ions using an enzyme self as a monolayer with surface plasmon resonance. Anal Chim Acta 2003; 500: 119-25.
[http://dx.doi.org/10.1016/S0003-2670(03)00943-7]

[62]   Doong RA, Tsai HC. Immobilization and characterization of sol-gel encapsulated acetylcholinesterase fiber-optic biosensor. Anal Chim Acta 2001; 434: 239-46.
[http://dx.doi.org/10.1016/S0003-2670(01)00853-4]

[63]   Evtugyn GA, Budnikov HC, Nikolskaya EB. Biosensors for the determination of environmental inhibitors of enzymes. Russ Chem Rev 1999; 68(12): 1041-64.
[http://dx.doi.org/10.1070/RC1999v068n12ABEH000525]

[64]   Zeng GM, Tang L, Shen GL, Huang GH, Niu CG. Determination of trace chromium (VI) by an inhibition-based enzyme biosensor incorporating an electropolymerized aniline membrane and ferrocene as an electron transfer mediator. Int J Environ Anal Chem 2004; 84(10): 761-74.
[http://dx.doi.org/10.1080/03067310410001730619]

[65]   Sánchez FG, Dıaz AN, Peinado MR, Belledone C. Free and sol-gel immobilized alkaline phosphatase based biosensor for the determination of pesticides and inorganic compounds. Anal Chim Acta 2003; 484(1): 45-51.

[66]   Renedo OD, Lomillo MAA, Martinez MJA. Optimisation procedure for the inhibitive determination of chromium (111) using an amperometric tyrosinase biosensor. Anal Chim Acta 2004; 521: 215-21.
[http://dx.doi.org/10.1016/j.aca.2004.06.026]

Recent Insights of Matrix Metalloproteinase Inhibitors in Therapeutic Applications

Ravichandran Rathna, Bethu Madhumitha, Ravichandran Viveka and **Ekambaram Nakkeeran**[*]

Research Laboratory, Department of Biotechnology, Sri Venkateswara College of Engineering (Autonomous), Sriperumbudur Tk - 602 117, Tamil Nadu, India

Abstract: Matrix metalloproteinases are proteolytic zinc-dependent enzymes that play a pivotal function in cell migration, proliferation, differentiation, programmed cell death, and other physiological processes. Recent studies demonstrated that the imbalance activation and inhibition of these enzymes resulted in unexpected physiological and pathological processes. Thus, it fueled the interest in matrix metalloproteinase and its inhibitors in medicinal and pharmaceutical chemists. This chapter discusses the therapeutic accomplishments of matrix metalloproteinase inhibitors in arthritis, autoimmune disease, inflammations, cancer, and cardiovascular disease. Further, the chapter discusses clinical trial implications, obstacles, and future research.

Keywords: Arthritis, Autoimmune disease, Classifications, Cancer, Cardiovascular disease, Clinical implications, Endogenous tissue inhibitors, Exogenous inhibitors, Functional roles, Inflammations, Matrix metalloproteinase, Matrix metalloproteinase inhibitors, Mechanism, Obstacles, Therapeutic accomplishments.

INTRODUCTION

In 1962, Gross and Lapiere investigated a compelling collagenolytic activity in small tissue fragments during metamorphosis in tadpoles [1]. The pioneer of interstitial collagenase from the amphibian, *Rana catesbeiana,* is a cornerstone for potent biological macromolecules, matrix metalloproteinases (MMPs). MMPs are also known as matrixins that belong to a metzincin superfamily, zinc-endopeptidases. It is a calcium-dependent zinc-containing endoproteases which are traditionally categorized into six types [2, 3] based on homology, substrate sp-

---

[*] **Corresponding author Ekambaram Nakkeeran:** Research Laboratory, Department of Biotechnology, Sri Venkateswara College of Engineering (Autonomous), Sriperumbudur Tk - 602 117, Tamil Nadu, India; E-mail: nakkeeran@svce.ac.in

G. Baskar, K. Sathish Kumar & K. Tamilarasan (Eds.)

ecificity and cellular localization (partly) as collagenases (MMP-1, -8, -13, -18), gelatinases (MMP-2, -9), stromelysins (MMP-3, -10, -11, -17), matrilysins (MMP-7, -26), membrane-type MMPs (MT-MMP-14, -15, -16, -17, -24, -25) and other MMPs (MMP-12, -19, -20, -21, -22, -23, -27, -28, -29). Recently, MMPs are classified into four types based on their domain structure as archetypal MMPs, matrilysins, gelatinases and furin-activatable MMPs [4]. All the MMPs consist of prodomain, a catalytic domain with the zinc metal active site, fibronectin-like domain, linker, and hemopexin domain that assist in the degradation of various extracellular matrix (ECM) and non-ECM [5, 6]. Further, they are released from fibroblasts [7], monocytes [8], macrophages [8], lymphocytes [9], epithelial cells [10], endothelial cells [11], polymorphonuclear leukocytes [12], and osteoblasts [13]. Catalytic domains of MMPs assist in cleaving the non-collagen substrates while N-terminal domains cleave the native fibrillar collagen molecules. In general, MMPs play a vital role in physiological or pathological processes and tissue remodeling *via* ECM protein turnover and degradation. This chapter discusses the therapeutic accomplishments of MMPI in arthritis, autoimmune disease, inflammations, cancer, and cardiovascular disease. Further, the chapter highlights the clinical trial implications, obstacles, and future researches.

MATRIX METALLOPROTEINASE

Archetypal MMPs consist of an inactive protein precursor (propeptide) with a critical cysteine residue and a zinc catalytic domain residing on the hemopexin domain. Interestingly, archetypal MMPs is sub-categorized into collagenases (MMP-1, -8, -13, -18), stromelysins (MMP-3, -10) and other archetypal MMPs (-12, -19, -20, -27). Collagenase was the first MMP discovered that is capable of cleaving triple helices collagen into fragments and essential for various biological functions such as for regulating proinflammatory factors [14], functions as an agonist of the non-ECM protein like protease-activated receptor-1 [15] and proteolysis insulin-like growth factor binding proteins [16]. Stromelysins are structurally the same as collagenase and have the capacity to degrade structural proteins of non-collagenous ECM like proteoglycans, glycoproteins, elastin, entactin, fibronectin and laminin and participate in proMMP activation. Jin *et al.* [17], reported the matriptase-dependent activation of MMP-3 increased the ECM degradation in the tumor cell microenvironment that was eventually promoting tumor development and angiogenesis. MMP-3 overexpression implicated in Osteoarthritis [18] while pulmonary hypertension associated with systemic sclerosis [19], esophageal cancer [20], and tumor progression [21] correlates to overexpression of MMP-10. MMP-12, -19, -20, and -27 are the four matrixins that are classified as other archetypal MMPs and have the same structure as archetypal MMPs but differ in their sequence and substrate specificity. MMP-12 is a macrophage metalloelastase capable of degrading elastin and also considered as a prime therapeutic target for chronic obstructive pulmonary complications [22] and cutaneous melanoma [23]. MMP-19 is also known as matrix metalloproteinase RASI-1 which exhibits potent basement membrane degradation [24] and overexpressed in gastrointestinal diseases [25] and stimulates proliferation and cell migration of tumors [26]. MMP-20 is commonly known as enamelysin, or enamel metalloproteinase plays a vital role in tooth enamel formation. In pancreatic ductal adenocarcinoma, overexpression of MMP-19 and MMP-20 results in the progression and prognosis of carcinoma tissues [27]. MMP-27 is unique and consists of intracellular retention motif and exhibits expression on macrophages during ovarian and peritoneal endometriotic lesions [28].

Matrilysins are the smallest MMPs among other MMPs since it lacks the hemopexin domain. MMP-7 actively degrades the ECM components like collagen IV, laminin, glycoprotein, and mucoproteins to form soluble ectodomains from the cell surface. However, MMP-7 is the sole MMP released by epithelial tumor cells [29] and MMP-26 expressed in multiple human cancer tissues and smooth muscle cells [30]. Gelatinase is a type IV collagenase which includes gelatinase A (MMP-2) and B (MMP-9). It contains three single repeats of fibronectin in its structure to facilitate the binding of gelatin and collagen. MMP-2 prevents the autolytic inactivation while binding to the intact collagen. However, MMP-2 is resistant to the native interstitial collagens. Besides, MMP-2 and -9 play a vital function in tissue homeostasis [31] and tumor-associated tissue remodeling [32].

Furin-activated MMPs are sub-divided as secreted MMPs (MMP-11, -21, -28); type-I transmembrane MMPs (MMP-14, -15, -16, -24), type-II transmembrane MMPs (MMP-23), and glycosylphosphatidylinositol (GPI)-anchored MMPs (MMP-17, -25). Formerly, MMP-11 is categorized under stromelysins based on the structural similarity, but intracellular activation is owing to the occurrence of furin-cleavage sites between propeptide and a catalytic domain. Hsin *et al*. [33], reported the expression of MMP-11 in the development and aggression of tumors through the focal adhesion kinase/Src kinase (FAK/Src) pathway. MMP-21 is implicated in embryogenesis and tumor progression [34] and expressed in macrophages and fibroblasts [35]. Lohi *et al*. [36], identified MMP-28 (Epilysin) expressed in normal and intact tissues like testis, intestine, lung, skin, and play an important role in tissue homeostasis, cutaneous wound repair, and tumor progression. MMP-14 is a multifunctional Membrane Type-1 MMP (MT1-MMP) associated with metastatic progression for its ability to cleave ECM. Further, low-level MT1-MMP mediated ECM degradation enhances cell migration and tumorigenesis, but high-level MT1-MMP expression does not augment tumor invasion and vascularization [37]. MMP-15 (MT2-MMP) assists inducing proteolysis and function as a capable mediator of epithelial-mesenchymal transition in carcinomas [38]. Further, the high-level expression of MT2-MMP is critical for tumor progression and angiogenesis in lung cancer [39], while esophageal and gastric related cancer are likely to cause angiogenesis with poor prognosis [40, 41]. MMP-16 (MT3-MMP) is identified as a novel molecular target and prognostic marker for gastric cancer [42]. MMP-24 (MT5-MMP) overexpression is associated with brain tumors [43], lung cancer metastasis [44] and ovarian cancer [45]. MMP-23 is a unique Membrane Type II MMP that plays a vital role in ovarian follicular development [46]. MMP-17 (MT4-MMP) and -25 (MT6-MMP) cell surface-attached through GPI-anchor. Yip *et al.* [47], reported MMP-17 as a potential biomarker during the pre-clinical study with breast cancer patients subjected to chemotherapy and erlotinib. Moreover, it plays a prime role in rheumatoid arthritis, osteoarthritis, inflammation and multiple human cancers [48]. MMP-25 is also known as serum metalloproteinase, leukolysin (MMP-25) primarily emancipate by neutrophils as an implication for chronic airways inflammation. Therefore, MT6-MMP could act as a dynamic metabolic biomarker for atopy-associated inflammation [49].

Membrane type-1 MMP's are the crucial regulator of fibronectin turnover and endocytosis [50]. Further, Philips *et al.* [51], described that MMP-1 and -2 cleave collagen fibers while elastases, MMP-2, and -9 cleave elastin fibers. Besides, the crucial accustomed process of MMPs such as tissue remodeling, cell migration, proliferation, angiogenesis, embryonic development, intercellular communication, and wound healing leads to a severe impediment during pathological conditions. For example, the typical role of MMP-2 and -9 is enhancing the cell migration; however, in the pathological process, especially in the case of tumors, it steers the metastatic pathway and acts as a potent determinant in tumor cell behavior [52]. MMPs also exhibit to inhibit the calcium mobilization mechanism of vascular smooth muscle contraction that fostered the aortic dilation [53]. Hence, MMP inhibition can have both favorable and detrimental effects. Since the 1990s, matrix metalloproteinase inhibitors (MMPI) as a broad scale therapeutic agent, but its bioavailability, efficacy, and lack of selective antagonist are the critical properties that make it unsuitable in clinical trials [54]. Classification, functional roles, and inhibitors of matrix metalloproteinases are shown in Table **1**.

MATRIX METALLOPROTEINASE INHIBITORS

Classification of MMP is quite capricious because the principal substrate is still a matter controvert. Further, studies suggested that stromal cells present in tumor microenvironment induce and regulates MMP but not the tumor cells [55]. Many pathophysiology conditions like rheumatoid arthritis, osteoarthritis, chronic obstructive pulmonary disease, osteoporosis, inflammatory bowel disease, psoriasis, multiple sclerosis, macular degeneration, and cancer are mainly due to the alteration/modification in ECM that significantly influence the fundamental cellular processes (cell proliferation, migration, and invasion). Thus, for the past few decades, inhibition of MMP is regarded as a fascinating target for therapeutic intervention. MMP inhibitors (MMPI) are used to modulate the MMP activity in different pathophysiological conditions. Naturally, MMPs are regulated at various levels of expression and activation with the help of proinflammatory cytokines and growth factors. Further, the cysteine switch mechanism influences the activation of MMP. Generally, regulation of MMPs ensues by post-translational modification, generation of zymogens (blocked active protein site) and simultaneous expression of tissue inhibitors of metalloproteinase (TIMPs) and MMPs [56]. MMPI is classified into endogenous and exogenous inhibitors (Fig. 1).

Table 1. Classification, functional roles, and inhibitors of matrix metalloproteinases.

Matrix Metalloproteinases (MMP)	Preferred Name	Category	Molecular Mass (kDa)	Chromosome Location	Function (process)	Inhibitors
MMP-1	Interstitial collagenase	Collagenase	54	11q22.2	Cellular protein metabolic process, collagen catabolic process, cytokine-mediated signaling pathway, ECM disassembly, leukocyte migration, proteolysis, viral process.	Minocycline, Marimastat, Ilomastat, Batimastat, BB-1101, MMI270B, Metastat, Doxycycline, FN-439
MMP-2	Type IV collagenase	Gelatinases	72	16q12.2	Metallopeptidase activity, protein and zinc ion binding, serine-type endopeptidase activity.	Minocycline, BB-2516, TIMP-4, BB-94, Doxycycline, BB-1101, MMI270B, Ilomastat

(Table 1) cont.....

Matrix Metalloproteinases (MMP)	Preferred Name	Category	Molecular Mass (kDa)	Chromosome Location	Function (process)	Inhibitors
MMP-3	Stromelysin-1	Stromelysins	54	11q22.3	Cellular response to nitric oxide, collagen catabolic process, cytokine-mediated signaling pathway, ECM disassembly, negative regulation of hydrogen peroxide metabolic process, positive regulation of oxidative stress-induced cell death, proteolysis.	BB-94, BB-1101, MMI270B, Doxycycline, FN-439, Ilomastat, BB-2516, Minocycline
MMP-7	Matrilysins	Matrilysins	29	11q22.2	The maternal process involved in female pregnancy, proteolysis, response to nutrient levels.	BB-94, BB-1101, Doxycycline, BB-2516, Minocycline
MMP-8	Neutrophil collagenase	Collagenases	52	11q22.3	Collagen catabolic process, proteolysis, endodermal cell differentiation, extracellular matrix disassembly, neutrophil degranulation.	TIMP-1, BB-94, BB-1101, MMI270B, CMT-3, Doxycycline, FN-439, Ilomastat, BB-2516
MMP-9	Matrix metalloproteinase-9	Gelatinases	92	20q13.12	-	TIMP-1, BB-94, BB-1101, MMI270B, FN-439, Ilomastat, BB-2516, Minocycline

(Table 1) cont.....

Matrix Metalloproteinases (MMP)	Preferred Name	Category	Molecular Mass (kDa)	Chromosome Location	Function (process)	Inhibitors
MMP-10	Stromelysin-2	Stromelysins	45	11q22.2	Development of embryos, tissue reconstruction, reproduction	-
MMP-11	Stromelysin-3	Furin-activated MMPs	55	22q11.23		-
MMP-12	Macrophage metalloelastase	Other Enzymes	54	11q22.2		BB-1101
MMP-13	Collagenase 3	Collagenases	60	11q22.2		Doxycycline, BB-1101, CMT-3, MMI270B
MMP-14	Matrix metalloproteinase-14	MT-MMP	66	14q11.2		Ilomastat, TIMP-1, TIMP-2, BB-1101, BB-2516
MMP-15	Matrix metalloproteinase-15	MT-MMP	76	16q21		-
MMP-16	Matrix metalloproteinase-16	MT-MMP	31	8q21.3		-
MMP-17	Matrix metalloproteinase-17	Stromelysins	125	12q24.33		TIMP1, TIMP2
MMP-18	Collagenase-4	Collagenases	53	-		-
MMP-19	RASI 1	Other Enzymes	57	12q13.2		-
MMP-20	Enamelysin	Other Enzymes	54	11q22.2		-
MMP-21	Matrix metalloproteinase-21	Other Enzymes	43	10q26.2		-
MMP-22	Matrix metalloproteinase-22	Other enzymes	-	-	-	-
MMP-23	Matrix metalloproteinase-23	Other Enzymes	-	1p36.33	Development of embryos, tissue reconstruction, reproduction	-
MMP-24	Matrix metalloproteinase-24	MT-MMP	-	-	-	-
MMP-25	Matrix metalloproteinase-25	MT-MMP	78	-	-	-
MMP-26	Matrix metalloproteinase-26	Matrilysins	28	11p15.4	Development of embryos, tissue reconstruction, reproduction	TIMP1, TIMP4
MMP-28	Epilysin	Other enzymes	56	-		-
MMP-29	Matrix metalloproteinase-28	Other enzymes	331	17q12		-

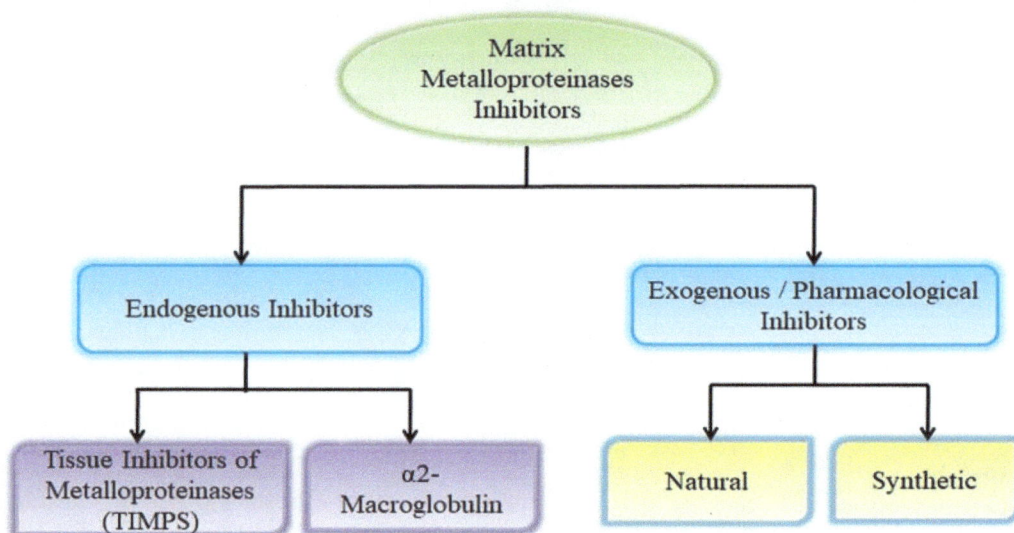

Fig. (1). Classification of matrix metalloproteinase inhibitors.

Endogenous Inhibitors

TIMPs and α-2-Macroglobulin are important endogenous inhibitors of MMP. TIMPs are naturally occurring inhibitors present in ECM and binds non-covalently with MMP in a 1:1 stoichiometric complex. As a result of the complex formation, the catalytic domain in the pro-peptide blocked for the access of substrates. Hence, TIMPs are important regulators of ECM turnover, tissue reconstruction, and cellular behavior. TIMPs family comprise of TIMP1, 2, 3, and 4 of molecular mass ranging from 21 to 27.5 kDa with a sequence homology ranging from 40-50% are identified in humans. Fig. (**2**) illustrates the characteristics of TIMPs. TIMPs contains two distinct domains, namely, N- (125 amino acid residues) and C-terminal domain (65 amino acid residues), where three disulfide bridges serve to stabilize each conformation [6]. TIMP1, 2, and 4 are secreted protein, while TIMP3 is a membrane-bound protein and confined to ECM. Further, TIMP1, TIMP2, TIMP3, and TIMP4 are potential biomarkers in tuberculosis [57], non-small cell lung carcinoma [58], oral cancer progression [59] and paroxysmal atrial fibrillation [60], respectively. Independent folding of the N-terminal domain could inhibit the MMP activity. All TIMPs are effective in inhibiting secreted and membrane-type MMPs except TIMP-1 while TIMP-2 is the most efficient inhibitor than others [61]. Lizarraga *et al.* [62], reported that TIMP-4 regulates carcinogenesis through the activation of apoptosis in cervical cancer cells. TIMP2 exhibits to control interleukin-10 (IL-10) expression and also impart anti-apoptotic effect [63]. α-2-Macroglobulin is one of the major proteins present in extravascular secretions like lymph. α-2-Macroglobulin when comp-lexed with MMP, acts as a biological marker for rheumatoid arthritis [64] and drug safety evaluation [65]. This protein activated by protease cleavage at the bait region; as a consequence, thiol ester bonds activation and thereby forms covalent attachment sites for proteinases [66]. α-2-Macroglobulin is a reserve proteinase inhibitor that gets activated during TIMPs deficiency [66].

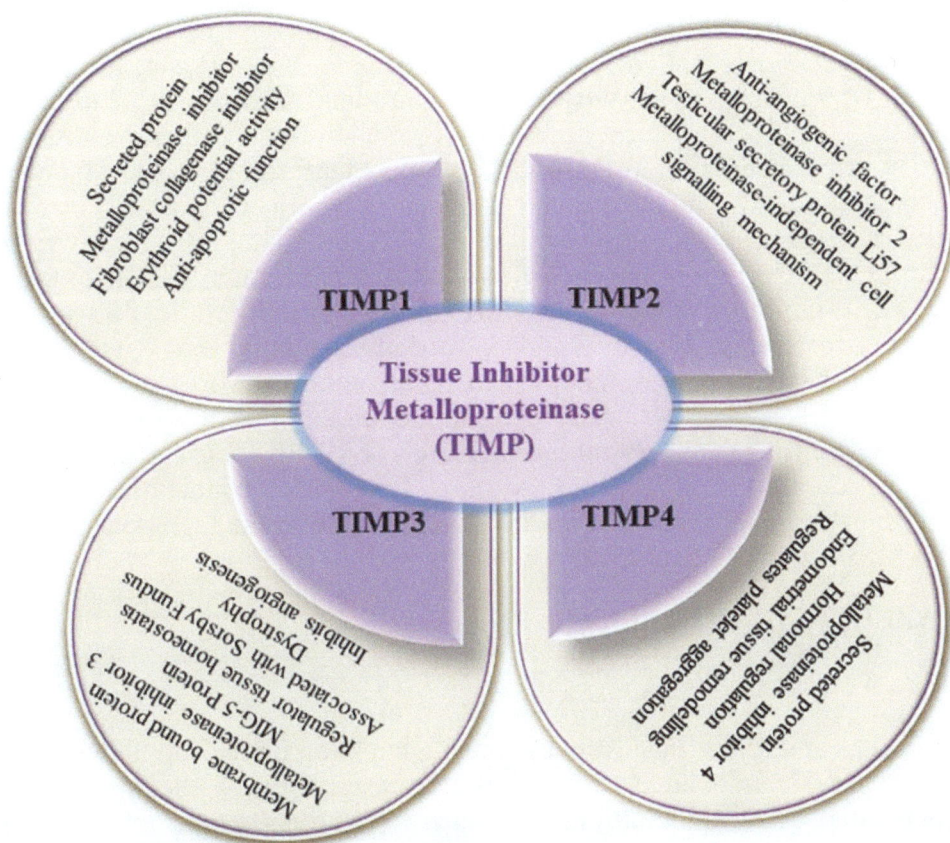

Fig. (2). Characteristics of tissue inhibitors of metalloproteinases (TIMPs).

Exogenous Inhibitors

Many MMP inhibitors evolved as a therapeutic tool for various diseases such as autoimmune disease, inflammation, and tumor growth. Exogenous MMPI can be natural or synthetically designed. For instance, natural polyphenols obtained from plant sources exhibited collagenase and elastase inhibitory activity [67] while collagen peptidomimetic, doxycycline, marimastat, and tetracycline derivatives are few synthetic inhibitors used for inhibiting MMPs. Generally, MMP degrades its substrate by catalytic zinc ions. Thereby, researchers identified zinc-binding groups (ZBG) like carboxylates, hydroxamic acids, thiols, and phosphonic, which are capable of halting the catalytic activity of zinc ions on the substrate and inactivate the enzyme [68]. The ZBG exhibits a broad spectrum of MMP inhibitory activity. The most popularly and commonly known ZBG MMPI is hydroxamic acids, capable of inhibiting MMP at lower concentrations [69]. Another promising MMPI is the chemically modified tetracycline compounds such as minocycline and doxycycline. In 1967, the Food and Drug Administration approved doxycycline. Minocycline is a potent MMPI with multiple actions to treat multiple sclerosis and inflammation. In Fragile X syndrome, minocycline inhibits MMP-9 gelatinase activity and alleviate the syndrome [70]. Researchers and pharmaceutical companies have discovered several synthetic MMPI and proved its effectiveness in pre-clinical studies. Unfortunately, all failed the clinical trials and exhibited a cytotoxic effect rather than cytostatic in cancer treatment [71], musculoskeletal syndrome [72] and lack of sensitivity and specificity [73]. The despondency of synthetic MMPI, such as designing the specific MMP-inhibitor complexes, MMP biology (inherent flexibility of the active site), metabolic activity, and hazardous potential has triggered the academicians or scientific researchers to focus on natural MMPI from marine and terrestrial species. Kumar *et al.* [74], reported curcumin could be a potential MMPI candidate for selectively inhibiting different MMPs. A few other examples of naturally derived MMPI are Omega 3 fatty acid, squalamine, catechin derivatives, betulinic acid, myricetin and so on that have the therapeutic activity to inhibit MMPI in the pathological conditions.

THERAPEUTIC ACCOMPLISHMENTS

Arthritis

Several studies and scientific advances contributed to an increasing understanding of the pathogenesis associated with arthritic diseases. Cartilage destruction leads to osteoarthritis (OA), which ultimately results in joint tissue destruction. Thus, it is imperative to get detailed knowledge about the events involved in the destruction process leading to arthritis disease for therapeutic developments. A biochemical study on the metalloproteinases (MMPs) stated that MMPs play a vital role in ECM turnover in physiology and pathology [75]. The relationship between the MMP expression and the development of various forms of arthritis focused on this chapter. Progressive degradation of ECM is the main feature of arthritic diseases and leads to permanent loss of function. Different mechanistic classes of proteinases have contributed to the degradation of connective tissue macromolecules and ECM processing, but recent studies state that the MMP family have a significant role in this process. Generally, MMPs are secreted by mesenchymal and hemopoietic cells, which active at neutral pH and possess the combined ability to degrade ECM components. Overexpression of MMPs involved in the development of pathogenesis in various diseases such as rheumatoid arthritis (RA) and OA [76 - 78]. Mainly, MMP-2 and -9 found to be overproduced in the tissues of patients and lead to rheumatoid arthritis. MMP-12 acted as a modulator by macrophages during inflammation and expressed its effect on matrix components like membrane proteins and elastin. MMP inhibitors like carboxylic acids, thiols, and various phosphorus-containing ligands are found valid. However, they require additional lipophilic interactions for compensating their lower binding affinities. Hydroxamic acids are the most potent inhibitors suitable for all molecular environments [79 - 81]. The expression of MMP-1 and -3 occurs in active inflammation of tissue microenvironment with the substantial expression in the rheumatoid nodule sites [82, 83].

Autoimmune Disease

Several proteins of MMPs have been identified for their involvement in many pathological processes such as embryonic development, reproduction, tissue remodeling, and physiological processes like cardiovascular diseases, inflammation, arthritis, and cancer. MMP-9 or Gelatinase B is mainly involved in the repair of tissues and physiological effects in autoimmune diseases such as multiple sclerosis, atherosclerosis. Specific cell types can produce MMP-9 on stimulation such as monocytes, keratinocytes and malignant cells [84]. However, MMP-9 is secreted mainly by granulocytes and alveolar macrophages. MMP-9, in particular, regulates the composition of cell-matrix and suppresses the autoimmune disease by various regulating molecules [85]. Henceforth, the role of MMP-9 in autoimmune diseases could be used as a potential target for immunomodulation.

Inflammations

MMPs are expressed in several tissue-specific functions related to inflammation, which includes activity and bioavailability of inflammatory cytokines and chemokines. Proteolytic activities of MMPs are important in the tumor microenvironment in the presence of specific inhibitors. Biologists later discovered the overexpression of proteinases (MMP) serves as the primary contributor for invasion and metastasis of cancer cells, naturally, MMP inhibitors (MMPIs) tries inhibits its action. After extensive preclinical trails, numerous MMPIs are used in inflammatory, cancer, and vascular diseases [71, 86]. During an inflammatory response, immune surveillance response through leukocytes traffic and cross the tissue barrier including basement membranes which leads to continuous remodeling of ECM. However, this is possible by equipping these cells with ECM degrading enzymes [87 - 90]. Most of the MMPs are secreted as pro-MMPs, which are activated by catalytic domains. Primary sources of MMPs for the inflammatory process include neutrophils, macrophages, cytokines and interleukins which stimulate MMP expression and activation. MMP-2 is a major MMP which showed a significant role in the vascular formation, and increased levels of MMP-2 inhibit the function of MMP by blocking the cleavage and active sites of the substrate. Examples of acute inflammatory reactions are myocardial infarction, multiple sclerosis, and stroke. Macromolecular inhibitors like natural TIMPs, monoclonal antibodies and tiny molecules such as synthetic and natural products have been used in therapies where overexpression of MMPs is involved [91].

Cancer

Several new anticancer agents identified for interfering with DNA synthesis, nucleotide turnover, replication, and other intracellular and extracellular processes. Cellular processes like abnormal growth, attack of neighboring tissues, replication, and formation of metastases observed in cancer cells. Nowadays, the major challenge faced in cancer therapy is the treatment of metastases. Treating the growth of malignant tumors without any change to the healthy tissues is of much importance. Thus, many anti-metastatic drugs have been developed and designed [71, 92]. Enzymes degrading the extracellular matrix (ECM) have been discovered. About 23 MMP genes have been implicated in humans for cancer. Extracellular matrix maintains the structure and also holds cell composition together within the body. Degradation of ECM by MMPs enhances tumor growth and its behavior, which leads to the progression of cancer. Numerous MMP inhibitors have been developed for the treatment of cancer [93]. Some of the inhibitors are tetracycline, hydroxamate derivatives and peptidomimetic inhibitors are extensively studied against the tumor progression in clinical and pre-clinical trials. Batimastat is a broad spectrum of MMPI, which shows a significant reduction in tumor growth in orthotopic colon, pancreatic, liver tumor models. Marimastat is a low molecular weight MMPI, which is closely related to batimastat with hydroxamate group used in clinical trials. This inhibitor shows no tumor regressions based on the rate of tumor marker levels during the chronic administration in Phase-I and -II clinical studies [94]. COL-3 is a tetracycline analog that mainly inhibits the activity and production of the collagenase enzyme, which downregulates the growth of primary and metastatic tumors. A preclinical study of BAY 12-9566 showed a selective oral biphenyl derivative, which inhibits the activity against MMP-3 and MMP-9 [95]. This inhibitor showed toxic effects in various tumor models in animals with an elevation level of transaminase, bilirubin elevations and erythropoiesis [96, 97]. Hence, there is a need for long term administration of MMPs as a target for cancer treatment and makes it compulsory to undergo novel clinical trial design strategies. Development of less toxic, novel, creative MMP inhibitors with conventional anticancer agents remains attractive in future efforts.

Cardiovascular Diseases

MMPs are a class of zinc and calcium-dependent enzymes that actively participate in the disintegration of ECM in many diseases such as cancer, arthritis and cardiovascular disorders [98]. Recently, the role of MMPs has the potential to act as diagnostic and therapeutic tools. Members of matrix metalloproteinases (MMPs) play an essential role in various pathological processes such as respiratory diseases, myocardial infection and inflammatory processes [99]. Due to its significance in vascular tissue remodeling, soluble MMPs act directly in the development of dilated cardiomyopathy and atherosclerotic plaque. Nowadays, tissue remodeling is practiced and divided into the two -prolonged process as a cellular component and ECM remodeling. Matrix metalloproteinases (MMPs) can degrade the ECM. Numerous MMPs are expressed and activated in the myocardium for atherosclerotic disease with greater alterations in the ischemic regions of arterial walls [100]. MMP-2 and -3 with its varied activities and increased plasma concentrations could reduce the chance of atherosclerosis in patients. MMP-9 is the most widely implicated MMP that regulates tissue remodeling by degrading the ECM proteins. Endogenous inhibitors of MMPs (TIMPs) have been identified with four sets of tiny molecules concerning cell viability and growth. Thus, TIMPs helps in understanding the mechanism between structure and the function of left ventricle remodeling [101, 102].

CLINICAL IMPLICATIONS AND OBSTACLES

The retroverted relationship of MMP expression and pathophysiological conditions makes it evident that inhibiting the performance of MMP leads to the development of the novel therapeutic intervention. However, the primary research focussed on inhibiting MMP activity has become the main target for treating the pathophysiological process. Based on this research, TIMPs and synthetic MMPI have been developed and resulted in a dramatic expansion in therapeutic intervention. In clinical trials, TIMPs exhibited poor bioavailability [103]. However, TIMPs showed appreciable results from multiple animal models but failed at clinical trials [71]. The first MMPI developed was Batimastat that suffered from poor bioavailability when subjected orally; therefore, this therapeutic compound was administered intrapleural and intraperitoneally in clinical trials. Batimastat still needs further development since it had an inadequate response. Marimastat showed better results in Phase-I/II study in treating tumor but the subsequent study failed to exhibit survival benefit and lead to musculoskeletal syndrome [104]. Wong *et al*. [105], reported that marimastat is not used clinically due to its unacceptable side effects. However, Walz and Cayabyab [106] reported that Batimastat and its derivatives could be used as an anti-inflammatory to counter neutrophil contribution to lung or cerebral cortex cavitation. Agouron Pharmaceuticals developed AG3340 for treating age-related macular degeneration and cancer [107]. However, a Phase-III study of Prinomastat has not improved the outcome of chemotherapy in advanced non-small-cell lung cancer [108]. For the patients with advanced solid cancer, the recommended Phase-II dosage of MMI270 or CGS27023A is 300 mg bid [109]. COL-3 or metastat is under Phase-I trial for recurrent high-grade gliomas [110] while Kaposi's sarcoma and advanced brain tumors are under Phase-II trials [104]. MMPIs and its clinical implications are shown in Table **2**.

Table 2. Matrix metalloproteinase inhibitors clinical implications.

Matrix Metalloproteinase Inhibitors	Molecular Weight (g/mol)	Clinical Phase	Diseases	Side Effects
Actinonin	385.50	2	Apoptosis	Loss in Cells
Batimastat	477.64	2	Metastasis	Bone Pain, Bowel Incontinence
CGS 27023A	393.46	Phase-II discontinued	Inflammatory Disorders, Demyelinating Disorders and Cancers	Erosion of cartilage matrix
Cipemastat	436.55	2	Rheumatoid arthritis	Breakdown of cartilage and bone
Col-3	371.35	2	Refractory metastatic cancer, brain tumors	Bone pain, weight loss
Doxycycline	444.44	4	Plague, tetanus	Severe headache, dizziness, blurred vision
Fluorouracil	130.08	2	Glioblastoma	Allergic Reaction
Gemcitabine	263.19	2	Pancreatic cancer, breast cancer	Nausea (mild), vomiting, poor appetite, skin rash
GI-129471	471.62	-	Respiratory disease	Heart-related chest pain, anemia
Ilomastat	388.47	-	Lung injury, trauma healing, and cornea repair	Adequate healing, stroke
KB-R-7785	349.43	-	Myocardial ischemia	Nausea and vomiting, sweating, fatigue

(Table 2) cont.....

Matrix Metalloproteinase Inhibitors	Molecular Weight (g/mol)	Clinical Phase	Diseases	Side Effects
Keracyanin Chloride	630.98	2	Diabetes mellitus	Allergic reaction, Trouble breathing
Marimastat (BB-2516)	331.41	3	Wound healing	Musculoskeletal pain, Inflammation.
Marimastat (TACE)	331.41	3	Human gastric cancer	Indigestion, constipation
Minocycline	457.48	4	Acute hepatitis-like syndrome	Mild nausea, mild diarrhea, stomach upset, joint or muscle pain
Prinomastat	423.50	3	Angiogenesis, tumor growth and invasion	Musculoskeletal Pain and Stiffness
Pyrimidine-2,4,6-trione	259.26	-	Amyotrophic lateral sclerosis	Respiratory problems and death
RO 28-2653	485.50	-	Prostate carcinoma	Dysuria or pain when urinating, bone pain, weight loss, Fatigue
SB-3CT	306.39	2	Human brain injury, cardiovascular diseases, inflammation	Skin or respiratory system irritation, inhalation, ingestion
Solimastat	408.49	Phase-I discontinued	Colorectal cancer	Decrease in blood cell counts, hair loss, mouth sores
Tanomastat	410.91	3	Pancreatic cancer, lung cancer, ovarian cancer, osteoarthritis	Tumour Growth, Infection, Aging
Temozolomide	194.15	1	Brain and central nervous system tumors	Low number of blood cells
Temozolomide	194.15	2	Malignant glioma	Liver damage
Vorinostat	264.33	1	Cutaneous manifestations in patients, Glioblastoma	Nausea, vomiting, diarrhoea, deep-vein thrombosis

CONCLUDING REMARKS AND FUTURE DIRECTIONS

Matrix metalloproteinases are considered as a promising target for various pathophysiological conditions due to its fundamental functional characteristic and up-regulation expression during the pathological process. Besides, MMPs could be used as a powerful prognostic marker in various pathophysiological conditions. However, MMPs are mainly used as a prospective target for treating cancer and its implication in malignant pathologies. Anti-MMP therapy is viable for a significant therapeutic benefit. Preclinical studies showed promising and favorable linear pharmacokinetics of MMPI in disease conditions, while similar results were not obtained in clinical studies. The prime cause for MMPI failure in clinical trials especially tumor treatment is that the compounds are investigated under advanced stages of pathological conditions, maybe, effective at early stages (cytostatic effect). Also, lack of knowledge on MMP biology, MMPI design, pathobiology of MMP and its inhibitors. Enhanced understanding of MMP on both molecular and clinical levels is needed for its potential applications in therapeutic intervention. The future of anti-MMP therapy remains in the responsibility of the pharmaceutical companies. Researchers and pharmaceutical companies have gained attention towards the identification and development of natural MMPI for treating diseases. These compounds are diet-based molecules, non-toxic and cost-effective with negligible adverse effects. The advent of genomics and proteomics would play a significant role in MMP substrates and functions since it remains unknown to date. Combined progress in molecular biology of the pathological condition and its targets will open the gateway for pioneering advanced and useful tools for new therapies and personalized medicine.

CONSENT FOR PUBLICATION

Not applicable.

CONFLICT OF INTEREST

The author(s) confirms that there is no conflict of interest.

ACKNOWLEDGEMENTS

Authors express gratitude to Prof. M. Sivanandham, Secretary, SVEHT and SVCE for the support.

REFERENCES

[1]   Gross J, Lapiere CM. Collagenolytic activity in amphibian tissues: a tissue culture assay. Proc Natl Acad Sci USA 1962; 48(6): 1014-22.
[http://dx.doi.org/10.1073/pnas.48.6.1014] [PMID: 13902219]

[2]   Cui N, Hu M, Khalil RA. Biochemical and biological attributes of matrix metalloproteinases. Prog Mol Biol Transl Sci 2017; 147: 1-73.
[http://dx.doi.org/10.1016/bs.pmbts.2017.02.005] [PMID: 28413025]

[3]   Beber AR, Polina ER, Biolo A, *et al.* Matrix metalloproteinase-2 polymorphisms in chronic heart failure: relationship with susceptibility and long-term survival. PLoS One 2016; 11(8)e0161666
[http://dx.doi.org/10.1371/journal.pone.0161666] [PMID: 27551966]

[4]   Fanjul-Fernández M, Folgueras AR, Cabrera S, López-Otín C. Matrix metalloproteinases: evolution,

gene regulation and functional analysis in mouse models. Biochim Biophys Acta 2010; 1803(1): 3-19.
[http://dx.doi.org/10.1016/j.bbamcr.2009.07.004] [PMID: 19631700]

[5]   Hadler-Olsen E, Fadnes B, Sylte I, Uhlin-Hansen L, Winberg JO. Regulation of matrix
       metalloproteinase activity in health and disease. FEBS J 2011; 278(1): 28-45.
       [http://dx.doi.org/10.1111/j.1742-4658.2010.07920.x] [PMID: 21087458]

[6]   Khalil RA. Matrix metalloproteinases and tissue remodeling in health and disease: Target tissues and
       therapy. Academic Press 2017.

[7]   Voronkina IV, Vakhromova EA, Kirpichnikova KM, Smagina LV, Gamaley IA. Matrix
       metalloproteinase activity in transformed cells exposed to an antioxidant. Cell Tissue Biol 2015; 9(1):
       16-23.
       [http://dx.doi.org/10.1134/S1990519X15010113]

[8]   Newby AC. Metalloproteinase expression in monocytes and macrophages and its relationship to
       atherosclerotic plaque instability. Arterioscler Thromb Vasc Biol 2008; 28(12): 2108-14.
       [http://dx.doi.org/10.1161/ATVBAHA.108.173898] [PMID: 18772495]

[9]   Abraham M, Shapiro S, Karni A, Weiner HL, Miller A. Gelatinases (MMP-2 and MMP-9) are
       preferentially expressed by Th1 vs. Th2 cells. J Neuroimmunol 2005; 163(1-2): 157-64.
       [http://dx.doi.org/10.1016/j.jneuroim.2005.02.001] [PMID: 15885317]

[10]   Tholozan FM, Gribbon C, Li Z, *et al*. FGF-2 release from the lens capsule by MMP-2 maintains lens
       epithelial cell viability. Mol Biol Cell 2007; 18(11): 4222-31.
       [http://dx.doi.org/10.1091/mbc.e06-05-0416] [PMID: 17699594]

[11]   Arenas IA, Xu Y, Lopez-Jaramillo P, Davidge ST. Angiotensin II-induced MMP-2 release from
       endothelial cells is mediated by TNF-α. Am J Physiol Cell Physiol 2004; 286(4): C779-84.
       [http://dx.doi.org/10.1152/ajpcell.00398.2003] [PMID: 14644777]

[12]   Fedorova NV, Ksenofontov AL, Serebryakova MV, Stadnichuk VI, *et al*. Neutrophils release
       metalloproteinases during adhesion in the presence of insulin, but Cathepsin G in the presence of
       glucagon. Mediators Inflamm 2018; 2018

[13]   Jonitz-Heincke A, Lochner K, Schulze C, *et al*. Contribution of human osteoblasts and macrophages to
       bone matrix degradation and proinflammatory cytokine release after exposure to abrasive
       endoprosthetic wear particles. Mol Med Rep 2016; 14(2): 1491-500.
       [http://dx.doi.org/10.3892/mmr.2016.5415] [PMID: 27357630]

[14]   Thirkettle S, Decock J, Arnold H, *et al*. Matrix metalloproteinase 8 (collagenase 2) induces the
       expression of interleukins 6 and 8 in breast cancer cells. J Biol Chem 2013; 288(23): 16282-94.
       [http://dx.doi.org/10.1074/jbc.M113.464230] [PMID: 23632023]

[15]   Boire A, Covic L, Agarwal A, *et al*. PAR1 is a matrix metalloprotease-1 receptor that promotes
       invasion and tumorigenesis of breast cancer cells. Cell 2005; 120(3): 303-13.
       [http://dx.doi.org/10.1016/j.cell.2004.12.018] [PMID: 15707890]

[16]   Lee SE, Han BD, Park IS, Romero R, Yoon BH. Evidence supporting proteolytic cleavage of insulin-
       like growth factor binding protein-1 (IGFBP-1) protein in amniotic fluid. J Perinat Med 2008; 36(4):
       316-23.
       [http://dx.doi.org/10.1515/JPM.2008.067] [PMID: 18598121]

[17]   Jin X, Yagi M, Akiyama N, *et al*. Matriptase activates stromelysin (MMP-3) and promotes tumor
       growth and angiogenesis. Cancer Sci 2006; 97(12): 1327-34.
       [http://dx.doi.org/10.1111/j.1349-7006.2006.00328.x] [PMID: 16999819]

[18]   Chao PZ, Hsieh MS, Cheng CW, Lin YF, Chen CH. Regulation of MMP-3 expression and secretion
       by the chemokine eotaxin-1 in human chondrocytes. J Biomed Sci 2011; 18(1): 86.
       [http://dx.doi.org/10.1186/1423-0127-18-86] [PMID: 22114952]

[19]   Avouac J, Guignabert C, Hoffmann-Vold AM, *et al*. Role of stromelysin 2 (Matrix Metalloproteinase
       10) as a novel mediator of vascular remodeling underlying pulmonary hypertension associated with

systemic sclerosis. Arthritis Rheumatol 2017; 69(11): 2209-21.
[http://dx.doi.org/10.1002/art.40229] [PMID: 28805015]

[20]   Juchniewicz A, Kowalczuk O, Milewski R, *et al.* MMP-10, MMP-7, TIMP-1 and TIMP-2 mRNA expression in esophageal cancer. Acta Biochim Pol 2017; 64(2): 295-9.
[http://dx.doi.org/10.18388/abp.2016_1408] [PMID: 28510611]

[21]   Zhang G, Miyake M, Lawton A, Goodison S, Rosser CJ. Matrix metalloproteinase-10 promotes tumor progression through regulation of angiogenic and apoptotic pathways in cervical tumors. BMC Cancer 2014; 14(1): 310.
[http://dx.doi.org/10.1186/1471-2407-14-310] [PMID: 24885595]

[22]   Lagente V, Le Quement C, Boichot E. Macrophage metalloelastase (MMP-12) as a target for inflammatory respiratory diseases. Expert Opin Ther Targets 2009; 13(3): 287-95.
[http://dx.doi.org/10.1517/14728220902751632] [PMID: 19236151]

[23]   Zhang Z, Zhu S, Yang Y, Ma X, Guo S. Matrix metalloproteinase-12 expression is increased in cutaneous melanoma and associated with tumor aggressiveness. Tumour Biol 2015; 36(11): 8593-600.
[http://dx.doi.org/10.1007/s13277-015-3622-9] [PMID: 26040769]

[24]   Stracke JO, Hutton M, Stewart M, *et al.* Biochemical characterization of the catalytic domain of human matrix metalloproteinase 19. Evidence for a role as a potent basement membrane degrading enzyme. J Biol Chem 2000; 275(20): 14809-16.
[http://dx.doi.org/10.1074/jbc.275.20.14809] [PMID: 10809722]

[25]   Cervinková M, Horák P, Kanchev I, *et al.* Differential expression and processing of matrix metalloproteinase 19 marks progression of gastrointestinal diseases. Folia Biol (Praha) 2014; 60(3): 113-22.
[PMID: 25056434]

[26]   Yu G, Herazo-Maya JD, Nukui T, *et al.* Matrix metalloproteinase-19 promotes metastatic behavior *in vitro* and is associated with increased mortality in non-small cell lung cancer. Am J Respir Crit Care Med 2014; 190(7): 780-90.
[http://dx.doi.org/10.1164/rccm.201310-1903OC] [PMID: 25250855]

[27]   Zhai LL, Wu Y, Cai CY, Huang Q, Tang ZG. High-level expression and prognostic significance of matrix metalloprotease-19 and matrix metalloprotease-20 in human pancreatic ductal adenocarcinoma. Pancreas 2016; 45(7): 1067-72.
[http://dx.doi.org/10.1097/MPA.0000000000000569] [PMID: 26692439]

[28]   Cominelli A, Gaide Chevronnay HP, Lemoine P, *et al.* Matrix metalloproteinase-27 is expressed in CD163+/CD206+ M2 macrophages in the cycling human endometrium and in superficial endometriotic lesions. Mol Hum Reprod 2014; 20(8): 767-75.
[http://dx.doi.org/10.1093/molehr/gau034] [PMID: 24810263]

[29]   Lynch CC, Vargo-Gogola T, Martin MD, *et al.* Matrix metalloproteinase 7 mediates mammary epithelial cell tumorigenesis through the ErbB4 receptor. Cancer Res 2007; 67(14): 6760-7.
[http://dx.doi.org/10.1158/0008-5472.CAN-07-0026] [PMID: 17638887]

[30]   Zhao YG, Xiao AZ, Ni J, Man YG, Sang QX. Expression of matrix metalloproteinase-26 in multiple human cancer tissues and smooth muscle cells. Chin J Cancer 2009; 28(11): 1168-75.
[http://dx.doi.org/10.5732/cjc.008.10768] [PMID: 19895737]

[31]   Jia Q, Liu S, Wen D, *et al.* Juvenile hormone and 20-hydroxyecdysone coordinately control the developmental timing of matrix metalloproteinase-induced fat body cell dissociation. J Biol Chem 2017; 292(52): 21504-16.
[http://dx.doi.org/10.1074/jbc.M117.818880] [PMID: 29118190]

[32]   DI Carlo A. Matrix metalloproteinase-2 and -9 and tissue inhibitor of metalloproteinase-1 and -2 in sera and urine of patients with renal carcinoma. Oncol Lett 2014; 7(3): 621-6.
[http://dx.doi.org/10.3892/ol.2013.1755] [PMID: 24520285]

[33] Hsin CH, Chou YE, Yang SF, *et al.* MMP-11 promoted the oral cancer migration and Fak/Src activation. Oncotarget 2017; 8(20): 32783-93.
[http://dx.doi.org/10.18632/oncotarget.15824] [PMID: 28427180]

[34] Ahokas K, Lohi J, Illman SA, *et al.* Matrix metalloproteinase-21 is expressed epithelially during development and in cancer and is up-regulated by transforming growth factor-β1 in keratinocytes. Lab Invest 2003; 83(12): 1887-99.
[http://dx.doi.org/10.1097/01.LAB.0000106721.86126.39] [PMID: 14691307]

[35] Skoog T, Ahokas K, Orsmark C, Jeskanen L, Isaka K, Saarialho-Kere U. MMP-21 is expressed by macrophages and fibroblasts in vivo and in culture. Exp Dermatol 2006; 15(10): 775-83.
[http://dx.doi.org/10.1111/j.1600-0625.2006.00460.x] [PMID: 16984259]

[36] Lohi J, Wilson CL, Roby JD, Parks WC. Epilysin, a novel human matrix metalloproteinase (MMP-28) expressed in testis and keratinocytes and in response to injury. J Biol Chem 2001; 276(13): 10134-44.
[http://dx.doi.org/10.1074/jbc.M001599200] [PMID: 11121398]

[37] Cepeda M. 2017.

[38] Liu Y, Sun X, Feng J, *et al.* MT2-MMP induces proteolysis and leads to EMT in carcinomas. Oncotarget 2016; 7(30): 48193-205.
[http://dx.doi.org/10.18632/oncotarget.10194] [PMID: 27374080]

[39] Chen L, Zhou Q, Xu B, *et al.* MT2-MMP expression associates with tumor progression and angiogenesis in human lung cancer. Int J Clin Exp Pathol 2014; 7(6): 3469-77.
[PMID: 25031779]

[40] Xu X, Chen L, Xu B, *et al.* Increased MT2-MMP expression in gastric cancer patients is associated with poor prognosis. Int J Clin Exp Pathol 2015; 8(2): 1985-90.
[PMID: 25973093]

[41] Chen L, Di D, Luo G, *et al.* Immunochemical staining of MT2-MMP correlates positively to angiogenesis of human esophageal cancer. Anticancer Res 2010; 30(10): 4363-8.
[PMID: 21036765]

[42] Cao L, Chen C, Zhu H, *et al.* MMP16 is a marker of poor prognosis in gastric cancer promoting proliferation and invasion. Oncotarget 2016; 7(32): 51865-74.
[http://dx.doi.org/10.18632/oncotarget.10177] [PMID: 27340864]

[43] Pullen N, Pickford A, Perry MM, Jaworski DM, *et al.* Current insights into Matrix metalloproteinases and glioma progression: Transcending the degradation boundary. Metalloproteinases Med 2018; 5: 13-30.
[http://dx.doi.org/10.2147/MNM.S105123]

[44] Okimoto RA, Breitenbuecher F, Olivas VR, *et al.* Inactivation of Capicua drives cancer metastasis. Nat Genet 2017; 49(1): 87-96.
[http://dx.doi.org/10.1038/ng.3728] [PMID: 27869830]

[45] Luo YP, Zhong M, Wang LP, Zhong M, Sun GQ, Li J. [Inhibitory effects of RNA interference on MMP-24 expression and invasiveness of ovarian cancer SKOV(3) cells]. Nan Fang Yi Ke Da Xue Xue Bao 2009; 29(4): 781-4.
[PMID: 19403421]

[46] Ohnishi J, Ohnishi E, Jin M, *et al.* Cloning and characterization of a rat ortholog of MMP-23 (matrix metalloproteinase-23), a unique type of membrane-anchored matrix metalloproteinase and conditioned switching of its expression during the ovarian follicular development. Mol Endocrinol 2001; 15(5): 747-64.
[http://dx.doi.org/10.1210/mend.15.5.0638] [PMID: 11328856]

[47] Yip C, Foidart P, Somja J, *et al.* MT4-MMP and EGFR expression levels are key biomarkers for breast cancer patient response to chemotherapy and erlotinib. Br J Cancer 2017; 116(6): 742-51.
[http://dx.doi.org/10.1038/bjc.2017.23] [PMID: 28196064]

[48] Murphy G, Lee MH. What are the roles of metalloproteinases in cartilage and bone damage? Ann Rheum Dis 2005; 64 (Suppl. 4): iv44-7.
[http://dx.doi.org/10.1136/ard.2005.042465] [PMID: 16239386]

[49] Blumenthal MN, Zhong W, Miller M, Wendt C, Connett JE, Pei D. Serum metalloproteinase leukolysin (MMP-25/MT-6): a potential metabolic marker for atopy-associated inflammation. Clin Exp Allergy 2010; 40(6): 859-66.
[http://dx.doi.org/10.1111/j.1365-2222.2010.03475.x] [PMID: 20337648]

[50] Shi F, Sottile J. MT1-MMP regulates the turnover and endocytosis of extracellular matrix fibronectin. J Cell Sci 2011; 124(Pt 23): 4039-50.
[http://dx.doi.org/10.1242/jcs.087858] [PMID: 22159414]

[51] Philips N, Conte J, Chen YJ, *et al.* Beneficial regulation of matrixmetalloproteinases and their inhibitors, fibrillar collagens and transforming growth factor-β by Polypodium leucotomos, directly or in dermal fibroblasts, ultraviolet radiated fibroblasts, and melanoma cells. Arch Dermatol Res 2009; 301(7): 487-95.
[http://dx.doi.org/10.1007/s00403-009-0950-x] [PMID: 19373483]

[52] Jarosz K, Chudzik R, Gołębiowska M, Gołębiowska B. TIMP way to stop metastasis? J Educ Health Sport 2017; 7(7): 598-604.

[53] Chew DK, Conte MS, Khalil RA. Matrix metalloproteinase-specific inhibition of Ca2+ entry mechanisms of vascular contraction. J Vasc Surg 2004; 40(5): 1001-10.
[http://dx.doi.org/10.1016/j.jvs.2004.08.035] [PMID: 15557917]

[54] Ramnath N, Creaven PJ. Matrix metalloproteinase inhibitors. Curr Oncol Rep 2004; 6(2): 96-102.
[http://dx.doi.org/10.1007/s11912-004-0020-7] [PMID: 14751086]

[55] Airola K, Fusenig NE. Differential stromal regulation of MMP-1 expression in benign and malignant keratinocytes. J Invest Dermatol 2001; 116(1): 85-92.
[http://dx.doi.org/10.1046/j.1523-1747.2001.00223.x] [PMID: 11168802]

[56] Tallant C, Marrero A, Gomis-Rüth FX. Matrix metalloproteinases: fold and function of their catalytic domains. Biochim Biophys Acta 2010; 1803(1): 20-8.
[http://dx.doi.org/10.1016/j.bbamcr.2009.04.003] [PMID: 19374923]

[57] Shapiro AM, Ricordi C, Hering BJ, *et al.* International trial of the Edmonton protocol for islet transplantation. N Engl J Med 2006; 355(13): 1318-30.
[http://dx.doi.org/10.1056/NEJMoa061267] [PMID: 17005949]

[58] Wang H, Zhan Y, Jin J, Zhang C, Li W. MicroRNA-15b promotes proliferation and invasion of non□small cell lung carcinoma cells by directly targeting TIMP2. Oncol Rep 2017; 37(6): 3305-12.
[http://dx.doi.org/10.3892/or.2017.5604] [PMID: 28498424]

[59] Su CW, Su BF, Chiang WL, Yang SF, Chen MK, Lin CW. Plasma levels of the tissue inhibitor matrix metalloproteinase-3 as a potential biomarker in oral cancer progression. Int J Med Sci 2017; 14(1): 37-44.
[http://dx.doi.org/10.7150/ijms.17024] [PMID: 28138307]

[60] Wakula P, Neumann B, Kienemund J, *et al.* CHA2DS2-VASc score and blood biomarkers to identify patients with atrial high-rate episodes and paroxysmal atrial fibrillation. Europace 2017; 19(4): 544-51.
[PMID: 28431065]

[61] Firestein GS, Budd R, Gabriel SE, Mcinnes IB, O'Dell JR. Kelley and Firestein's Textbook of Rheumatology E-Book. 10th ed., Elsevier 2017.

[62] Pereira JL. Relevance of circulating markers of inflammation and metastization in the NSCLC landscape. The use of Lymphoblastoid Cell Lines from patients in 2016.

[63] Guedez L, Mansoor A, Birkedal-Hansen B, *et al.* Tissue inhibitor of metalloproteinases 1 regulation of interleukin-10 in B-cell differentiation and lymphomagenesis. Blood 2001; 97(6): 1796-802.

[http://dx.doi.org/10.1182/blood.V97.6.1796] [PMID: 11238122]

[64]   Tchetverikov I, Lard LR, DeGroot J, *et al.* Matrix metalloproteinases-3, -8, -9 as markers of disease activity and joint damage progression in early rheumatoid arthritis. Ann Rheum Dis 2003; 62(11): 1094-9.
[http://dx.doi.org/10.1136/ard.62.11.1094] [PMID: 14583574]

[65]   Anadón A, Castellano V, Martinez-Larrañaga MR. Biomarkers in drug safety evaluation.Biomarkers in Toxicology. San Diego, CA, USA: Elsevier/Academic Press 2014; pp. 923-45.
[http://dx.doi.org/10.1016/B978-0-12-404630-6.00055-5]

[66]   Mocchegiani E, Costarelli L, Giacconi R, Cipriano C, Muti E, Malavolta M. Zinc-binding proteins (metallothionein and α-2 macroglobulin) and immunosenescence. Exp Gerontol 2006; 41(11): 1094-107.
[http://dx.doi.org/10.1016/j.exger.2006.08.010] [PMID: 17030107]

[67]   Horng CT, Wu HC, Chiang NN, *et al.* Inhibitory effect of burdock leaves on elastase and tyrosinase activity. Exp Ther Med 2017; 14(4): 3247-52.
[http://dx.doi.org/10.3892/etm.2017.4880] [PMID: 28912875]

[68]   Rao BG. Recent developments in the design of specific Matrix Metalloproteinase inhibitors aided by structural and computational studies. Curr Pharm Des 2005; 11(3): 295-322.
[http://dx.doi.org/10.2174/1381612053382115] [PMID: 15723627]

[69]   Verma RP. Hydroxamic Acids as Matrix Metalloproteinase Inhibitors.Matrix Metalloproteinase Inhibitors Experientia Supplementum. Basel: Springer 2012; Vol. 103: pp. 137-76.
[http://dx.doi.org/10.1007/978-3-0348-0364-9_5]

[70]   Siller SS, Broadie K. Matrix metalloproteinases and minocycline: therapeutic avenues for fragile X syndrome. Neural Plast 2012; 2012
[http://dx.doi.org/10.1155/2012/124548]

[71]   Coussens LM, Fingleton B, Matrisian LM. Matrix metalloproteinase inhibitors and cancer: trials and tribulations. Science 2002; 295(5564): 2387-92.
[http://dx.doi.org/10.1126/science.1067100] [PMID: 11923519]

[72]   Herszényi L, Hritz I, Lakatos G, Varga MZ, Tulassay Z. The behavior of matrix metalloproteinases and their inhibitors in colorectal cancer. Int J Mol Sci 2012; 13(10): 13240-63.
[http://dx.doi.org/10.3390/ijms131013240] [PMID: 23202950]

[73]   Glassmire DM, Kinney DI, Greene RL, Stolberg RA, Berry DT, Cripe L. Sensitivity and specificity of MMPI-2 neurologic correction factors: receiver operating characteristic analysis. Assessment 2003; 10(3): 299-309.
[http://dx.doi.org/10.1177/1073191103256129] [PMID: 14503653]

[74]   Kumar D, Kumar M, Saravanan C, Singh SK. Curcumin: a potential candidate for matrix metalloproteinase inhibitors. Expert Opin Ther Targets 2012; 16(10): 959-72.
[http://dx.doi.org/10.1517/14728222.2012.710603] [PMID: 22913284]

[75]   Davidson RK, Waters JG, Kevorkian L, *et al.* Expression profiling of metalloproteinases and their inhibitors in synovium and cartilage. Arthritis Res Ther 2006; 8(4): R124.
[http://dx.doi.org/10.1186/ar2013] [PMID: 16859525]

[76]   Cathcart JM, Cao J, Brook S. MMP Inhibitors: Past, present and future. Front Biosci 2015; 20: 1164-78.
[http://dx.doi.org/10.2741/4365] [PMID: 25961551]

[77]   Eckhard U, Huesgen PF, Schilling O, *et al.* Active site specificity profiling of the matrix metalloproteinase family: Proteomic identification of 4300 cleavage sites by nine MMPs explored with structural and synthetic peptide cleavage analyses. Matrix Biol 2016; 49: 37-60.
[http://dx.doi.org/10.1016/j.matbio.2015.09.003] [PMID: 26407638]

[78]   Van Doren SR. Matrix metalloproteinase interactions with collagen and elastin. Matrix Biol 2015; 44-

46: 224-31.
[http://dx.doi.org/10.1016/j.matbio.2015.01.005] [PMID: 25599938]

[79]   Muri EM, Nieto MJ, Williamson JS. Hydroxamic acids as pharmacological agents: An update. Med Chem Res 2004; 1(4): 385-94.

[80]   Bencsik P, Kupai K, Görbe A, *et al.* Development of Matrix Metalloproteinase-2 Inhibitors for Cardioprotection. Front Pharmacol 2018; 9: 296.
[http://dx.doi.org/10.3389/fphar.2018.00296] [PMID: 29674965]

[81]   Nikolaou A, Ninou I, Kokotou MG, *et al.* Hydroxamic acids constitute a novel class of autotaxin inhibitors that exhibit *in vivo* efficacy in a pulmonary fibrosis model. J Med Chem 2018; 61(8): 3697-711.
[http://dx.doi.org/10.1021/acs.jmedchem.8b00232] [PMID: 29620892]

[82]   Harikrishnan A, Veena V. Therapeutic molecules for fumigating inflammatory tumor environment. Curr Signal Transduct Ther 2018; 13(2): 129-52.
[http://dx.doi.org/10.2174/1574362413666180402145336]

[83]   Kraniak JM, Mattingly RR, Sloane BF. Roles of pericellular proteases in tumor angiogenesis: Therapeutic implications. extracellular targeting of cell signaling in cancer: strategies directed at MET and RON receptor tyrosine kinase pathways 2018; 411-6.

[84]   Chakraborti S, Mandal M, Das S, Mandal A, Chakraborti T. Regulation of matrix metalloproteinases: an overview. Mol Cell Biochem 2003; 253(1-2): 269-85.
[http://dx.doi.org/10.1023/A:1026028303196] [PMID: 14619979]

[85]   Ram M, Sherer Y, Shoenfeld Y. Matrix metalloproteinase-9 and autoimmune diseases. J Clin Immunol 2006; 26(4): 299-307.
[http://dx.doi.org/10.1007/s10875-006-9022-6] [PMID: 16652230]

[86]   Overall CM, López-Otín C. Strategies for MMP inhibition in cancer: innovations for the post-trial era. Nat Rev Cancer 2002; 2(9): 657-72.
[http://dx.doi.org/10.1038/nrc884] [PMID: 12209155]

[87]   Kelley LC, Lohmer LL, Hagedorn EJ, Sherwood DR. Traversing the basement membrane *in vivo*: a diversity of strategies. J Cell Biol 2014; 204(3): 291-302.
[http://dx.doi.org/10.1083/jcb.201311112] [PMID: 24493586]

[88]   Rohl J. Intracellular trafficking and secretion of matrix metalloproteinases during macrophage migration (Doctoral dissertation, Queensland University of Technology)
[http://dx.doi.org/10.5204/thesis.eprints.102374]

[89]   Van Den Steen PE, Wuyts A, Husson SJ, Proost P, Van Damme J, Opdenakker G. Gelatinase B/MMP-9 and neutrophil collagenase/MMP-8 process the chemokines human GCP-2/CXCL6, ENA-78/CXCL5 and mouse GCP-2/LIX and modulate their physiological activities. Eur J Biochem 2003; 270(18): 3739-49.
[http://dx.doi.org/10.1046/j.1432-1033.2003.03760.x] [PMID: 12950257]

[90]   McQuibban GA, Gong JH, Tam EM, McCulloch CA, Clark-Lewis I, Overall CM. Inflammation dampened by gelatinase A cleavage of monocyte chemoattractant protein-3. Science 2000; 289(5482): 1202-6.
[http://dx.doi.org/10.1126/science.289.5482.1202] [PMID: 10947989]

[91]   Kim YS, Joh TH. Matrix metalloproteinases, new insights into the understanding of neurodegenerative disorders. Biomol Ther (Seoul) 2012; 20(2): 133-43.
[http://dx.doi.org/10.4062/biomolther.2012.20.2.133] [PMID: 24116286]

[92]   Overall CM, Kleifeld O. Tumour microenvironment - opinion: validating matrix metalloproteinases as drug targets and anti-targets for cancer therapy. Nat Rev Cancer 2006; 6(3): 227-39.
[http://dx.doi.org/10.1038/nrc1821] [PMID: 16498445]

[93]   Zucker S, Cao J, Chen WT. Critical appraisal of the use of matrix metalloproteinase inhibitors in

cancer treatment. Oncogene 2000; 19(56): 6642-50.
[http://dx.doi.org/10.1038/sj.onc.1204097] [PMID: 11426650]

[94]   Germanov E, Berman JN, Guernsey DL. Current and future approaches for the therapeutic targeting of metastasis (review). Int J Mol Med 2006; 18(6): 1025-36.
[http://dx.doi.org/10.3892/ijmm.18.6.1025] [PMID: 17089005]

[95]   Pavlaki M, Zucker S. Matrix metalloproteinase inhibitors (MMPIs): the beginning of phase I or the termination of phase III clinical trials. Cancer Metastasis Rev 2003; 22(2-3): 177-203.
[http://dx.doi.org/10.1023/A:1023047431869] [PMID: 12784996]

[96]   Sanz-Moreno V, Gadea G, Ahn J, *et al.* Rac activation and inactivation control plasticity of tumor cell movement. Cell 2008; 135(3): 510-23.
[http://dx.doi.org/10.1016/j.cell.2008.09.043] [PMID: 18984162]

[97]   Wolf K, Mazo I, Leung H, *et al.* Compensation mechanism in tumor cell migration: mesenchymal-amoeboid transition after blocking of pericellular proteolysis. J Cell Biol 2003; 160(2): 267-77.
[http://dx.doi.org/10.1083/jcb.200209006] [PMID: 12527751]

[98]   Liu P, Sun M, Sader S. Matrix metalloproteinases in cardiovascular disease. Can J Cardiol 2006; 22 (Suppl. B): 25B-30B.
[http://dx.doi.org/10.1016/S0828-282X(06)70983-7] [PMID: 16498509]

[99]   Brinckerhoff CE, Matrisian LM. Matrix metalloproteinases: a tail of a frog that became a prince. Nat Rev Mol Cell Biol 2002; 3(3): 207-14.
[http://dx.doi.org/10.1038/nrm763] [PMID: 11994741]

[100] Spinale FG, Villarreal F. Targeting matrix metalloproteinases in heart disease: lessons from endogenous inhibitors. Biochem Pharmacol 2014; 90(1): 7-15.
[http://dx.doi.org/10.1016/j.bcp.2014.04.011] [PMID: 24780447]

[101] Lindsey ML, Zamilpa R. Temporal and spatial expression of matrix metalloproteinases and tissue inhibitors of metalloproteinases following myocardial infarction. Cardiovasc Ther 2012; 30(1): 31-41.
[http://dx.doi.org/10.1111/j.1755-5922.2010.00207.x] [PMID: 20645986]

[102] Frangogiannis NG. Regulation of the inflammatory response in cardiac repair. Circ Res 2012; 110(1): 159-73.
[http://dx.doi.org/10.1161/CIRCRESAHA.111.243162] [PMID: 22223212]

[103] Skiles JW, Gonnella NC, Jeng AY. The design, structure, and clinical update of small molecular weight matrix metalloproteinase inhibitors. Curr Med Chem 2004; 11(22): 2911-77.
[http://dx.doi.org/10.2174/0929867043364018] [PMID: 15544483]

[104] Avendaño C, Menendez JC. Medicinal chemistry of anticancer drugs. Elsevier 2015.

[105] Wong MS, Sidik SM, Mahmud R, Stanslas J. Molecular targets in the discovery and development of novel antimetastatic agents: current progress and future prospects. Clin Exp Pharmacol Physiol 2013; 40(5): 307-19.
[http://dx.doi.org/10.1111/1440-1681.12083] [PMID: 23534409]

[106] Walz W, Cayabyab FS. Neutrophil infiltration and matrix metalloproteinase-9 in lacunar infarction. Neurochem Res 2017; 42(9): 2560-5.
[http://dx.doi.org/10.1007/s11064-017-2265-1] [PMID: 28417261]

[107] Griffioen AW. AG-3340 (Agouron Pharmaceuticals Inc). IDrugs 2000; 3(3): 336-45.
[PMID: 16103944]

[108] Bissett D, O'Byrne KJ, von Pawel J, *et al.* Phase III study of matrix metalloproteinase inhibitor prinomastat in non-small-cell lung cancer. J Clin Oncol 2005; 23(4): 842-9.
[http://dx.doi.org/10.1200/JCO.2005.03.170] [PMID: 15681529]

[109] Davis DW, McConkey DJ, Abbruzzese JL, Herbst RS. Surrogate markers in antiangiogenesis clinical trials. Br J Cancer 2003; 89(1): 8-14.

[http://dx.doi.org/10.1038/sj.bjc.6601035] [PMID: 12838293]

[110]   Hagemann C, Anacker J, Ernestus RI, Vince GH. A complete compilation of matrix metalloproteinase expression in human malignant gliomas. World J Clin Oncol 2012; 3(5): 67-79.
[http://dx.doi.org/10.5306/wjco.v3.i5.67] [PMID: 22582165]

CHAPTER 7

Current Potentialities and Perspectives of Enzyme Inhibition Based Biosensor

Senthil Nagappan[1], Rajvikram Madurai Elavarasan[2], Nilavunesan Dhandapani[1] and Ekambaram Nakkeeran[1,*]

[1] *Department of Biotechnology, Sri Venkateswara College of Engineering (Autonomous), Sriperumbudur Tk - 602 117, Tamil Nadu, India*

[2] *Department of Electrical and Electronics Engineering, Sri Venkateswara College of Engineering (Autonomous), Sriperumbudur Tk - 602 117, Tamil Nadu, India*

Abstract: A biosensor is an electrical device encompassing biological components, with the intent to detect and measure the concentration of the analyte in a sample. The applications of the biosensor are numerous including, but not limited to food, healthcare and environmental sectors. In particular, biosensors based on enzymatic inhibition are highly sensitive in monitoring the inhibitory analytes affecting the catalytic activity of enzymes. A careful selection of transducer, a biosensor component for transforming the biochemical signal to an electrical signal along with the choice of enzymes is crucial in determining the commercial viability of enzymatic inhibitory biosensor. This chapter highlights the recent studies as well as products available in the market related to biosensors based on enzymatic inhibition. Besides, the economic analysis of existing and futuristic biosensors is also discussed.

Keywords: Analyte, Analytes, Allergens, Biosensor, Biochemicals, Detection, Enzymes, Environmental, Enzyme inhibition, Enzyme inhibition, Food industry, Healthcare, Heavy metal, Inhibition-based biosensors, Nerve agents, Pharmaceuticals, Potentiometric, Pesticide, Screening, Toxins, Transducer.

INTRODUCTION

Biosensors are finding applications in a wide range of fields such as healthcare, environment, food industry, *etc.* for the detection of the desired analyte. The popularity of biosensor can be mainly attributed to its characteristics of simplicity, affordability and rapid detection. Enzymatic inhibition based biosensor is a class of biosensor developed first in the 1960s by Guilbault for the detection of nerve agents [1]. The principle of enzymatic inhibition based biosensor is the correlation

---

[*] **Corresponding author Ekambaram Nakkeeran:** Research Laboratory, Department of Biotechnology, Sri Venkateswara College of Engineering (Autonomous), Sriperumbudur Tk - 602 117, Tamil Nadu, India; E-mail: nakkeeran@svce.ac.in

G. Baskar, K. Sathish Kumar & K. Tamilarasan (Eds.)

of inhibiting analyte concentration with the degree of enzyme inhibition. The parts of a typical biosensor are substrate, bioreceptor, transducer, and signal measuring device. The additional component in an enzyme inhibition biosensor is the inhibitor. Bioreceptors can be an enzyme, antibody, nucleic acid or the whole cell itself. The signal transducer senses chemical and physical cues arising from enzyme inhibition activity such as electroactive material, pH change, heat, light and mass change with the help of either electrode or optical sensing. Finally, an electrical signal is read from the transducer through signal conducting circuits like amplifiers, filters, multiplexers, analog to digital converters, linearizers and compressors.

The major advantages of enzymatic inhibition based biosensor are as follows:

- Shorter response time
- A wider range of detection
- Sensitivity
- Robustness

Furthermore, the enzyme immobilization is reported to improve response time, reusability and stability of inhibition biosensor [2]. Due to the above-said features, this class of biosensor is used for the detection of a variety of compounds such as pesticides, insecticides, surfactants, heavy metals, toxins, nerve agents, glycoalkaloids, pharmaceutical drugs, nicotine, fluoride, benzoic acid, *etc*. For detection by inhibitory effect, different enzymes are reported. This includes acetylcholinesterase, butyrylcholinesterase, acid phosphatase, alkaline phosphatase, ascorbic oxidase, catalase, chymotrypsin, urease, tyrosinase, protease, peroxidase, lipoxygenase, elastase, laccase, glucose oxidase, invertase, *etc*. The above-mentioned enzymes, although display effective inhibition at a lower concentration of analyte they suffer from the problem of selectivity. To overcome the problem of selectivity, pretreatment of samples to remove the inhibitory compounds of non-interest and protein engineering techniques have been employed. This chapter discusses the recent studies on biosensor based on inhibition of the enzyme.

FUNDAMENTALS OF ENZYME INHIBITION BASED BIOSENSOR

Enzyme Inhibitors

Enzyme inhibitors are of two types - reversible and irreversible inhibitors. Irreversible inhibitors form a stable binding with enzyme resulting in chemical modification of enzyme's active site. In the above case, modification of the enzyme's active site is irreversible in nature and thus subsequent substrate or inhibitor binding is impossible unless the active site is reactivated. In contrast, reversible inhibitors bind to the enzyme through weak interaction such as hydrogen bond, hydrophobic interaction, ionic bonds and van der Waals bond. The weak interaction allows the enzyme's active site to be reused for subsequent substrate or inhibitor binding. As a result, biosensors based on reversible inhibitors will have an advantage over one based on irreversible inhibitors in terms of reusability [2].

The mode of inhibition can be predicted Reversible inhibitors can be classified into four types - competitive, uncompetitive, non-competitive and mixed reversible inhibitors. Competitive inhibitors compete with the substrate for binding to the enzyme's active site. This class of inhibitor has similar structural confirmation to that of the substrate. Competitive inhibition happens only when the concentration of inhibitors is higher than the substrate [3]. Unlike competitive inhibitors, uncompetitive inhibitors bind to the enzyme-substrate complex rather than the active site of the enzyme. In such a case, there is no competition between substrate and inhibitor and thus the required inhibitor concentration for uncompetitive inhibition is lower than competitive inhibition. Non-competitive inhibitors bind both to enzyme and enzyme-substrate complex with the same affinity. They are not substrate analogues and bind to the enzyme on a site other than the active site. Similar to uncompetitive inhibitors, the required inhibitor concentration for non-competitive inhibition is lower than competitive inhibition. Mixed inhibitors bind to both enzyme and enzyme-substrate complex but with different affinity. Also, as in both cases of uncompetitive and non-competitive inhibitors, the site of binding in the enzyme is not the active site. A biosensor based on uncompetitive, non-competitive or mixed inhibition has the advantage of lower detection limits in comparison to a competitive inhibition biosensor [2]. Therefore understanding inhibition mechanism through enzyme inhibition kinetics is essential for constructing a biosensor.

Enzyme Inhibitors Kinetics

The mode of inhibition can be predicted by methods such as double reciprocal plots [4], Dixon plots [5], Cornish-Bowden plots [6]. Fig. (**1**) shows the double reciprocal plot for (a) competitive, (b) non-competitive (c) uncompetitive inhibition, and Dixon plot for (d) mixed inhibition. Fig. (**1A**) shows the characteristic pattern of competitive inhibition plot in which the lines representing enzymatic activity in absence of inhibitor (I=0) and enzymatic activity in the presence of inhibitor (I>0) intercept at a single point on y-axis, above the x-axis. Furthermore, dissociation constants of enzyme-substrate (Km) and enzyme-inhibitor (Km, app) can be estimated from the x-intercepts of their respective lines. Similarly, characteristic patterns are shown in Fig. (**1B**), (**1C**) and (**1D**) represent the other three types of inhibition while x-intercepts in the plots are used for estimation of the kinetic parameters.

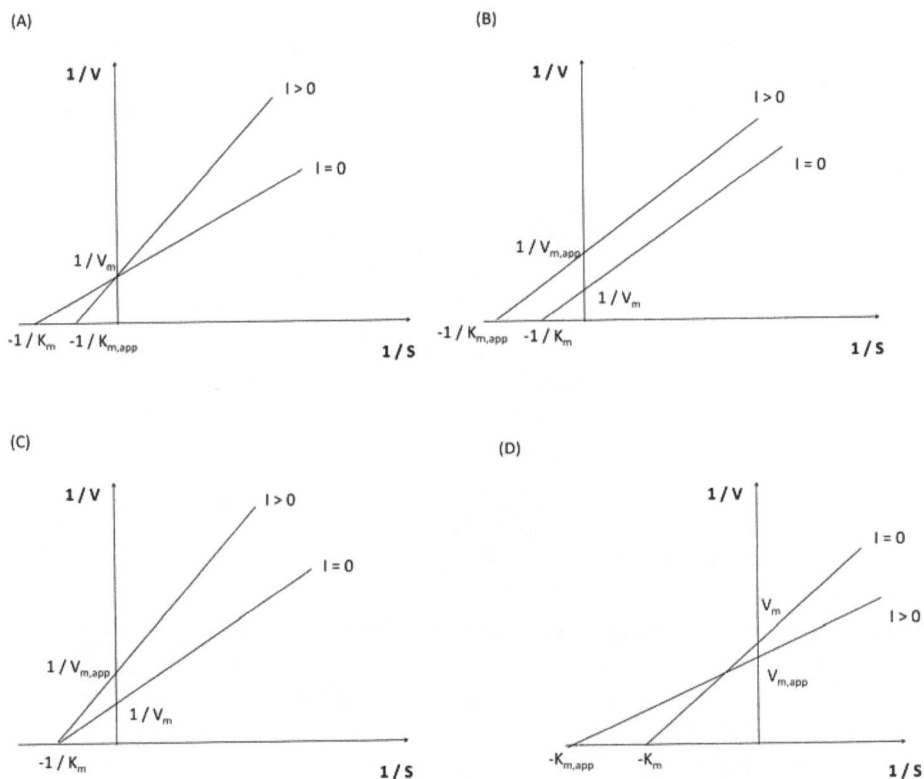

Fig. (1). Double reciprocal plots for competitive inhibition (A), non-competitive inhibition (B), uncompetitive inhibition (C) and Dixon plot for mixed inhibition (D).

Components of Biosensor

Transducer

A transducer generates an electrical output in accordance with either physical or chemical change resulting from the enzymatic activity. Transducers can be classified into (1) electrochemical, (2) optical, (3) piezoelectric transducers. Electrochemical transducers comprise of amperometric, potentiometric, field-effect transistors, and conductometric transducers. An amperometric transducer measures change in current arising from the enzymatic activity. A potentiometric transducer measures the accumulated charge or potential in the electrode generated by the activity of the enzyme. A field-effect transistor measures change in current through the transistor due to enzymatic activity. The conductometric transducer measures the change in conductance of the medium; here, the enzyme-inhibitor interaction results in a change in concentration of ionic species in the medium which in turn changes the conductance.

The pH based enzyme biosensor s are widely studied due to ease in the detection of H^+ ions and also the methods of determination are well established for H^+ ions [7]. The acetylcholinesterase-acetylcholine and butyrylcholinesterase-butyrylcholine are the most preferred enzyme utilized in pH based transducer since H+ ions are formed as a result of the enzymatic activity [7]. The optical transducer measures the change in photon as a result of enzyme inhibition. They are classified into two categories (1) Labelled, *e.g.*, fluorescence and (2) Label-free, *e.g.*, surface plasmon resonance. The piezoelectric transducer consists of a sensor that detects a change in mass of piezoelectric crystal occurring while inhibitor binds enzyme. When enzyme acetylcholinesterase is used along with thiocholine and acetate salts, piezoelectric and amperometric transducers are the preferred respectively [7]. This is because Thiokol produced from thiocholine is electroactive while precipitation of product after enzymatic activity by acetate salts changes the mass. In general, the optical biosensor is more sensitive than electrochemical biosensors [8]. The choice of the transducer for the appropriate biological element is vital for a commercially viable biosensor.

Substrate

Enzymes in inhibition based enzymatic biosensor are not single substrate-specific but act on multiple different substrates. For example, acetylcholinesterase act on many substrates like acetylcholine, acetylthiocholine, Indoxylacetate, 3-indoxyl acetate, *etc* [9 - 12]. The enzyme polyphenol oxidase act on substrates such as catechol, Tyrosine, p-cresol, m-cresol, Phenol, p-chlorophenol [13 - 20]. Hydrogen peroxidase act on hydroquinone, hydrogen dioxide, *etc* [21 - 26]. Phosphatase act on mono fluoro phosphate, p-nitrophenyl phosphate, catechol monophosphate, phenyl phosphate, Riboflavin-5-monophosphate, 2-phospho-L-ascorbic acid [27 - 33].

The type of substrate selected for a particular enzyme can influence the mode of detection. For example, the typical substrate associated with acetylcholinesterase is acetylcholine and the mode of detection for this enzyme-substrate complex is usually an electrochemical method. However, when indoxyl acetate is used for acetylcholinesterase it produces a colour change that could be detected by an optical transducer [11]. The following two studies indicate the role of substrate in biosensor in terms of binding affinity and inhibitor concentration relation. A substrate with a low affinity for the enzyme (high K_m) generates a better signal than a substrate with high affinity [34]. A higher concentration of inhibitor than substrate was observed to produce a better signal in an inhibition based enzyme biosensor [35].

Enzymes

Acetylcholinesterase is the most popular choice of the enzyme when it comes to enzyme inhibition based biosensor. Acetylcholinesterase has been successfully used for the detection of (1) pesticides and insecticides such as paraoxon, dichlorvos, malathion, carbofuran, aldicarb, *etc* [36 - 42], (2) surfactants like sodiumdodecylsulfate and benzalkonium chloride [26, 41], (3) heavy metals [41, 43] (4) nerve agents [11], (5) toxins like aflatoxin, mycotoxins [9, 44], (6) drugs like Codeine, Neostigmine, *etc* [42, 45].

Many other enzymes are also employed in biosensor for detection of analyte inhibitors. These include horse radish peroxidase [21, 23 - 25, 46], glucose oxidase [47 - 51], laccase [52 - 55], acid phosphatise [29], alkaline phosphatise [27, 30, 32, 33], bromelain [56], tyrosinase [14, 16, 18, 20, 57, 58], butyrlcholinestserase [59 - 64], catalase [12], urease [9, 65 - 67], xanthine oxidase [34].

Immobilization

Immobilization of enzymes can be performed by techniques such as physical entrapment, covalent cross-linking and adsorption. The advantages of enzyme immobilization are

- Enzymes can be reused especially in cases of reversible inhibitors where a simple washing procedure renders enzyme reusability.
- Stability of biosensor *i.e.* prolongs the shelf life
- Reduces response time
- Increases the sensitivity of the enzyme

However, immobilization can sometimes affect the performance of biosensors. It decreases the sensitivity of the enzyme as reported in the detection of okadaic acid by immobilized phosphatase [68]. Immobilization can also change the type of inhibition as shown in a study involving the detection of heavy metal and polyphenols by immobilized acetylcholinesterase and polyphenol oxidase [19, 69].

In recent years, biosensor's sensitivity was drastically improved with nanotechnology. Various nanomaterials are reported for use in biosensor. These include carbon nanotubes [15, 21, 38, 51, 70, 71], graphene [36, 55, 66, 72 - 74], gold nanoparticles [22, 31, 36, 75, 76], CdTe semiconductor [37, 77], Silicon oxide nanosheets [78], magnetic nanobeads [67, 79, 80].

ENZYME INHIBITION BASED BIOSENSOR IN ENVIRONMENTAL POTENTIALITIES

Heavy Metals

The presence of heavy metals above the permissible limit is a serious threat to the environment. Enzyme inhibition based biosensors offer rapid and sensitive detection of these high-density pollutants. Although heavy metals act as a cofactor for enzymes, it also inhibits the activity of enzymes. This inhibitory activity has been successfully used for the detection of various heavy metals. The inhibition of alkaline phosphatase, immobilized on a graphene oxide nanofiber coated electrode by Hg^{2+}, Pb^{2+}, and Cd^{2+} was used for the determination of these metals at concentrations as low as 0.0075, 0.015 and 0.0312 µg L^{-1} [73]. In another study, alkaline phosphatase was immobilized on gold nanorods containing electrode for the determination of heavy metals [31]. The Cu^{2+}, Pb^{2+}, and Cd^{2+} were determined by photoluminescence measurement of the urease and glucose oxidase, immobilized on porous silica [81]. Horseradish peroxidase, immobilized on a maize tassel-carbon nanotube was used for the determination of Cu^{2+} and Pb^{2+} with limits of detection of 4.2 and 2.5 µg L^{-1} respectively [21]. The cross-linking of glucose oxidase with cobalt hexacyanoferrate deposited carbon film electrode was used for the measurement of Cu^{2+}, Co^{2+}, and Cd^{2+} [82]. A range of 60-120 µg L^{-1} was achieved as the limit of detection for Ag+, Pb^{2+}, and Cu^{2+} using optical biosensing of the horseradish peroxidase inhibition [78]. This enzyme was immobilized on a nanostructured porous silica-coated electrode. Urease immobilized on magnetic particles was used for the measurement of Hg^{2+} and Cu^{2+} [67]. Amperometric measurement of inhibition of catalase, immobilized on a glass electrode achieved the lowest limit of detection for Hg^{2+} at 0.000018 µg L^{-1} [83]. Tyrosinase and glucose oxidase immobilized on an electrode was used for the determination of Cr(III) and Cr(VI) with high sensitivity and good stability [17]. The inhibition of glucose oxidase immobilized on an electrode coated with a nanocomposite of manganese dioxide allowed the determination of a variety of heavy metals using a chronoamperometric transducer [84]. The cross-linking of β-galactosidase with a glass electrode was used for measurement of Cd(II) and Cr(VI) achieving limits of detection of 6950 and 91700 µg L^{-1}, respectively [85]. A reduced-graphene-oxide based field-effect transistor containing urease was used for the determination of Cu^{2+} with a limit of detection of 0.01 µg L^{-1} [66]. A 25 fold improvement in the limit of detection of heavy metal over conventional amperometry was achieved by a novel accumulation mode sensing of inhibitory activity of oxidoreductase [86]. Heavy metals such as Cu^{2+}, Hg^{2+}, Cd^{2+} and Pb^{2+} at limits of detection of 0.079, 0.025, 0.024 and 0.044 µg L^{-1}, respectively were achieved using glucose oxidase immobilized on ultrathin polypyrrole electrode [87]. The reported limits of detection for these metals are more than 60 fold higher than the previous studies. The inhibition of alkaline phosphatase immobilized on a poly(neutral red) carbon film electrode by Cr(III) and Cr(VI) was used for the amperometric determination of these metals [46]. The inhibition of alkaline phosphatase by Hg^{2+}, Cd^{2+}, Pb^{2+}, Zn^{2+}, and Cu^{2+} was used for the chronoamperometric determination of these metals [88]. The change in the ratio of bioluminescence intensities at two different wavelengths due to inhibition of Firefly luciferases by Hg^{2+} was used for its determination at a limit of detection of 10000 µg L^{-1} [89]. These studies demonstrate the robustness of various enzymes for the determination of heavy metals.

Pesticides

The prevalent use of pesticides has boosted food production around the world. Nevertheless, even trace amounts of these toxic compounds in consumables create acute toxicity in humans. Two major classes of pesticides are organophosphates and carbamates. Organophosphates include malathion, paraxon, parathion, dichlorvos, fenitrothion, *etc* and carbamates include aldicarb, carbaryl, carbofuran, *etc*. Inhibition based biosensor is constructed based on the ability of both pesticides to inhibit the activity of enzymes. Lu and co-workers [90] reported a highly sensitive and stable biosensor based on acetylcholinesterase immobilized on nanorod containing palladium and gold for the detection of paraxon [90]. The above biosensor was able to detect paraxon at concentrations as low as 3.6×10^{-6} µg L^{-1}. Fluorescence quenching biosensor based on immobilization of acetylcholinesterase on CdTe quantum dots was able to measure paraxon and parathion at lower limits of detection of 4.30×10^{-6} µg L^{-1} and 2.47×10^{-6} µg L^{-1} [77]. A highly sensitive biosensor based on immobilization of acetylcholinesterase on the tert-butyllithium exfoliated transition metal dichalcogenides electrode was able to detect fenitrothion at a lower limit of detection of 0.0029 µg L^{-1} [91]. In another study acetylcholinesterase immobilized on electrode containing nanocomposite mixture of reduced graphene oxide, nano-gold, cyclodextrin, Prussian blue and chitosan was used to measure malathion and carbaryl with detection limits of 0.004 pg mL^{-1} and 0.001 µg L^{-1}, respectively [74]. A biosensor based on ratiometric fluorescent quantum dots was used to measure an organophosphate at a very low detection limit of 0.018 µg L^{-1} [92]. Acetylcholinesterase immobilized on graphitic carbon nitride functionalized with cyclodextrin was able to measure organophosphates with high sensitivity by electrochemiluminescence [93]. Lipase immobilized on gold nanoparticles was used to measure diazinon, methyl parathion and methyl paraoxon at detection limits of 60, 26 and 25 µg L^{-1}, respectively [94]. The above biosensor was selective for organophosphates and insensitive to carbamate contamination. A biosensor based on immobilization of acetylcholinesterase and choline oxidase on graphene quantum dots, measured dichlorvos with a detection limit of 172 µg L^{-1} [72]. The immobilization of acetylcholinesterase on a porous carbon composite matrix of ionic liquid and gold nanoparticle was able to detect dichlorvos at a detection limit of 6.61×10^{-5} µg L^{-1} [95]. A biosensor based on acetylcholinesterase immobilized on carbon electrode containing Fe_2O_3 nanoparticle was able to measure paraoxon with a limit of detection of 1.2×10^{-8} µg L^{-1} [96]. A highly sensitive, novel biosensor based on acetylcholinesterase catalyzed the activation of DNA probes was used to measure aldicarb with a detection limit of 3.3 µg L^{-1} [97]. The novel structure of reticulated hollow spheres made of $NiCo_2S_4$ was crosslinked with acetylcholinesterase for detection of methyl parathion and paraoxon at concentrations as low as 4.2×10^{-4} µg L^{-1} and 3.5×10^{-4} µg L^{-1}, respectively [98]. Esterase immobilized on halloysite nanotubes was used for the measurement of carbofuran pesticide with a detection limit of 1.69 µg L^{-1} [99]. These studies demonstrate that enzyme inhibition based biosensor capability of sensitive detection of pesticides.

Toxins

Toxins especially microbe derived toxins have an adverse effect on human health and environmental ecology. For example, microcystin produced by certain algae species causes liver damage once ingested by a human. Timely detection of these compounds can check the spread of their source, aquatic organisms. A novel biosensor based on immobilization of recombinant protein phosphatase on screen-printed electrode containing Cobalt-Phthalocyanine was used for the detection of microcystin with a detection limit of 0.93 µg L^{-1} [100]. Marine toxin okadaic acid was detected using protein phosphatase immobilized onto sepharose beads activated by cyanogen bromide with detection limits of 1 µg L^{-1} [28]. Table **1** represents Environmental applications of enzyme inhibition based biosensor.

Table 1. Environmental applications of enzyme inhibition based biosensor.

Enzyme	Inhibitor/Analyte	Immobilization/Nanomaterial	Transducer	Detection Level	Study
Alkaline phosphatase	Hg^{2+}, Pb^{2+}, Cd^{2+}	Graphene oxide nanofibers	Electrochemical	0.0075, 0.015, 0.0312 µg L^{-1}	[73]
Horseradish peroxidase	Cu^{2+}, Pb^{2+}	Maize tassel -multiwalled carbon nanotube	Electrochemical	4.2, 2.5 µg L^{-1}	[21]
Horseradish peroxidase	Ag^+, Pb^{2+}, Cu^{2+}	Nanostructured porous Si	Optical	60 - 120 µg L^{-1}	[78]
Catalase	Hg^{2+}	Cross-linking with glutaraldehyde	Electrochemical	1.8×10^{-5} µg L^{-1}	[83]
β -galactosidase	Cd^{2+} and Cr^{6+}	Cross-linking with glutaraldehyde	Electrochemical	6.95 µg L^{-1}, 0.092 µg L^{-1}	[85]

(Table 1) cont.....

Enzyme	Inhibitor/Analyte	Immobilization/Nanomaterial	Transducer	Detection Level	Study
Urease	Cu^{2+}	Reduced-graphene-oxide based field-effect transistors	Electrochemical	10×10^3 µg L^{-1}	[66]
Oxidoreductase	As^{5+}	Immobilized with a redox polymer	Electrochemical	4.7×10^3 µg L^{-1}	[86]
Glucose oxidase	Cu^{2+}, Hg^{2+}, Cd^{2+} and Pb^{2+}	Ultrathin polypyrrole	Electrochemical	0.079 µg L^{-1} Cu^{2+}, 0.025 µg L^{-1} Hg^{2+}, 0.024 µg L^{-1} Pb^{2+} and 0.044 µg L^{-1} Cd^{2+}	[87]
Firefly luciferases	Heavy metal	Cross-linking with glutaraldehyde	Optical	0.10×10^3 µg L^{-1}	[89]
Acetylcholinesterase	Paraoxon	Pd and Au core-shell nanorods	Electrochemical	3.6×10^{-6} µg L^{-1}	[90]
Hydrogen peroxidase	Paraoxon and parathion	CdTe quantum dots	Optical	$4.30 \times 10^{-6} \times 10^3$ µg L^{-1} for paraoxon and $2.47 \times 10^{-6} \times 10^3$ µg L^{-1} for parathion	[77]
Acetylcholinesterase	Fenitrothion	Tert-butyllithium exfoliated transition metal dichalcogenides	Electrochemical	2.86×10^3 µg L^{-1}	[91]
Acetylcholinesterase	Malathion and carbaryl	Reduced graphene oxide-Au nanoparticles-β-cyclodextrin/Prussian blue-chitosan nanocomposites	Electrochemical	0.00414 µg L^{-1} for malathion and 0.00115 µg L^{-1} for carbaryl	[74]
Lipase	Diazinon, methyl parathion and methyl paraoxon	Gold nanoparticles	Electrochemical	60 µg L^{-1} for diazinon, 26×10^3 µg L^{-1} for methyl parathion and 25 µg L^{-1} for methyl paraoxon	[94]
Acetylcholinesterase and choline oxidase	Dichlorvos	Graphene quantum dots	Optical	172 µg L^{-1}	[72]
Acetylcholinesterase	Dichlorvos	Iionic liquids-Au nanoparticle - porous carbon composite matrix	Electrochemical	6.61×10^{-5} µg L^{-1}	[95]
Acetylcholinesterase	Paraoxon	Fe_2O_3 nanoparticle	Electrochemical	1.2×10^{-8} µg L^{-1}	[96]
Acetylcholinesterase	Aldicarb	DNA conformational switch	Optical	3.3 µg L^{-1}	[97]
Acetylcholinesterase	Methyl parathion and Araoxon	Reticulated hollow spheres structure $NiCO_2S_4$	Electrochemical	4.2×10^{-4} µg L^{-1} for methyl parathion and 3.5×10^{-5} µg L^{-1} for paraoxon	[98]
Esterase	Carbofuran	Halloysite nanotubes	Electrochemical	1.69 µg L^{-1}	[99]
Acetylcholinesterase	Malathion and Acephate	Carbon paste electrode	Electrochemical	58 µg L^{-1} for Malathion, 44 µg L^{-1} for Acephate	[101]
Recombinant protein phosphatase	Microcystin	Cobalt-Phtalocyanine modified electrode	Electrochemical	0.93 µg L^{-1}	[100]

ENZYME INHIBITION BASED BIOSENSOR IN PHARMACEUTICALS POTENTIALITIES

The detection of a pharmaceutical drug in our body is crucial for determining the effectiveness of the treatment. With the advent of new drugs in the market, the detection of these compounds by a fast, cheap and sensitive method is essential. Enzyme inhibition based biosensors have been addressing these issues in recent times. Methimazole, an antithyroid drug was measured using a biosensor based on immobilization of tyrosine and nanocomposite of iridium oxide onto a screen-printed electrode [79]. A highly sensitive biosensor was developed for the detection of analgesic drug acetaminophen by horseradish peroxidase immobilized on a gold electrode containing ZrO and Fe_3O_4 nanoparticles with detection limits of 0.01 µg L^{-1} [102]. Horse-radish peroxidase cross-linked with graphite electrode containing gold nanorods was able to detect anti-HIV replication drug deferiprone with a detection limit as low as 0.005 µg L^{-1} [75]. A high concentration of uric acid in our blood sample indicates the diabetic condition and kidney-related problems. A biosensor containing uricase immobilized on an electrode containing ZnO nanosheets was able to measure uric acid with a limit of the detection value of 0.019 µg L^{-1} [103]. The drug atrazine was detected by tyrosinase immobilized on graphite, graphene and multiwalled carbon nanotube screen-printed electrode with a detection limit of 300 µg L^{-1} [104]. A biosensor based on horse-radish peroxidase immobilized on electrode containing nanocomposite of polyhydroxyalkanoate polymer and gold nanoparticle was used for measurement of antimalarial drug artemisinin with a limit of detection of 3.5 µg L^{-1} [76]. Anticancer drugs are probed by the enzyme, glutathione-s-transferase; a biosensor containing the above enzyme immobilized on the carbon paste electrode was able to detect anticancer drugs cisplatin with a very low detection limit of 8.8 µg L^{-1} [105]. A biosensor based on tyrosinase immobilized on a multiwalled carbon nanotube electrode containing Nafion–cysteamine was used for measurement of dopamine at a concentration as low as 1 µg L^{-1} [15]. Harmane, norharmane, and harmaline β-carbolines were detected by monoamine oxidase -A immobilized on screen-printed electrodes containing Prussian blue and copper with detection limits of 5, 2.5 and 2.5 µg L^{-1} respectively [106]. Table **2** represents Pharmaceutical application of enzyme inhibition based biosensor.

Table 2. Pharmaceutical application of enzyme inhibition based biosensor.

Enzyme	Inhibitor/Analyte	Immobilization	Transducer	Detection Level	Study
Tyrosinase	Antithyroid drug - methimazole	The nanocomposite of magnetic nanoparticles with iridium oxide nanoparticles	Electrochemical	0.006 µg L^{-1}	[79]
Horseradish peroxidase	Acetaminophen-analgesic drug	Core-shell zro@Fe3O$_4$ nanoparticles	Electrochemical	0.01 µg L^{-1}	[102]
Horse radish peroxidase	Anti-HIV replication drug - deferiprone	Gold nanorods	Electrochemical	0.005 µg L^{-1}	[75]

(Table 1) cont.....

Enzyme	Inhibitor/Analyte	Immobilization	Transducer	Detection Level	Study
Uricase	Uric acid	Zno nanosheets	Electrochemical	0.019 µg L⁻¹	[103]
Tyrosinase	Atrazine	Grapheme and multiwalled carbon nanotubes	Electrochemical	300 µg L⁻¹	[104]
Horse-radish peroxidase	Antimalarial drug - artemisinin	Nanocomposite polyhydroxyalkanoate and gold nanoparticle	Electrochemical	3.5 µg L⁻¹	[76]
Glutathione-s-transferase	Cisplatin – Anticancer drug	Carbon paste electrodes	Electrochemical	8.8 µg L⁻¹	[105]
Tyrosinase	Dopamine	Multiwalled carbon nanotube–Nafion–cysteamine	Electrochemical	1 µg L⁻¹	[15]
Monoamine oxidase-A	B-carbolines: harmane, norharmane, and harmaline in drug	Prussian blue and copper	Electrochemical	5.0 µg L⁻¹ for harmane, 2.5 µg L⁻¹ for both harmaline and norharmane	[106]

ENZYME INHIBITION BASED BIOSENSOR IN FOOD INDUSTRY

The application of biosensor in the food industry fulfills the purpose of (1) food safety by detection of contaminants, pathogens, allergens in food, (2) food quality by measurement of levels of alcohol, amino acids, sugars, sweeteners, additives and flavours in food. Food contaminants display an inhibitory effect on some enzymes and biosensors based on this property have been reported. Toxic compounds malaoxon and chlorpyriphos-oxon in milk samples were rapidly detected with high sensitivity using genetically modified acetylcholinesterase immobilized on screen-printed electrodes [107]. Malathion in lettuce was measured using acetylcholinesterase immobilized on a multiwalled carbon nanotube electrode containing Poly(3,4-ethylene dioxythiophene) with a very low detection limit of 11×10^{-9} µg L⁻¹ [70]. Carbosulfan in rice was measured using a biosensor having acetylcholinesterase immobilized on platinum electrode containing zinc oxide nanocuboids with detection limits of 0.24×10^3 µg L⁻¹ [108]. A biosensor based on immobilization of catalase on a magnetic nanocomposite of Fe_3O_4 and chitosan was used for the detection of herbicide, 2,4-dichloroph-enoxyacetic acid in food sample with a limit of detection of 0.02 µg L⁻¹ [80]. Food allergens putrescine and histamine were detected using a biosensor based on immobilization of histamine dehydrogenase and putrescine oxidase on a disposable electrode with detection limits of 10 and 8.1 µg L⁻¹, respectively [109]. Toxic formetanate hydrochloride in mango and grapes was measured using a biosensor based on laccase immobilized onto an electrode containing gold nanoparticle with a limit of detection of 9.5×10^{-2} µg L^{-1} [110]. Paraoxon in milk and tap water was detected using butyrylcholinesterase cross-linked with graphite electrode containing silver nanowires and conjugated polymer with a detection limit of 0.212 µg L^{-1} [59]. Fungal aflatoxins, commonly associated with crops such as peanuts, corn, *etc* was detected by a pH-sensitive field effective transistor-based biosensor containing acetylcholinesterase [41]. Mycotoxin patulin found in rotten apples was detected by a biosensor containing urease immobilized onto a gold electrode [9]. Food preservative, benzoic acid in a mayonnaise sauce and soft drinks were detected by a biosensor based on tyrosinase inhibitory activity with a detection limit of 0.9 µg L^{-1} [111]. Table **3** represents Food industry applications of enzyme inhibition based biosensor.

Table 3. Food industry applications of enzyme inhibition based biosensor.

Enzyme	Inhibitor/Analyte	Immobilization/Nanomaterials	Transducer	Detection Level	Study
Acetylcholinesterase	Aflatoxin B1 in food sample	Cross-linking with bovine serum albumin and glutaraldehyde	Electrochemical (Field-effect transistors)	200 µg L^{-1}	[41]
Genetically modified Acetylcholinesterase	Chlorpyriphos-oxon and malaoxon in milk	Entrapment in a polymeric matrix	Electrochemical	2.58×10^{-5} µg L^{-1} for Chlorpyriphos-oxon, 4.25×10^{-4} µg L^{-1} for malaoxon	[107]
Acetylcholinesterase	Malathion in lettuce	Nanocomposites of Poly(3,4-ethylene dioxythiophene) and Multiwalled carbon nanotube	Electrochemical	1×10^{-9} µg L^{-1}	[70]
Acetylcholinesterase	Carbosulfan in rice	ZnO nanocuboids	Electrochemical	0.24×10^{3} µg L^{-1}	[108]
Catalase	2,4-dichlorophenoxyacetic acid	Magnetic Fe_3O_4-chitosan nanocomposite	Electrochemical	0.02 µg L^{-1}	[80]
Histaminedehydrogenase and putrescine oxidase	Allergens Histamine and Putrescine in food samples	Cross-linking with bovine serum albumin and glutaraldehyde	Electrochemical	8.1 µg L^{-1} for Histamine and 10 µg L^{-1} for Putrescine	[109]
Laccase	Formetanate hydrochloride in mango and grapes	Gold nanoparticles	Electrochemical	9.5×10^{-2} µg L^{-1}	[110]
Butyrylcholinesterase	Paraoxon in tap water and milk	Silver nanowires	Electrochemical	0.212 µg L^{-1}	[59]
Superoxide dismutase	Antioxidant, superoxide in beverages	Multiwalled carbon nanotubes with poly(3,4-ethylene dioxythiophene)	Electrochemical	1 µg L^{-1}	[112]
Protein phosphatase	Okadaic acid	CNBr-activated Sepharose beads	Electrochemical	1 µg L^{-1}	[28]

APPLICATIONS IN OTHER MAJOR FIELDS

This section reviews enzyme inhibition based biosensors for applications in sectors rather than environment, food and pharmaceuticals. Cyanide is an important substance utilized in a wide range of industries however it causes acute toxicity at even trace amount. A biosensor based on horseradish peroxidase immobilized onto electrode containing gold sononanoparticles was used for the detection of cyanide with a detection limit of 0.03 µg L^{-1} [22]. In an earlier study, cyanide was measured using horseradish peroxidase cross-linked with the electrode by chitosan and acrylamide with a detection limit of 0.43 µg L^{-1} [113]. A biosensor containing polyphenol oxidase immobilized along with Zn− Al layered double hydroxides was used for the detection of cyanide with a limit of detection of 0.1 ×10^3 µg L^{-1} [13]. Sodium azide is used in automobile airbags and also in hospitals and laboratories as a preservative. This highly acute toxic compound was detected using a biosensor containing laccase immobilized onto an electrode containing redox-active layered double hydroxides with a limit of detection of 5.5 ×10^3 µg L^{-1} [52]. Sodium azide was also detected by the inhibition of catalase [114]. The amount of superoxide, reactive oxygen species in a sample indicate the antioxidant levels. A biosensor based on superoxide dismutase immobilized onto a multiwalled carbon nanotube electrode using poly (3,4-ethylene dioxythiophene) - a conducting polymer was used for determination of superoxide with a detection limit of 1 µg L^{-1} [112]. Immobilizations of glucose oxidase and choline oxidase onto carbon paste electrodes were reported for the detection of nitric oxide and nicotine with a limit of detection of 250 µg L^{-1} and 10 µg L^{-1} [115, 116]. Table **4** mentions the Other applications of enzyme inhibition based biosensor.

Table 4. Other applications of enzyme inhibition based biosensor.

Enzyme	Inhibitor/Analyte	Immobilization	Transducer	Detection Level	Study
Horseradish peroxidase	Cyanide	Gold sononanoparticles	Electrochemical	0.03 µg L^{-1}	[22]
Horseradish peroxidase	Cyanide	Chitosan and acrylamide	Electrochemical	0.43 µg L^{-1}	[113]
Polyphenol oxidase	Cyanide	Zn - Al layered double hydroxides	Electrochemical	0.1 ×10^3 µg L^{-1}	[13]
Laccase	Azide, fluoride, and cyanide	Redox-active layered double hydroxides	Electrochemical	5.5 ×10^3 µg L^{-1} for azide, 6.9 ×10^3 µg L^{-1} for fluoride, and 6.2 ×10^3 µg L^{-1} for cyanide	[52]
Glucose oxidase	Nitrous oxide	Carbon paste electrode	Electrochemical	250 µg L^{-1}	[115]
Choline oxidase	Nicotine	Carbon paste electrode	Electrochemical	10 µg L^{-1}	[116]

FUTURE PERSPECTIVES

Commercially available biosensors are based mainly on direct enzymatic activity rather than enzyme inhibition assay. Some of the major manufacturers in the biosensor market are Yellow springs instruments, Abbott, Daiichi, Roche, Eppendorf, Bayer, Medline, Nova Biomedical, United Surgical & Diagnostics, Nippon Laser and Electronics Lab, Texas Instruments, Abtech Scientific, Inc, Biacore AB-Sweeden. Much of the biosensors have been developed for glucose detection [117]. However, in the coming days, it is expected that enzyme inhibition based biosensors will play a major role in the expansion of the market in other areas also.

The future biosensors should meet the criteria of low cost, short response time, high sensitivity and selectivity. In this regard, Table **5** presents an economical overview by categorizing biosensor components in terms of cost [71, 118 - 120]. In recent times, drastic improvements in response time and sensitivity were achieved using novel transducers and nanomaterials but the selectivity remains an obstacle. Genetic engineering and bioinformatics can play a major role in addressing this problem. Recent genetic tools have gained the ability to modify active sites or other binding sites of the enzyme by altering the amino acids responsible for sensitivity and selectivity. This was demonstrated in a recent study involving site-directed mutagenesis of firefly's luciferase, in which histidine and glutamic acid at position 310 and 354 were identified as key residues for metal sensitivity and subsequent engineering improved the sensitivity towards mercury, zinc, and nickel [89].

Table 5. Economy based categorization of biosensor components.

Low-Cost Components	High-Cost Components
• Screen-printed electrodes • Photovoltaics • Electronic paper -paper-microfluidics • Organic light-emitting diodes • Radio-frequency identification tags • Organic thin-film transistors • Poly (brilliant cresyl blue) electrode • Physical adsorption Immobilization • Entrapment Immobilization • Surface plasmon resonance • Microcantilever • Quartz crystal microbalances • Molecularly imprinted polymers	• Enzyme purification • Optical detection by fluorescence • Silicon-based transistors • Platinum electrode • Cross-linking immobilization • Covalent binding immobilization

CONCLUDING REMARKS

This chapter discusses recent works on enzyme inhibition based biosensor. Various enzymes, substrates and immobilizing material used in inhibition based biosensor were reviewed. This review points out that by changing the substrate for enzyme inhibition based biosensor, one can change the transducer and thereby achieving differential sensitivity and response time. The identification of the inhibition type for the construction of an effective biosensor is emphasized. The role of enzyme kinetics using plots such as Lineweaver-Burk, Dixon, *etc* in predicting the type of inhibition is highlighted. Finally, this chapter concludes by summarizing enzyme inhibition based biosensors on three main streams of application namely environmental, food industry and pharmaceuticals.

CONSENT FOR PUBLICATION

Not applicable.

CONFLICT OF INTEREST

The author(s) confirms that there is no conflict of interest.

ACKNOWLEDGEMENTS

Authors thank Prof. M. Sivanandham, Secretary, SVEHT and SVCE for their support and encouragement.

REFERENCES

[1]   Guilbault G, Tyson B, Kramer D, Cannon P. Electrochemical determination of glucose oxidase using diphenylamine sulfonic acid as a potential poiser. Anal Chem 1963; 35: 582-6.
[http://dx.doi.org/10.1021/ac60197a014]

[2]   Amine A, Arduini F, Moscone D, Palleschi G. Recent advances in biosensors based on enzyme inhibition. Biosens Bioelectron 2016; 76: 180-94.
[http://dx.doi.org/10.1016/j.bios.2015.07.010] [PMID: 26227311]

[3]   Shuler ML, Kargi F, DeLisa M. Bioprocess engineering: basic concepts: Prentice Hall Englewood Cliffs, NJ. 2017.

[4]   Lineweaver H, Burk D. The determination of enzyme dissociation constants. J Am Chem Soc 1934; 56: 658-66.
[http://dx.doi.org/10.1021/ja01318a036]

[5]   Dixon M. The determination of enzyme inhibitor constants. Biochem J 1953; 55(1): 170-1.
[http://dx.doi.org/10.1042/bj0550170] [PMID: 13093635]

[6]   Cornish-Bowden A. A simple graphical method for determining the inhibition constants of mixed, uncompetitive and non-competitive inhibitors. Biochem J 1974; 137(1): 143-4.
[http://dx.doi.org/10.1042/bj1370143] [PMID: 4206907]

[7]   Rajangam B, Daniel DK, Krastanov AI. Progress in enzyme inhibition based detection of pesticides. Eng Life Sci 2018; 18: 4-19.

[http://dx.doi.org/10.1002/elsc.201700028]

[8]   Rodríguez-Delgado MM, Alemán-Nava GS, Rodríguez-Delgado JM, Dieck-Assad G, *et al.* Laccase-based biosensors for detection of phenolic compounds. TrAC. Trends Analyt Chem 2015; 74: 21-45.
[http://dx.doi.org/10.1016/j.trac.2015.05.008]

[9]   Soldatkin OO, Stepurska K, Arkhypova V, Soldatkin A, *et al.* Conductometric enzyme biosensor for patulin determination. Sens Actuators B Chem 2017; 239: 1010-5.
[http://dx.doi.org/10.1016/j.snb.2016.08.121]

[10]   Ding J, Qin W. Current-driven ion fluxes of polymeric membrane ion-selective electrode for potentiometric biosensing. J Am Chem Soc 2009; 131(41): 14640-1.
[http://dx.doi.org/10.1021/ja906723h] [PMID: 19785410]

[11]   Pohanka M, Adam V, Kizek R. An acetylcholinesterase-based chronoamperometric biosensor for fast and reliable assay of nerve agents. Sensors (Basel) 2013; 13(9): 11498-506.
[http://dx.doi.org/10.3390/s130911498] [PMID: 23999806]

[12]   Chen H, Mousty C, Chen L, Cosnier S. A new approach for nitrite determination based on a HRP/catalase biosensor. Mater Sci Eng C 2008; 28: 726-30.
[http://dx.doi.org/10.1016/j.msec.2007.10.015]

[13]   Shan D, Mousty C, Cosnier S. Subnanomolar cyanide detection at polyphenol oxidase/clay biosensors. Anal Chem 2004; 76(1): 178-83.
[http://dx.doi.org/10.1021/ac034713m] [PMID: 14697048]

[14]   Vidal JC, Esteban S, Gil J, Castillo JR. A comparative study of immobilization methods of a tyrosinase enzyme on electrodes and their application to the detection of dichlorvos organophosphorus insecticide. Talanta 2006; 68(3): 791-9.
[http://dx.doi.org/10.1016/j.talanta.2005.06.038] [PMID: 18970392]

[15]   Canbay E, Akyilmaz E. Design of a multiwalled carbon nanotube-Nafion-cysteamine modified tyrosinase biosensor and its adaptation of dopamine determination. Anal Biochem 2014; 444: 8-15.
[http://dx.doi.org/10.1016/j.ab.2013.09.019] [PMID: 24090870]

[16]   Asav E, Yorganci E, Akyilmaz E. An inhibition type amperometric biosensor based on tyrosinase enzyme for fluoride determination. Talanta 2009; 78(2): 553-6.
[http://dx.doi.org/10.1016/j.talanta.2008.12.010] [PMID: 19203623]

[17]   Liu L, Kang X, Chen C, Zhang H, Chen C, Xie Q. L-tyrosine polymerization-based ultrasensitive multi-analyte enzymatic biosensor. Talanta 2018; 179: 803-9.
[http://dx.doi.org/10.1016/j.talanta.2017.12.014] [PMID: 29310310]

[18]   Campanella L, Lelo D, Martini E, Tomassetti M. Organophosphorus and carbamate pesticide analysis using an inhibition tyrosinase organic phase enzyme sensor; comparison by butyrylcholinesterase+choline oxidase opee and application to natural waters. Anal Chim Acta 2007; 587(1): 22-32.
[http://dx.doi.org/10.1016/j.aca.2007.01.023] [PMID: 17386749]

[19]   Narli I, Kiralp S, Toppare L. Preventing inhibition of tyrosinase with modified electrodes. Anal Chim Acta 2006; 572(1): 25-31.
[http://dx.doi.org/10.1016/j.aca.2006.04.088] [PMID: 17723457]

[20]   Sima VH, Patris S, Aydogmus Z, Sarakbi A, Sandulescu R, Kauffmann JM. Tyrosinase immobilized magnetic nanobeads for the amperometric assay of enzyme inhibitors: application to the skin whitening agents. Talanta 2011; 83(3): 980-7.
[http://dx.doi.org/10.1016/j.talanta.2010.11.005] [PMID: 21147347]

[21]   Moyo M, Okonkwo JO, Agyei NM. An amperometric biosensor based on horseradish peroxidase immobilized onto maize tassel-multi-walled carbon nanotubes modified glassy carbon electrode for determination of heavy metal ions in aqueous solution. Enzyme Microb Technol 2014; 56: 28-34.
[http://dx.doi.org/10.1016/j.enzmictec.2013.12.014] [PMID: 24564899]

[22]   Attar A, Cubillana-Aguilera L, Naranjo-Rodríguez I, de Cisneros JLH-H, Palacios-Santander JM, Amine A. Amperometric inhibition biosensors based on horseradish peroxidase and gold sononanoparticles immobilized onto different electrodes for cyanide measurements. Bioelectrochemistry 2015; 101: 84-91.
[http://dx.doi.org/10.1016/j.bioelechem.2014.08.003] [PMID: 25179932]

[23]   Oliveira GC, Moccelini SK, Castilho M, *et al.* Biosensor based on atemoya peroxidase immobilised on modified nanoclay for glyphosate biomonitoring. Talanta 2012; 98: 130-6.
[http://dx.doi.org/10.1016/j.talanta.2012.06.059] [PMID: 22939138]

[24]   Huang J, Huang W, Wang T. Catalytic and inhibitory kinetic behavior of horseradish peroxidase on the electrode surface. Sensors (Basel) 2012; 12(11): 14556-69.
[http://dx.doi.org/10.3390/s121114556] [PMID: 23202175]

[25]   Moyo M, Okonkwo JO. Horseradish peroxidase biosensor based on maize tassel–MWCNTs composite for cadmium detection. Sens Actuators B Chem 2014; 193: 515-21.
[http://dx.doi.org/10.1016/j.snb.2013.11.086]

[26]   Ko S, Takahashi Y, Fujita H, Tatsuma T, *et al.* Peroxidase-modified cup-stacked carbon nanofiber networks for electrochemical biosensing with adjustable dynamic range. RSC Advances 2012; 2: 1444-9.
[http://dx.doi.org/10.1039/C1RA00649E]

[27]   Samphao A, Suebsanoh P, Wongsa Y, Pekec B, *et al.* Alkaline phosphatase inhibition-based amperometric biosensor for the detection of carbofuran. Int J Electrochem Sci 2013; 8: 3254-64.

[28]   Allum LL, Mountfort DO, Gooneratne R, Pasco N, Goussain G, Hall EA. Assessment of protein phosphatase in a re-usable rapid assay format in detecting microcystins and okadaic acid as a precursor to biosensor development. Toxicon 2008; 52(7): 745-53.
[http://dx.doi.org/10.1016/j.toxicon.2008.08.010] [PMID: 18812183]

[29]   Sanllorente-Méndez S, Domínguez-Renedo O, Arcos-Martínez MJ. Development of acid phosphatase based amperometric biosensors for the inhibitive determination of As(V). Talanta 2012; 93: 301-6.
[http://dx.doi.org/10.1016/j.talanta.2012.02.037] [PMID: 22483914]

[30]   Alvarado-Gámez AL, Alonso-Lomillo MA, Domínguez-Renedo O, Arcos-Martínez MJ. A disposable alkaline phosphatase-based biosensor for vanadium chronoamperometric determination. Sensors (Basel) 2014; 14(2): 3756-67.
[http://dx.doi.org/10.3390/s140203756] [PMID: 24569772]

[31]   Homaei A. Immobilization of *Penaeus merguiensis* alkaline phosphatase on gold nanorods for heavy metal detection. Ecotoxicol Environ Saf 2017; 136: 1-7.
[http://dx.doi.org/10.1016/j.ecoenv.2016.10.023] [PMID: 27810575]

[32]   Akyilmaz E, Turemis M. An inhibition type alkaline phosphatase biosensor for amperometric determination of caffeine. Electrochim Acta 2010; 55: 5195-9.
[http://dx.doi.org/10.1016/j.electacta.2010.04.038]

[33]   Tekaya N, Saiapina O, Ben Ouada H, Lagarde F, Ben Ouada H, Jaffrezic-Renault N. Ultra-sensitive conductometric detection of heavy metals based on inhibition of alkaline phosphatase activity from Arthrospira platensis. Bioelectrochemistry 2013; 90: 24-9.
[http://dx.doi.org/10.1016/j.bioelechem.2012.10.001] [PMID: 23174485]

[34]   Shan D, Wang Y, Zhu M, Xue H, Cosnier S, Wang C. Development of a high analytical performance-xanthine biosensor based on layered double hydroxides modified-electrode and investigation of the inhibitory effect by allopurinol. Biosens Bioelectron 2009; 24(5): 1171-6.
[http://dx.doi.org/10.1016/j.bios.2008.07.023] [PMID: 18760589]

[35]   Benilova IV, Arkhypova VN, Dzyadevych SV, Jaffrezic-Renault N, *et al.* Kinetics of human and horse sera cholinesterases inhibition with solanaceous glycoalkaloids: Study by potentiometric biosensor. Pestic Biochem Physiol 2006; 86: 203-10.

[http://dx.doi.org/10.1016/j.pestbp.2006.04.002]

[36]   Yang Y, Asiri AM, Du D, Lin Y. Acetylcholinesterase biosensor based on a gold nanoparticle-polypyrrole-reduced graphene oxide nanocomposite modified electrode for the amperometric detection of organophosphorus pesticides. Analyst (Lond) 2014; 139(12): 3055-60.
[http://dx.doi.org/10.1039/c4an00068d] [PMID: 24770670]

[37]   Liang H, Song D, Gong J. Signal-on electrochemiluminescence of biofunctional CdTe quantum dots for biosensing of organophosphate pesticides. Biosens Bioelectron 2014; 53: 363-9.
[http://dx.doi.org/10.1016/j.bios.2013.10.011] [PMID: 24184599]

[38]   Ding J, Zhang H, Jia F, Qin W, Du D. Assembly of carbon nanotubes on a nanoporous gold electrode for acetylcholinesterase biosensor design. Sens Actuators B Chem 2014; 199: 284-90.
[http://dx.doi.org/10.1016/j.snb.2014.04.012]

[39]   Dutta RR, Puzari P. Amperometric biosensing of organophosphate and organocarbamate pesticides utilizing polypyrrole entrapped acetylcholinesterase electrode. Biosens Bioelectron 2014; 52: 166-72.
[http://dx.doi.org/10.1016/j.bios.2013.08.050] [PMID: 24041663]

[40]   Evtugyn GA, Shamagsumova RV, Padnya PV, Stoikov II, Antipin IS. Cholinesterase sensor based on glassy carbon electrode modified with Ag nanoparticles decorated with macrocyclic ligands. Talanta 2014; 127: 9-17.
[http://dx.doi.org/10.1016/j.talanta.2014.03.048] [PMID: 24913851]

[41]   Stepurska KV, Soldatkin OO, Arkhypova VM, *et al.* Development of novel enzyme potentiometric biosensor based on pH-sensitive field-effect transistors for aflatoxin B1 analysis in real samples. Talanta 2015; 144: 1079-84.
[http://dx.doi.org/10.1016/j.talanta.2015.07.068] [PMID: 26452930]

[42]   Du D, Chen S, Cai J, Song D. Comparison of drug sensitivity using acetylcholinesterase biosensor based on nanoparticles–chitosan sol–gel composite. J Electroanal Chem (Lausanne Switz) 2007; 611: 60-6.
[http://dx.doi.org/10.1016/j.jelechem.2007.08.007]

[43]   Barquero-Quirós M, Domínguez-Renedo O, Alonso-Lomillo MA, Arcos-Martínez MJ. Acetylcholinesterase inhibition-based biosensor for aluminum(III) chronoamperometric determination in aqueous media. Sensors (Basel) 2014; 14(5): 8203-16.
[http://dx.doi.org/10.3390/s140508203] [PMID: 24811076]

[44]   Puiu M, Istrate O, Rotariu L, Bala C. Kinetic approach of aflatoxin B1-acetylcholinesterase interaction: a tool for developing surface plasmon resonance biosensors. Anal Biochem 2012; 421(2): 587-94.
[http://dx.doi.org/10.1016/j.ab.2011.10.035] [PMID: 22093609]

[45]   Turan J, Kesik M, Soylemez S, Goker S, *et al.* Development of an amperometric biosensor based on a novel conducting copolymer for detection of anti-dementia drugs. J Electroanal Chem (Lausanne Switz) 2014; 735: 43-50.
[http://dx.doi.org/10.1016/j.jelechem.2014.10.007]

[46]   Attar A, Emilia Ghica M, Amine A, Brett CM. Poly(neutral red) based hydrogen peroxide biosensor for chromium determination by inhibition measurements. J Hazard Mater 2014; 279: 348-55.
[http://dx.doi.org/10.1016/j.jhazmat.2014.07.019] [PMID: 25080156]

[47]   Ghica ME, Carvalho RC, Amine A, Brett CM. Glucose oxidase enzyme inhibition sensors for heavy metals at carbon film electrodes modified with cobalt or copper hexacyanoferrate. Sens Actuators B Chem 2013; 178: 270-8.
[http://dx.doi.org/10.1016/j.snb.2012.12.113]

[48]   Chey CO, Ibupoto ZH, Khun K, Nur O, Willander M. Indirect determination of mercury ion by inhibition of a glucose biosensor based on ZnO nanorods. Sensors (Basel) 2012; 12(11): 15063-77.
[http://dx.doi.org/10.3390/s121115063] [PMID: 23202200]

[49] Samphao A, Rerkchai H, Jitcharoen J, Nacapricha D, *et al.* Indirect determination of mercury by inhibition of glucose oxidase immobilized on a carbon paste electrode. Int J Electrochem Sci 2012; 7: 1001-10.

[50] Ghica ME, Brett CM. Glucose oxidase inhibition in poly (neutral red) mediated enzyme biosensors for heavy metal determination. Mikrochim Acta 2008; 163: 185-93.
[http://dx.doi.org/10.1007/s00604-008-0018-1]

[51] Yang Q, Qu Y, Bo Y, Wen Y, Huang S. Biosensor for atrazin based on aligned carbon nanotubes modified with glucose oxidase. Mikrochim Acta 2010; 168: 197-203.
[http://dx.doi.org/10.1007/s00604-009-0272-x]

[52] Mousty C, Vieille L, Cosnier S. Laccase immobilization in redox active layered double hydroxides: a reagentless amperometric biosensor. Biosens Bioelectron 2007; 22(8): 1733-8.
[http://dx.doi.org/10.1016/j.bios.2006.08.020] [PMID: 17023155]

[53] Deng L, Chen C, Zhou M, Guo S, Wang E, Dong S. Integrated self-powered microchip biosensor for endogenous biological cyanide. Anal Chem 2010; 82(10): 4283-7.
[http://dx.doi.org/10.1021/ac100274s] [PMID: 20402491]

[54] Zapp E, Brondani D, Vieira IC, Scheeren CW, *et al.* Biomonitoring of methomyl pesticide by laccase inhibition on sensor containing platinum nanoparticles in ionic liquid phase supported in montmorillonite. Sens Actuators B Chem 2011; 155: 331-9.
[http://dx.doi.org/10.1016/j.snb.2011.04.015]

[55] Oliveira TM, Fátima Barroso M, Morais S, *et al.* Laccase-Prussian blue film-graphene doped carbon paste modified electrode for carbamate pesticides quantification. Biosens Bioelectron 2013; 47: 292-9.
[http://dx.doi.org/10.1016/j.bios.2013.03.026] [PMID: 23587791]

[56] Gorodkiewicz E, Breczko J, Sankiewicz A. Surface Plasmon Resonance Imaging biosensor for cystatin determination based on the application of bromelain, ficin and chymopapain. Folia Histochem Cytobiol 2012; 50(1): 130-6.
[http://dx.doi.org/10.5603/FHC.2012.0019] [PMID: 22532148]

[57] Amine A, El Harrad L, Arduini F, Moscone D, Palleschi G. Analytical aspects of enzyme reversible inhibition. Talanta 2014; 118: 368-74.
[http://dx.doi.org/10.1016/j.talanta.2013.10.025] [PMID: 24274310]

[58] Shim J, Woo J-J, Moon S-H, Kim G-Y. A preparation of a single-layered enzyme-membrane using asymmetric pBPPO base film for development of pesticide detecting biosensor. J Membr Sci 2009; 330: 341-8.
[http://dx.doi.org/10.1016/j.memsci.2009.01.013]

[59] Turan J, Kesik M, Soylemez S, Goker S, Coskun S, Unalan HE, *et al.* An effective surface design based on a conjugated polymer and silver nanowires for the detection of paraoxon in tap water and milk. Sens Actuators B Chem 2016; 228: 278-86.
[http://dx.doi.org/10.1016/j.snb.2016.01.034]

[60] Mitsubayashi K, Nakayama K, Taniguchi M, Saito H, Otsuka K, Kudo H. Bioelectronic sniffer for nicotine using enzyme inhibition. Anal Chim Acta 2006; 573-574: 69-74.
[http://dx.doi.org/10.1016/j.aca.2006.01.091] [PMID: 17723507]

[61] Espinoza MA, Istamboulie G, Chira A, *et al.* Detection of glycoalkaloids using disposable biosensors based on genetically modified enzymes. Anal Biochem 2014; 457: 85-90.
[http://dx.doi.org/10.1016/j.ab.2014.04.005] [PMID: 24747413]

[62] Arduini F, Amine A, Moscone D, Ricci F, Palleschi G. Fast, sensitive and cost-effective detection of nerve agents in the gas phase using a portable instrument and an electrochemical biosensor. Anal Bioanal Chem 2007; 388(5-6): 1049-57.
[http://dx.doi.org/10.1007/s00216-007-1330-z] [PMID: 17508205]

[63] Arduini F, Ricci F, Tuta CS, *et al.* Detection of carbamic and organophosphorous pesticides in water

samples using a cholinesterase biosensor based on Prussian Blue-modified screen-printed electrode. Anal Chim Acta 2006; 580(2): 155-62.
[http://dx.doi.org/10.1016/j.aca.2006.07.052] [PMID: 17723768]

[64]   Khaled E, Kamel MS, Hassan HN, Abdel-Gawad H, Aboul-Enein HY. Performance of a portable biosensor for the analysis of ethion residues. Talanta 2014; 119: 467-72.
[http://dx.doi.org/10.1016/j.talanta.2013.11.001] [PMID: 24401442]

[65]   Kuralay F, Özyörük H, Yıldız A. Inhibitive determination of Hg2+ ion by an amperometric urea biosensor using poly (vinylferrocenium) film. Enzyme Microb Technol 2007; 40: 1156-9.
[http://dx.doi.org/10.1016/j.enzmictec.2006.08.025]

[66]   Piccinini E, Bliem C, Reiner-Rozman C, Battaglini F, Azzaroni O, Knoll W. Enzyme-polyelectrolyte multilayer assemblies on reduced graphene oxide field-effect transistors for biosensing applications. Biosens Bioelectron 2017; 92: 661-7.
[http://dx.doi.org/10.1016/j.bios.2016.10.035] [PMID: 27836616]

[67]   Pogorilyi RP, Melnyk IV, Zub YL, Seisenbaeva GA, Kessler VG. Enzyme immobilization on a nanoadsorbent for improved stability against heavy metal poisoning. Colloids Surf B Biointerfaces 2016; 144: 135-42.
[http://dx.doi.org/10.1016/j.colsurfb.2016.04.003] [PMID: 27085045]

[68]   Campàs M, Marty J-L. Enzyme sensor for the electrochemical detection of the marine toxin okadaic acid. Anal Chim Acta 2007; 605(1): 87-93.
[http://dx.doi.org/10.1016/j.aca.2007.10.036] [PMID: 18022415]

[69]   Stoytcheva M. Electrochemical evaluation of the kinetic parameters of a heterogeneous enzyme reaction in presence of metal ions. Electroanalysis. An International Journal Devoted to Fundamental and Practical Aspects of Electroanalysis 2002; 14: 923-7.

[70]   Kaur N, Thakur H, Prabhakar N. Conducting polymer and multi-walled carbon nanotubes nanocomposites based amperometric biosensor for detection of organophosphate. J Electroanal Chem (Lausanne Switz) 2016; 775: 121-8.
[http://dx.doi.org/10.1016/j.jelechem.2016.05.037]

[71]   Barsan MM, Ghica ME, Brett CM. Electrochemical sensors and biosensors based on redox polymer/carbon nanotube modified electrodes: a review. Anal Chim Acta 2015; 881: 1-23.
[http://dx.doi.org/10.1016/j.aca.2015.02.059] [PMID: 26041516]

[72]   Sahub C, Tuntulani T, Nhujak T, Tomapatanaget B. Effective biosensor based on graphene quantum dots via enzymatic reaction for directly photoluminescence detection of organophosphate pesticide. Sens Actuators B Chem 2018; 258: 88-97.
[http://dx.doi.org/10.1016/j.snb.2017.11.072]

[73]   Mishra RK, Nawaz MH, Hayat A, Nawaz MAH, et al. Electrospinning of graphene-oxide onto screen printed electrodes for heavy metal biosensor. Sens Actuators B Chem 2017; 247: 366-73.
[http://dx.doi.org/10.1016/j.snb.2017.03.059]

[74]   Zhao H, Ji X, Wang B, et al. An ultra-sensitive acetylcholinesterase biosensor based on reduced graphene oxide-Au nanoparticles-β-cyclodextrin/Prussian blue-chitosan nanocomposites for organophosphorus pesticides detection. Biosens Bioelectron 2015; 65: 23-30.
[http://dx.doi.org/10.1016/j.bios.2014.10.007] [PMID: 25461134]

[75]   Narang J, Malhotra N, Singh G, Pundir CS. Electrochemical impediometric detection of anti-HIV drug taking gold nanorods as a sensing interface. Biosens Bioelectron 2015; 66: 332-7.
[http://dx.doi.org/10.1016/j.bios.2014.11.038] [PMID: 25437372]

[76]   Phukon P, Radhapyari K, Konwar BK, Khan R. Natural polyhydroxyalkanoate-gold nanocomposite based biosensor for detection of antimalarial drug artemisinin. Mater Sci Eng C 2014; 37: 314-20.
[http://dx.doi.org/10.1016/j.msec.2014.01.019] [PMID: 24582254]

[77]   Xue G, Yue Z, Bing Z, Yiwei T, Xiuying L, Jianrong L. Highly-sensitive organophosphorus pesticide

biosensors based on CdTe quantum dots and bi-enzyme immobilized eggshell membranes. Analyst (Lond) 2016; 141(3): 1105-11.
[http://dx.doi.org/10.1039/C5AN02163D] [PMID: 26688862]

[78] Shtenberg G, Massad-Ivanir N, Segal E. Detection of trace heavy metal ions in water by nanostructured porous Si biosensors. Analyst (Lond) 2015; 140(13): 4507-14.
[http://dx.doi.org/10.1039/C5AN00248F] [PMID: 25988196]

[79] Kurbanoglu S, Mayorga-Martinez CC, Medina-Sánchez M, Rivas L, Ozkan SA, Merkoçi A. Antithyroid drug detection using an enzyme cascade blocking in a nanoparticle-based lab-on-a-chip system. Biosens Bioelectron 2015; 67: 670-6.
[http://dx.doi.org/10.1016/j.bios.2014.10.014] [PMID: 25459057]

[80] Wang H, Wang J, Wang J, Zhu R, *et al.* Spectroscopic method for the detection of 2, 4-dichlorophenoxyacetic acid based on its inhibitory effect towards catalase immobilized on reusable magnetic Fe3O4-chitosan nanocomposite. Sens Actuators B Chem 2017; 247: 146-54.
[http://dx.doi.org/10.1016/j.snb.2017.02.175]

[81] Syshchyk O, Skryshevsky VA, Soldatkin OO, Soldatkin AP. Enzyme biosensor systems based on porous silicon photoluminescence for detection of glucose, urea and heavy metals. Biosens Bioelectron 2015; 66: 89-94.
[http://dx.doi.org/10.1016/j.bios.2014.10.075] [PMID: 25460887]

[82] Attar A, Ghica ME, Amine A, Brett CM. Comparison of cobalt hexacyanoferrate and poly (neutral red) modified carbon film electrodes for the amperometric detection of heavy metals based on glucose oxidase enzyme inhibition. Anal Lett 2015; 48: 659-71.
[http://dx.doi.org/10.1080/00032719.2014.952372]

[83] Elsebai B, Ghica ME, Abbas MN, Brett CMA. Catalase based hydrogen peroxide biosensor for mercury determination by inhibition measurements. J Hazard Mater 2017; 340: 344-50.
[http://dx.doi.org/10.1016/j.jhazmat.2017.07.021] [PMID: 28728113]

[84] El-Haleem HSA, Hefnawy A, Hassan RY, Badawi AH, El-Sherbiny IM. Manganese dioxide-core–shell hyperbranched chitosan (MnO 2–HBCs) nano-structured screen printed electrode for enzymatic glucose biosensors. RSC Advances 2016; 6: 109185-91.
[http://dx.doi.org/10.1039/C6RA24419J]

[85] Fourou H, Zazoua A, Braiek M, Jaffrezic-Renault N. An enzyme biosensor based on beta-galactosidase inhibition for electrochemical detection of cadmium (II) and chromium (VI). Int J Environ Anal Chem 2016; 96: 872-85.
[http://dx.doi.org/10.1080/03067319.2016.1209659]

[86] Oja SM, Feldman B, Eshoo MW. Method for Low Nanomolar Concentration Analyte Sensing Using Electrochemical Enzymatic Biosensors. Anal Chem 2018; 90(3): 1536-41.
[http://dx.doi.org/10.1021/acs.analchem.7b04075] [PMID: 29265807]

[87] Ayenimo JG, Adeloju SB. Inhibitive potentiometric detection of trace metals with ultrathin polypyrrole glucose oxidase biosensor. Talanta 2015; 137: 62-70.
[http://dx.doi.org/10.1016/j.talanta.2015.01.006] [PMID: 25770607]

[88] Islam MS, Sazawa K, Hata N, Sugawara K, Kuramitz H. Determination of heavy metal toxicity by using a micro-droplet hydrodynamic voltammetry for microalgal bioassay based on alkaline phosphatase. Chemosphere 2017; 188: 337-44.
[http://dx.doi.org/10.1016/j.chemosphere.2017.09.008] [PMID: 28888859]

[89] Gabriel GV, Viviani VR. Engineering the metal sensitive sites in Macrolampis sp2 firefly luciferase and use as a novel bioluminescent ratiometric biosensor for heavy metals. Anal Bioanal Chem 2016; 408(30): 8881-93.
[http://dx.doi.org/10.1007/s00216-016-0011-1] [PMID: 27815607]

[90] Lu X, Tao L, Song D, Li Y, Gao F. Bimetallic Pd@ Au nanorods based ultrasensitive acetylcholinesterase biosensor for determination of organophosphate pesticides. Sens Actuators B

Chem 2018; 255: 2575-81.
[http://dx.doi.org/10.1016/j.snb.2017.09.063]

[91]   Nasir MZM, Mayorga-Martinez CC, Sofer Z, Pumera M. Two-Dimensional 1T-Phase Transition Metal Dichalcogenides as Nanocarriers To Enhance and Stabilize Enzyme Activity for Electrochemical Pesticide Detection. ACS Nano 2017; 11(6): 5774-84.
[http://dx.doi.org/10.1021/acsnano.7b01364] [PMID: 28586194]

[92]   Yan X, Li H, Han X, Su X. A ratiometric fluorescent quantum dots based biosensor for organophosphorus pesticides detection by inner-filter effect. Biosens Bioelectron 2015; 74: 277-83.
[http://dx.doi.org/10.1016/j.bios.2015.06.020] [PMID: 26143468]

[93]   Wang B, Wang H, Zhong X, Chai Y, Chen S, Yuan R. A highly sensitive electrochemiluminescence biosensor for the detection of organophosphate pesticides based on cyclodextrin functionalized graphitic carbon nitride and enzyme inhibition. Chem Commun (Camb) 2016; 52(28): 5049-52.
[http://dx.doi.org/10.1039/C5CC10491B] [PMID: 26987783]

[94]   Zehani N, Kherrat R, Dzyadevych SV, Jaffrezic-Renault N. A microconductometric biosensor based on lipase extracted from Candida rugosa for direct and rapid detection of organophosphate pesticides. Int J Environ Anal Chem 2015; 95: 466-79.
[http://dx.doi.org/10.1080/03067319.2015.1036864]

[95]   Wei M, Wang J. A novel acetylcholinesterase biosensor based on ionic liquids-AuNPs-porous carbon composite matrix for detection of organophosphate pesticides. Sens Actuators B Chem 2015; 211: 290-6.
[http://dx.doi.org/10.1016/j.snb.2015.01.112]

[96]   Wei W, Dong S, Huang G, Xie Q, Huang T. MOF-derived $Fe_2 O_3$ nanoparticle embedded in porous carbon as electrode materials for two enzyme-based biosensors. Sens Actuators B Chem 2018; 260: 189-97.
[http://dx.doi.org/10.1016/j.snb.2017.12.207]

[97]   Wang X, Hou T, Dong S, Liu X, Li F. Fluorescence biosensing strategy based on mercury ion-mediated DNA conformational switch and nicking enzyme-assisted cycling amplification for highly sensitive detection of carbamate pesticide. Biosens Bioelectron 2016; 77: 644-9.
[http://dx.doi.org/10.1016/j.bios.2015.10.034] [PMID: 26492468]

[98]   Peng L, Dong S, Wei W, Yuan X, Huang T. Synthesis of reticulated hollow spheres structure $NiCo_2S_4$ and its application in organophosphate pesticides biosensor. Biosens Bioelectron 2017; 92: 563-9.
[http://dx.doi.org/10.1016/j.bios.2016.10.059] [PMID: 27836591]

[99]   Grawe GF, de Oliveira TR, de Andrade Narciso E, et al. Electrochemical biosensor for carbofuran pesticide based on esterases from Eupenicillium shearii FREI-39 endophytic fungus. Biosens Bioelectron 2015; 63: 407-13.
[http://dx.doi.org/10.1016/j.bios.2014.07.069] [PMID: 25127475]

[100] Catanante G, Espin L, Marty J-L. Sensitive biosensor based on recombinant PP1α for microcystin detection. Biosens Bioelectron 2015; 67: 700-7.
[http://dx.doi.org/10.1016/j.bios.2014.10.030] [PMID: 25459056]

[101] Raghu P, Madhusudana Reddy T, Reddaiah K, Kumara Swamy BE, Sreedhar M. Acetylcholinesterase based biosensor for monitoring of Malathion and Acephate in food samples: a voltammetric study. Food Chem 2014; 142: 188-96.
[http://dx.doi.org/10.1016/j.foodchem.2013.07.047] [PMID: 24001830]

[102] Narang J, Malhotra N, Singh S, Singh G, Pundir C. Monitoring analgesic drug using sensing method based on nanocomposite. RSC Advances 2015; 5: 2396-404.
[http://dx.doi.org/10.1039/C4RA11255E]

[103] Ahmad R, Tripathy N, Jang NK, Khang G, Hahn Y-B. Fabrication of highly sensitive uric acid biosensor based on directly grown ZnO nanosheets on electrode surface. Sens Actuators B Chem 2015; 206: 146-51.

[http://dx.doi.org/10.1016/j.snb.2014.09.026]

[104] Tortolini C, Bollella P, Antiochia R, Favero G, Mazzei F. Inhibition-based biosensor for atrazine detection. Sens Actuators B Chem 2016; 224: 552-8.
[http://dx.doi.org/10.1016/j.snb.2015.10.095]

[105] Materon EM, Jimmy Huang PJ, Wong A, Pupim Ferreira AA, Sotomayor MdelP, Liu J. Glutathione--transferase modified electrodes for detecting anticancer drugs. Biosens Bioelectron 2014; 58: 232-6.
[http://dx.doi.org/10.1016/j.bios.2014.02.070] [PMID: 24657642]

[106] Radulescu M-C, Bucur M-P, Bucur B, Radu GL. Biosensor based on inhibition of monoamine oxidases A and B for detection of β-carbolines. Talanta 2015; 137: 94-9.
[http://dx.doi.org/10.1016/j.talanta.2015.02.013] [PMID: 25770611]

[107] Mishra RK, Alonso GA, Istamboulie G, Bhand S, Marty J-L. Automated flow based biosensor for quantification of binary organophosphates mixture in milk using artificial neural network. Sens Actuators B Chem 2015; 208: 228-37.
[http://dx.doi.org/10.1016/j.snb.2014.11.011]

[108] Nesakumar N, Sethuraman S, Krishnan UM, Rayappan JBB. Electrochemical acetylcholinesterase biosensor based on ZnO nanocuboids modified platinum electrode for the detection of carbosulfan in rice. Biosens Bioelectron 2016; 77: 1070-7.
[http://dx.doi.org/10.1016/j.bios.2015.11.010] [PMID: 26562329]

[109] Henao-Escobar W, Del Torno-de Román L, Domínguez-Renedo O, Alonso-Lomillo MA, Arcos-Martínez MJ. Dual enzymatic biosensor for simultaneous amperometric determination of histamine and putrescine. Food Chem 2016; 190: 818-23.
[http://dx.doi.org/10.1016/j.foodchem.2015.06.035] [PMID: 26213043]

[110] Ribeiro FWP, Barroso MF, Morais S, *et al.* Simple laccase-based biosensor for formetanate hydrochloride quantification in fruits. Bioelectrochemistry 2014; 95: 7-14.
[http://dx.doi.org/10.1016/j.bioelechem.2013.09.005] [PMID: 24161938]

[111] Morales MD, Morante S, Escarpa A, González MC, Reviejo AJ, Pingarrón JM. Design of a composite amperometric enzyme electrode for the control of the benzoic acid content in food. Talanta 2002; 57(6): 1189-98.
[http://dx.doi.org/10.1016/S0039-9140(02)00236-9] [PMID: 18968725]

[112] Braik M, Barsan MM, Dridi C, Ali MB, Brett CM. Highly sensitive amperometric enzyme biosensor for detection of superoxide based on conducting polymer/CNT modified electrodes and superoxide dismutase. Sens Actuators B Chem 2016; 236: 574-82.
[http://dx.doi.org/10.1016/j.snb.2016.06.032]

[113] Ghanavati M, Azad RR, Mousavi SA. Amperometric inhibition biosensor for the determination of cyanide. Sens Actuators B Chem 2014; 190: 858-64.
[http://dx.doi.org/10.1016/j.snb.2013.09.055]

[114] Sezgintürk MK, Göktuğ T, Dinçkaya E. A biosensor based on catalase for determination of highly toxic chemical azide in fruit juices. Biosens Bioelectron 2005; 21(4): 684-8.
[http://dx.doi.org/10.1016/j.bios.2005.01.007] [PMID: 16202884]

[115] Kilinc E, Ozsoz M, Sadik OA. Electrochemical detection of NO by inhibition on oxidase activity. Electroanalysis. An International Journal Devoted to Fundamental and Practical Aspects of Electroanalysis 2000; 12: 1467-71.

[116] Yang Y, Yang M, Wang H, Tang L, *et al.* Inhibition biosensor for determination of nicotine. Anal Chim Acta 2004; 509: 151-7.
[http://dx.doi.org/10.1016/j.aca.2003.12.028]

[117] Bahadır EB, Sezgintürk MK. Applications of commercial biosensors in clinical, food, environmental, and biothreat/biowarfare analyses. Anal Biochem 2015; 478: 107-20.
[http://dx.doi.org/10.1016/j.ab.2015.03.011] [PMID: 25790902]

[118] Sokolov AN, Roberts ME, Bao Z. Fabrication of low-cost electronic biosensors. Mater Today 2009; 12: 12-20.
[http://dx.doi.org/10.1016/S1369-7021(09)70247-0]

[119] Verma N, Bhardwaj A. Biosensor technology for pesticides--a review. Appl Biochem Biotechnol 2015; 175(6): 3093-119.
[http://dx.doi.org/10.1007/s12010-015-1489-2] [PMID: 25595494]

[120] Ahmed MU, Hossain MM, Safavieh M, *et al.* Toward the development of smart and low cost point-o--care biosensors based on screen printed electrodes. Crit Rev Biotechnol 2016; 36(3): 495-505.
[PMID: 25578718]

Frontiers in Enzyme Inhibition, 2020, *Vol. 1*, 122-147

CHAPTER 8

Product Inhibition in Bioethanol Fermentation: An Overview

E. Raja Sathendra[1,*]**, G. Baskar**[2] **and R. Praveen Kumar**[1]

[1] *Department of Biotechnology, Arunai Engineering College, Tiruvannamalai-606603, India*
[2] *Department of Biotechnology, St. Joseph's College of Engineering, Chennai-600119, India*

Abstract: Many organisms are used for alcohol production in an industry that includes *Saccharomyces cerevisiae, Zymomonas mobilis*, and *Clostridium sp* are better microbes with respect to ethanol production and ethanol tolerance, Zymomonas mobilis can use glucose, sucrose, and fructose through Entner-Deodoroff pathway. In bioprocessing product inhibition is undesirable that limits the product's final titer and volumetric productivity more precisely known as product toxicity those utilizing the whole cell as biocatalyst. During the ethanol fermentation the yield of cell mass decreases gradually as the ethanol concentration increases progressively indicating product inhibition. The decrease in cell mass concentration is the accumulation of ethanol in the fermentation broth beyond the limits. This is because the increase in alcohol concentration during fermentation destroys the microorganism lipid bilayer membrane and denatures the enzymatic protein thereby creating instability conditions. The product inhibition is very well studied only in the batch reactor. Conventionally maximum ethanol concentration of 7-8% (v/v) is achieved in the time frame of 50-70 hr with the operating temperature of the 32-34°C and stirring rate of 180rpm during fermentation. To overcome this problem the continuous product removal solves the product inhibition through maintaining the ethanol concentration below the inhibitory limit.

Keywords: Bioethanol, Cell mass, Ethanol fermentation, Ethanol concentration, *Escherichia coli*, Growth kinetics, Hinshelwood model, Lignocelluloses, *Lactis aerogenes*, Monod's model, Product inhibition, *Saccharomyces cerevisiae*, Substrate, Specific growth rate, Toxin concentration, *Zymomonas mobilis*.

INTRODUCTION

Bioethanol produced through biological processes is mostly used in pharmaceutical, biofuel, fuel additive and food industries. Plants are the main sources for bioethanol production (*e.g.* Cellulose and starch). These polysacc-

* **Corresponding author E. Raja Sathendra:** Department of Biotechnology, Arunai Engineering College, Tiruvannamalai-606603, India; E-mail: e.rajasathendra@gmail.com

G. Baskar, K. Sathish Kumar & K. Tamilarasan (Eds.)

harides can be hydrolyzed to release sugar for bioethanol fermentation. Lignocellulose is the most abundant and cheap source of fermentable sugars. A large volume of agricultural crop residues and non-edible plant biomass mainly contains 10–40% hemicellulose, 40–75% cellulose and 15–35% lignin, depending on the source. Hemicellulose and cellulose fraction of plant biomass can be used for bioethanol production. Lignocellulosic materials are highly recalcitrant nature and they need extensive pre-treatment to release sugars [1 - 3]. Tubers, grains, and roots are the main source of starch materials for bioethanol production [4]. Grains contain 62–89% fermentable starch typically.

LIGNOCELLULOSIC PRETREATMENT

Pretreatment changes the structural and physicochemical property of original lignocellulose by opening the biomass recalcitrance structure and then renders the direct microbial or enzymatic hydrolysis of cellulose to produce monomers [5]. Hemicellulose can be easily hydrolyzed using acids, alkalis and enzymes to release sugars for fermentation. The hydrolytic process mainly releases pentose sugars ($C_6H_{10}O_5$) *e.g.* rhamnose, arabinose and xylose and some hexose sugars ($C_6H_{10}O_6$) *e.g.* galactose and mannose. Hemicelluloses can release major parts of chemical constituents like monomers on thermochemical degradation *i.e.* xylose, mannose, acetic acid, galactose, and glucose, in conjunction with several inhibitors that are toxic to the fermenting microorganism. The important inhibitors generated on hydrolysis that include furans phenolics, (furfurals and 5-Hydroxy methyl furfural (5-HMF)), weak acids (acetic acid). Among inhibitors [6 - 9] Hibbert's ketones have also been noticed in the acid hydrolysate of pine and spruce [10, 11]. Fungi like yeast cells are highly exposed to growth inhibition during fermentation through such toxic compounds or at higher concentration of ethanol formed and it is the major drawback of using lignocellulosic material in converting to ethanol. During the bioethanol fermentation, microorganisms capable of tolerating the lignocellulose ethanol production by maintaining the high metabolic activity are desirable.

Pretreatment Strategy

Pre-treatment strategy plays an important part in the lignocellulose processing for bioethanol production. Extensive researches are made to improve the breaking down of lignocellulose material. Pre-treatment changes the structural and physicochemical property of original lignocellulose by opening the biomass recalcitrance structure and then renders the direct enzymatic hydrolysis of cellulose to produce monomers [12]. In the native form of C5 and C6 sugar-based hemicelluloses the major part of the component is the lignin in an intertwined and complicated form [13] shown in Fig. (**1**).

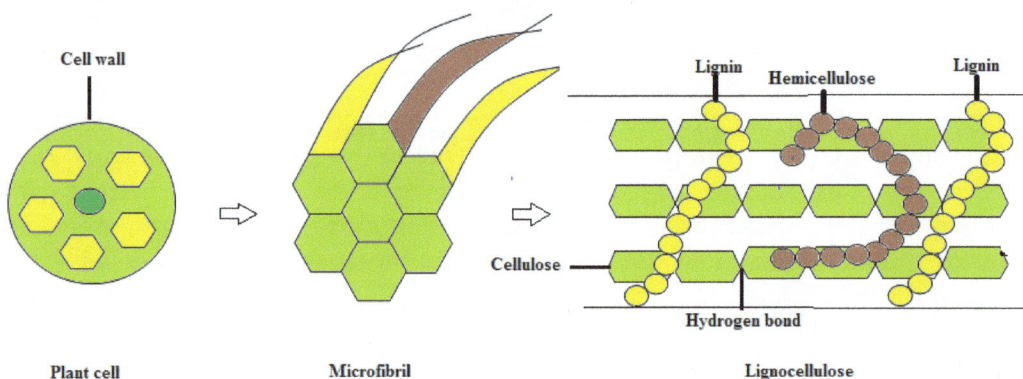

Fig. (1). showing the various internal components in the plant cell wall.

Lignocellulosic materials mostly contain a mixture of carbohydrate polymers (cellulose and hemicellulose), lignin, and ashes. The "holocellulose" is a term used to describe the total carbohydrate contained within a plant. Lignocellulosic materials are therefore comprised of cellulose and hemicellulose together referred to as Holocellulose. Cellulose, an unbranched linear polymer and hemicelluloses are a group of heterogeneous polysaccharides. The dry weight of the lignocellulosic material contains between 11% and 37% of hemicelluloses, can be easily hydrolyzed to their components of monomer consisting of xylose, mannose, glucose, galactose, arabinose, and small amounts of rhamnose, glucuronic acid, methyl glucuronic acid, and galacturonic acid by acids [14, 15]. Furthermore, among the various substrates lignocellulosic materials hold a good promise for biofuel production.

The non-fossil fluid fuel at present has an impact on a global scale is biofuels, including biodiesel and bioethanol. Utilization of bioethanol as a transportation fuel and as a gasoline supplement has been proved to be more environmentally friendly. The vehicle and industries are one of the major potential contributors to global warming by the release of carbon dioxide (CO_2) gas. The development of alternative energy sources such as biofuels becomes important to reduce these problems. The production of bioethanol from renewable sources such as plant biomass can reduce urban air pollution and the accumulation of carbon dioxide (CO_2), the so-called effect of greenhouse gases (GHG). Bioethanol is a complete and clean-burning of high octane fuel that can readily alter gasoline and its combustion results in significant reductions of toxic gas emissions such as 1-3 butadiene, benzene, and formaldehyde, while mixing ethanol with gasoline can enhance the octane of the mixture and can decrease carbon monoxide (CO) emissions by 10 ~ 30% [16]. Ethanol with gasoline, derived from renewable biomass feedstocks that seize carbon-dioxide growth, is expected to reduce the emissions by 90 ~ 100% (ASTMD 1993). Furthermore, the development of biofuels is predicted to assure the accessibility of new and renewable energy resources that increase the economic price of forests. It is evident that the potential for rivalry with food production for both the starch feed-stocks and sugar that prime agricultural lands normally required for producing crops should not be diverted for the production of fuel. The bio-transformation of plant lignocellulosic biomass into bioethanol is significant to be developed since this resource is economical and easy availability. Many sugar crops such as sugar cane; sugar beets and also starch crops, including wheat, potatoes and sweet potatoes are currently focused on, bioethanol production. Biomass resources obtained from lignocellulosic materials such as municipal solid waste, agricultural and forestry residues and large waste from industrial are not well utilized still, hence creating discarding problems. These residues can be set up easily for the production of bioethanol. Besides, herbaceous energy crops and woody can be planted and underutilized land can be in use to maintain native production of such forms of biomass. Apart from renewable, biofuels can also decrease the emission of gases which can cause global warming [17]. The biofuels from the second generation may consist of the fuels produced from mixed paper waste which is separated from the municipal solid waste, Cotton, Maize, Honge, cash crops Jatropha, *etc.* can be utilized for bioethanol production. The biofuels from the third generation can be produced from Algae of micro-organisms mainly. The biofuels from the fourth generation are produced from genetically engineered microbes.

Pre-treatment is mainly done to enhance the digestibility of cellulose by increasing enzyme accessibility. To maximize the positive potential of biofuels and minimize their negative impacts the proper deployment of a consistent scientific framework is required. The pre-treatment step in bioethanol production is the most crucial step when processing plant biomass. This step represents a physiochemical, chemical or thermochemical breakdown of the biomass so that the effect of the lignin is reduced and the enzymatic hydrolysis is improved. To achieve this, many pre-treatment techniques have been developed by the scientific community, some of which are currently commercial. In many cases, the techniques developed are difficult for scale-up and lack the necessary data for a proper economic analysis therefore this study focuses on those that are practical, show great potential and are well documented in the literature.

```
                        ┌─────────────────────┐
                        │   Lignocellulosic   │
                        │      Material       │
                        └──────────┬──────────┘
                                   │
                                   ▼
                        ┌─────────────────────┐
                        │  Crushing and Size  │
                        │      reduction      │
                        └──────────┬──────────┘
                                   │
                                   ▼
                        ┌─────────────────────┐
                        │   Pre-treatment     │
                        │     (Refining)      │
                        └─────────────────────┘
```

Chemical	Physical	Biological
1. Catalyzed steam-explosion 2. Dilute acid 3. Conc. Acid 4. CO_2 explosion 5. Dilute Alkaline 6. Ionic liquid 7. Ozonolysis 8. Ammonia fiber/freeze explosion (AFEX) 9. Organosolv 10. pH-controlled liquid hot water	1. Uncatalyzed steam-explosion 2. Liquid hot water (LHW) 3. Mechanical comminution 4. High energy radiation 5. Wet oxidation 6. Hot air oven 7. Pyrolysis	1. Microorganism 2. Enzyme

```
                        ┌─────────────────────┐
                        │   Detoxification    │
                        └──────────┬──────────┘
                                   │
                                   ▼
                        ┌─────────────────────┐
                        │    Fermentation     │
                        └──────────┬──────────┘
                                   │
                                   ▼
        ┌─────────────────┐   ┌─────────────────────┐
        │  Distillation   │──▶│     Dehydration     │
        └─────────────────┘   │ Pure Ethanol > 99%  │
                              └─────────────────────┘
```

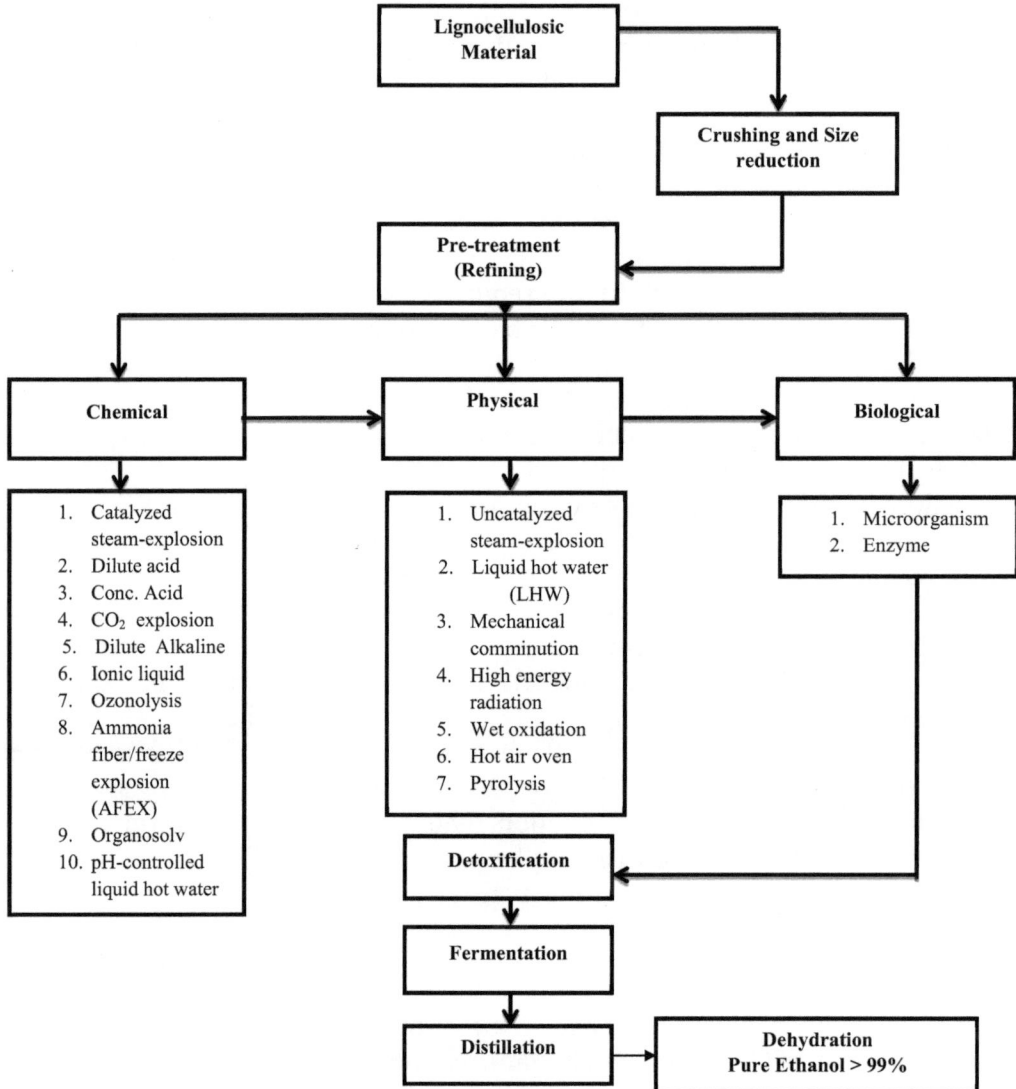

Flowsheet: Ethanol Production from various Lignocellulosic pre-treatment methods.

Detoxication of Lignocellulosic Hydrolysate

The carbohydrate fraction of the plant cell wall can be converted into fermentable monomeric sugars through acidic and/or enzymatic (hemicellulase/cellulase) reactions, which have been exploited to produce ethanol. Pre-treatment is unavoidable, which has been examined and employed extensively in the past [18]. The acidic pretreatment of lignocellulosic hydrolyzes the hemicellulose fraction, enabling subsequent enzymatic digestion of the cellulose in fermentation reaction [19]. However, the non-specificity of acidic treatment led to the formation of complex sugars and compounds inhibitory to the microorganisms for ethanol production [20].

Thermo-chemical degradation of hemicellulose liberates the majority of sugar monomers *i.e.* xylose, mannose, acetic acid, galactose, and glucose, in conjunction with several inhibitors toxic to the fermenting microorganism. These inhibitors include furans (furfurals and 5-Hydroxy methyl furfural (5-HMF)) phenolics, weak acids (acetic acid). Among inhibitors [19] Hibbert's ketone has also been noticed in the acid hydrolysate of pine and spruce [20].

To overcome the fermentation inhibitors, there are several detoxification methods such as physical (evaporation [21], membrane-mediated detoxification [22]), chemical (neutralization [19], calcium hydroxide over-liming [23]), activated charcoal treatment [24], ion exchange resins, and extraction with ethyl acetate [23] and biological detoxification (enzymatic mediated using laccase [25], lignin peroxidase, detoxification, *etc*). Every method represents its specificity to eliminate a particular inhibitor from the hydrolysate. The microbial mediated pre-treatment resulted in the maximum de-polymerisation of carbohydrate polymers into a mixture of simple sugars with fewer fermentation inhibitors. The pre-treated lignocellulose substrate when hydrolysed leads to maximum sugar recovery with minimum inhibitors in a short period by eliminating the requirement of a detoxification step [26].

MICROORGANISM FOR ETHANOL FERMENTATION

An ideal microorganism for ethanol production should have [27]:

1. The rapid growth rate during fermentation.
2. Optimum growth of microorganisms during fermentation for maximum ethanol production.
3. High tolerance to ethanol concentration produced in fermentation.
4. Ability to ferment easily available substrate [28].
5. Ability to tolerate growth inhibitors produced during hydrolysis [29].
6. High fermentation rate with maximum ethanol yield.
7. Ability to tolerate extreme chemical condition *e.g.* low pH.

The various microorganisms used in ethanol fermentation with the different lignocellulosic substrate are listed below in Table **1**:

Table 1. Showing bioethanol production using various microorganisms and substrate pre-treatment method under optimized condition.

S.No	Strain/Media Formulation	Raw Materials	Pre-treatment Method	Fermentation Condition	Yield	Ref
1	*Kluyveromyces marxianus*	Crude whey, 3.45% lactose, 0.1% yeast, 0.1% malt extract, tracer elements.	-	Temperature maintained at 34°C, pH of 4.5 and 6N of NaOH at 500 rpm.	2.10g/L	[30]
2	*T. reesei, A. niger, T. longibrachiatum*	Wheat straw	Physical, Physico-chemical, Chemical and biological treatment	Temperature maintained from 40-50°C, pH of 5.0	higher than 4% (w/w)	[31]
3	Immobilized *Saccharomyces cerevisiae* ATCC 24553	Pineapple cannery waste	pH is adjusted to 4.5	Temperature maintained at 30 +/- 0.2°C, pH is 4.5	37g/L	[32]
4	*Saccharomyces cerevisiae, cytophagahutchisonni*	Waste newspaper	Dust-free and fungus free state and dried. Hydrolysis of the pretreated substrate.	pH of 4.6 and temperature of 36°C	6.91% from pure culture, 6.12% from an isolated organism	[33]
5	Immobilized cells of *Pachysolen tannophilus*	Crude glycerol	Solvent extraction	Aerobic condition, the temperature of 28°C, pH is 5.6, yeast extract of 5g/L, tryptone of 10g/L	83%	[34]
6	*Zymomonas mobilis* ATCC29191	Cellulosic feedstock Bagasse hydrolysate	Treated with concentrated phosphoric acid, centrifugation, 2 stage saccharification process	SSF was operated with an agitation rate of 100 rpm and a temperature-controlled at 30°C.	2.01g.hr/L	[35]
7.	*S. cerevisiae*	Corn stem ground tissue	Cleaned and cut into small pieces of equal size	It is kept in a rotary shaker at 120 rpm for72 hrs and maintained at 4.5 pH.	4.91% v/v	[36]
8.	*S. cerevisiae* NCIM 3176	Grapefruits bunch	Washed with diluted liquid detergent labolene for 2-3 times, finely grinded.	Incubated at 30°C for 7 days. It is kept in shaft agitator at 50 rpm and maintained at 6.0 pH	7.2% a. b. w 9.05% a. b. v	[37]

(Table 1) cont.....

S.No	Strain/Media Formulation	Raw Materials	Pre-treatment Method	Fermentation Condition	Yield	Ref
9.	*S. cerevisiae*	Pteris (biomass)	Chemical and mechanical method	1g of Pteris is maintained in the temperature of 298k (25°c) at the pH 7	0.33mg/ l	[38]
10.	*S. cerevisiae*	Wheat	Milling, heating, and addition of enzymes and water	Temperature is maintained at -28°C.	25.46 v/v 99.7vol%	[39]
11.	*E.aerogens* HU-101	Glycerol waste	Diluted with distilled water	pH is maintained at 6.8	1.7g/ l glycerol	[40]
12.	*Saccharomyces cerevisiae, Zymomonas mobilis*	Fresh madhuca flowers	Dried, mixed thoroughly, grinded	It is incubated at 96 hrs and maintained in temperature of -30+/- 2°C at pH 5.5-6.5	21.2%	[41]
13.	*Saccharomyces cerevisiae CCUB*	Roots of carrot	Plant leaves were cut and whole roots were washed, dried	Batch fermentations were developed at 28± 1°C, in a 500 mL stirred tank bioreactor	35.3 g /l	[42]
14.	*Pichia stipitis* NRRL Y-7124	Wheat straw hemicellulose	Straw was hydrolysed at 90°C with 1.85% (w:v) sulfuric acid for 18 h, with an initial liquid to the solid ratio of 20:1.	It is stirred at 350 rpm and temperature is of 28 +/- 0.5°C at pH 6.5 +/- 0.1	80.4+/- 0.55%	[43]
15.	*Zymomonas mobilis* and *Saccharomyces cerevisiae* 3319	Simple sugar, starch, cellulose, sweet potato	Physical and chemical method	For *Zymomonos mobilis* pH is at 4.5 in *Saccharomyces cerevisiae* it is at Ph 5.0 and temperature is maintained at 27-30°C.	Yield in *Zymomonas mobilis* is 3.63g/ l and yield in *Saccharomyces cerevisiae* is 3.03 g/l	[44]
16.	*Zymomonas mobilis* ATCC 29191	Molasses	Hydrolysed	It is maintained at 32.4 c and 4.93 of pH	53.3g/L	[45]
17.	*Zymomonas mobilis* 3881, *Saccharomyces cerevisiae safdistil c-70*	Sugar beet pulp and raw juice	-	It is taken in the ratio of 1:1 and 1:0.75 of the substrate and raw juice. It is maintained at 30°C and pH 5.5	*Z.mobiles* showed a lower yield of 30% than *S.cerevisae*	[46]

(Table 1) cont.....

S.No	Strain/Media Formulation	Raw Materials	Pre-treatment Method	Fermentation Condition	Yield	Ref
18.	*Saccharomyces cerevisiae*	Cellulose	Thermo-chemical, physical and enzymatic pretreatment were carried out	-	3g/L within 40 hours	[47]
19.	Enzymatic hydrolysis	Sugarcane Bagasse	It includes physical, chemical and enzymatic treatment.	It is maintained at 30°C at 5.5 pH for the concentration of various clarified liquor	For 80% brix it is 1.15% (v/v) and for 50% brix it is 0.579% (v/v)	[48]
20.	The enzyme from *Aspergillus niger, Trichoderma viride*	Various agricultural raw materials	Hydrolysed and filtered	It is maintained at 37°C	13.8% (w/w)	[49]
21.	*Zymomonas mobilis*	Rice straw, wheat straw, rice husk.		It is done in SDDL media with the temperature maintained at 30°C and pH of 5.5-6.0	25%	[50]
22.	-	Lignocellulose	Physical, physicochemical, chemical and biological pretreatment is carried out.	The temperature of 30-35°C was maintained	34.0g/L	[51]
23.	*Saccharomyces cerevisiae* MTC173, *Zymomonas mobilis* MTCC2427	Agave leaves	Enzymatic hydrolysis	It is maintained at 80°C in YEPD, MA RM medium	Amylase yielded 4.0% and cellulose yielded 5.0%.	[52]
24.	*Zymomonas mobilis* 2427, *Saccharomyces cerevisiae*	Sunflower head	Powdered and treated with sulphuric acid and sodium hydroxide	It is maintained at a temperature of 30°C and pH of 5.0	*Zymomonas mobilis yielded* 21.89g/L and *Saccharomyces cerevisiae yielded* 20.52g/L	[53]
25.	-	Lignocellulose	Physical, physicochemical, chemical and biological pretreatment	-	60g/L	[54]
26.	Engineered *Zymomonas mobilis* and *Saccharomyces cerevisiae*	Lignocellulosic Biomass	Physical, chemical, solvent and biological pretreatment	Temperature is maintained 30-35°C.	5-7% (v/v)	[55]

Saccharification and Fermentation

Saccharification process is carried out on 250 ml flask and kept in orbital shaker containing supplemented nutrient and pretreated substrate and the pH is adjusted to 4.3 then sterilize at for 15 min at 121°C when the medium comes into room temperature *Trichoderma reesei* MTCC 4876 was inoculated for the conversion of cellulose into fermentable sugar. The enzyme cellulase secreted by *T. reesei* was determined using FPU method [56]. The saccharification process was measured in the interval time. Samples are centrifuged for 10 min at 5000 rpm and analyzed the reducing sugar content in the hydrolysate was estimated using the DNS method. After the saccharification, the fermentation broth was filtrated to remove the cell mass, and then inoculated with *Kluveromyces marxianus* MTCC 1389 along with fresh nutrient medium and incubated in orbital shaper at 120 rpm for 7 days for ethanol production. The bioethanol amount was determined using the potassium dichromate method for every interval of 12 h [58].

Distillation

Distillation is the most dominant and recognized industrial purification technique of ethanol. It utilizes the differences of volatilities of components in a mixture. The basic principle is that by heating a mixture, low boiling point components are concentrated in the vapour phase. By condensing this vapour, more concentrated less volatile compounds are obtained in the liquid phase. Distillation is one of the most efficient separation techniques. The fermented broth was distilled and 50 ml of the distillate was collected for analysing the ethanol concentrated. Ethanol assay was done by standard dichromate test also by the spectrophotometric method [59].

Dehydration

The large-scale production is still dominated by the extractive and azeotropic distillation despite recent advances in pervaporation and adsorption with molecular sieves. The large-scale production of bioethanol fuel requires energy demanding distillation steps to concentrate the diluted streams from the fermentation step and to overcome the azeotropic behaviour of the ethanol-water mixture. The conventional separation sequence consists of three distillation columns performing several tasks with high energy penalties: first a column for the pre-concentration of ethanol, second a column for extractive distillation and the third column for solvent recovery [60].

INHIBITION OF ETHANOLIC FERMENTATION

The main advantage of using *S. cerevisiae* for ethanol production is the extremely high tolerance to concentration shows respect to other microorganisms. To understand the inhibitory effect of various agents in the metabolism of *S. cerevisiae* in the ethanol fermentation is very important to enhance the technology of biofuel ethanol. For example, *S. cerevisiae* can tolerate ethanol concentrations up to between 115 and 200 gL^{-1} [61] compared to *Zymomonas mobilis* and *Escherichia coli have* maximum tolerances around 60–127 gL^{-1} [62, 63]. A high concentration of butanol, ethanol impairs the cellular wall permeability, signaling function and also increases cell size by causing a delay in the cell cycle [64, 65]. The important factors for growth inhibition are the osmotic pressure through sucrose concentration in the medium and ethanol formed via fermentation, pH affecting the proton pump and high temperature affecting the cell membrane. Trehalose pathway is a storage carbohydrate and one important function is the protection of yeast under stress such as toxicity of ethanol, cellular dehydration, high temperature and increased osmotic pressure. This carbohydrate is accumulated in the presence of oxygen at low concentrations of sugars when there is the exhaustion of glucose in the medium during the diauxic growth [66 - 68]. The inhibitory effect of organic acids on yeast is because of the strain of *S. cerevisiae* used in the fermentation process, acid concentration, the osmotic pressure of the medium and the synergism with other inhibitors [69]. The models of [70 - 76] only account for the effect of product (ethanol) inhibition.

GROWTH KINETICS

The microbial growth during bioconversion is a complex process. The Monod equation is usually used to relate the specific growth rate to the concentration of the limiting substrate.

$$\mu = \mu max \frac{s}{Ks+s} \tag{1}$$

The kinetics model was implemented to analyze and optimize the biological processes to understand and quantify [77].

The kinetics model describe the biological system behaviour, it can reduce the number of experiments performed to eliminate the extreme things and provide mathematical expressions that quantitatively explain the mechanism of the process required to control and optimize [78, 79]

Batch culture techniques under anaerobic conditions (the medium was deaerated) were used and high concentrations of ethanol were produced by the yeast itself. The cell can multiply rapidly, the number of living cells or the cell mass, exponentially increase with respect to time.

$$\frac{dx}{dt} = \mu x \qquad (2)$$

The equations describe the increase in cell numbers during $(x=x_o, t=t_o)$ period. Thus the rate of increase in x is proportional to x from the integrated form of eqn. 1.

$$ln = \frac{x}{xo} = \mu(t = tlag) \qquad (3)$$

From this, we can deduce that time interval t_d required to double the population is given by

$$td = \frac{ln2}{\mu} \qquad (4)$$

As is the case for steady state growth in a continuous stir tank reactor, only a single parameter μ or t_d is required to characterize the population during exponential batch growth. To a reasonable extent, growth is balanced during this stage of batch cultivation. Consequently, useful balanced growth kinetic data may be obtained from batch experiments provided attention is confined to the exponential growth phase.

Deviation from exponential growth eventually arises when some significant variation such as toxin concentration or nutrient level achieves a value that can no longer support the maximum growth rate. Exhaustion of a particular critical nutrient may appear rather sharply at a given time since the cells are rapidly increasing the total rate of nutrient consumption in the exponential growth phase. To formulate a rough analysis of this event, we suppose that the rate of nutrient consumption is proportional to the cell mass concentration of living cells until the stationary phase is reached:

$$\frac{da}{dt} = -kax \qquad (5)$$

We assume that exponential growth continues unabated until the stationary phase is reached, and we consider the time when exponential growth begins as time zero. Then

$$x = xoe^{kt} \qquad (6)$$

Where x_o is the cell mass concentration of living cells when exponential growth starts. If the concentration of A at time zero is a_o, we can determine from equations. 5 and 6 that A is completely consumed when:

$$ao = \frac{Ka}{\mu} (Xs - Xo) \qquad (7)$$

Where x_s is the mass of the population when A is exhausted and the population enters the stationary phase. Consequently, x_s is the maximum population size achieved during the batch culture. Rearranging the eqn. 7 we can find the maximum population in the case of nutrient depletion to be given by

$$Xs = Xo + \frac{\mu}{Ka} ao \qquad (8)$$

Linear dependence of x_s on the initial nutrient level has been observed experimentally in many cases. The other factors can influence the onset of the stationary phase and the size of the maximum population.

If toxin accumulates which allows the slows the rate of growth from exponential, an equation of the form

$$\frac{dx}{dt} = kx[1 - f(\text{toxin concentration})$$

is found useful. In the particular case where toxin linearly decreases the growth rate, we have a specific growth rate

$$\frac{1}{x}\frac{dx}{dt} = k(1 - bCt) \tag{9}$$

Where C_t is toxin concentration and b is a constant. A plausible assumption is that the rate of toxin production depends only on x proportional to it

$$\frac{dct}{dt} = qx \tag{10}$$

So that $Ct = q\int_{o}^{x} xdt$ (assuming ct =0 at t=0) and the growth equation becomes

$$\frac{dx}{dt} = kx(1 - bq \int_{o}^{t} xdt) \tag{11}$$

The instantaneous value of the effective specific growth rate μ_{eff}

$$\mu eff = \frac{1}{x}\frac{1dx}{dt} = (1 - bq \int_{0}^{t} xdt) \tag{12}$$

diminishes more and more rapidly with time *i.e.,*

$$\frac{d\mu eff}{dt} = kbqx < 0) \tag{13}$$

Growth halts when

$$\frac{1}{bq} = \int_{o}^{t} xdt) \tag{14}$$

Equation 9 indicates that growth ceases only when C_t reaches a particular level $c_t=1/b$.

PRODUCT INHIBITION VALIDATION

During fermentation of alcohol accumulation continues to persistent still termination of cell growth at a level of about 10^8 cells/ml. The ethanol high concentration inhibits the metabolic path activity of the yeast cells that can be determined only after the qualitative manipulation.

In the relatively low concentration of ethanol the inhibitory effect on cell growth is apparent during fermentation, *i.e.*, above the minimal percent of ethanol, while the cells fermentation activity seems to have tolerance until the ethanol concentration leading to about 20% for example. The ethanol inhibition on the metabolism of the yeast cell and its effect on inhibition can be described in terms of the quantitative aspect of kinetics.

The experiment on bacteria *lactis aerogenes* by Hinshelwood derived the equation that describes the growth of cell culture is affected by the addition of alcohol [76].

$$\mu = \mu m(1 - ap) \tag{15}$$

provided μm = maximum value of **the** specific rate of growth; p = alcohol concentration; a = empirical constant, on the species that is in alcohol addition most probably.

The studies on the growth of yeast cells on ethanol production formation during the fermentation of grape juice by Holzberg *et al.* [80] in the stationary phase of cell growth and they determined the following empirical formula:

$$\frac{dp}{dt} = b(pm - p) \tag{16}$$

Where, B = empirical constant; p_m = the concentration of ethanol attainable at maximum value.

The Hinshelwood kinetic model developed and adopted for investigating cell growth during fermentation.

$$\mu = \mu max \frac{s}{Ks} \left(1 - \frac{p}{Pm}\right) \tag{17}$$

Where, *μmax* = maximum rate of specific growth of biomass, *μ* = growth at a specific rate, k_s = the affinity of substrate constant and P_m = maximum concentration of ethanol above which cell growth ceases.

The gradual decrease in the substrate concentration on fermentation with respect to time, the data observed from that shows substrate metabolized by microorganism *Kluveromyces marxianus* during fermentation to yield ethanol. This is reflected in the increase in biomass yield this is because the cells utilize the substrate to enhance its growth and ethanol production. This is confirmed and reported [81, 82].

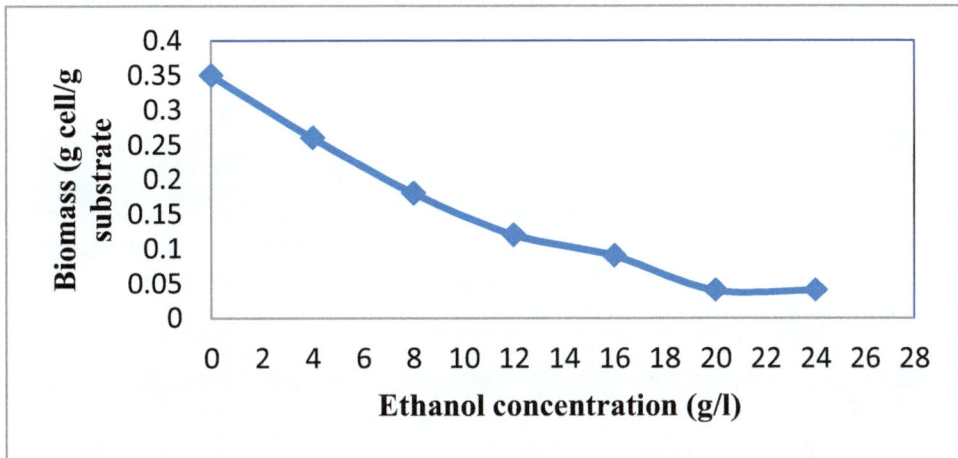

Fig. (2). Concentration of ethanol affects the yield of the fermenting organism was observed through Hinshelwood models in a hyperbolic manner.

The effect of ethanol inhibition was evaluated. This was determined by ethanol concentration that effecting the important variable during fermentation on the yield of fermenting microorganisms. This is because the cell growth gradually decreases and increases in the ethanol concentration showing the relationship between the yield of biomass and inhibition on ethanol product.

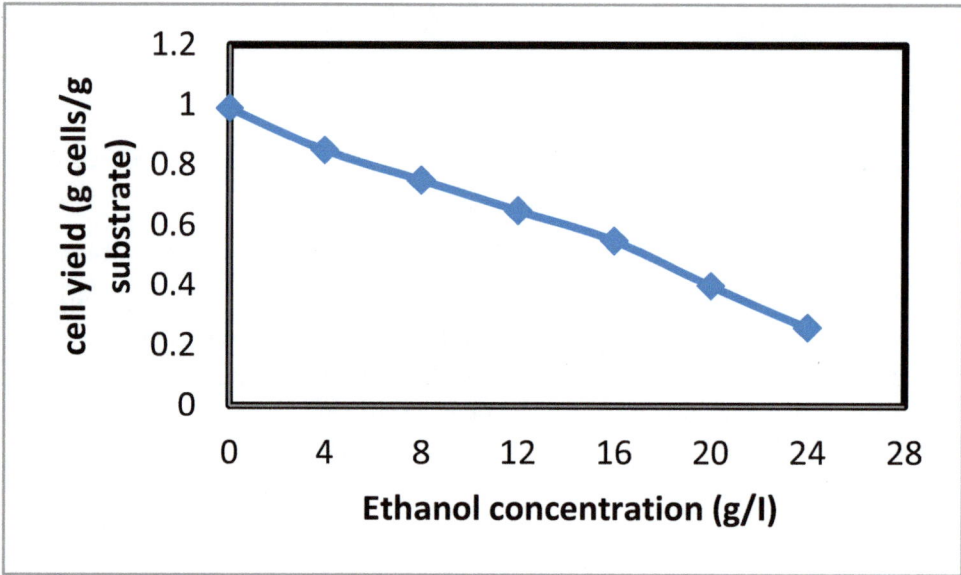

Fig. (3). The effect of concentration of ethanol on fermenting organism yield through hyperbolic and Hinshelwood models.

The gradual decrease in the fermenting organism was observed as a result of ethanol accumulation in the fermentation broth. Warren and Taylor reported the same decrease in the yield of biomass with a progressive increase in the concentration of ethanol [83, 84].

The yield produced in the fermentation is assumed to be constant for biomass and ethanol. In little cases, some parameters like maintenance and cell death are represented to include product formation and biomass production yield changes. The concentration of ethanol and yield of ethanol effects are shown in Fig. (4).

The increase in the ethanol concentration and the ethanol yield are predicted very well and it seems to be constant which is determined through quantification in the fermentation broth. The ethanol yield becomes almost constant until the termination period, this is because when the inhibition takes place the cells are no longer formed in the fermentation medium. This result was confirmed with Warren [85] showing ethanol yield as a function of the concentration of ethanol.

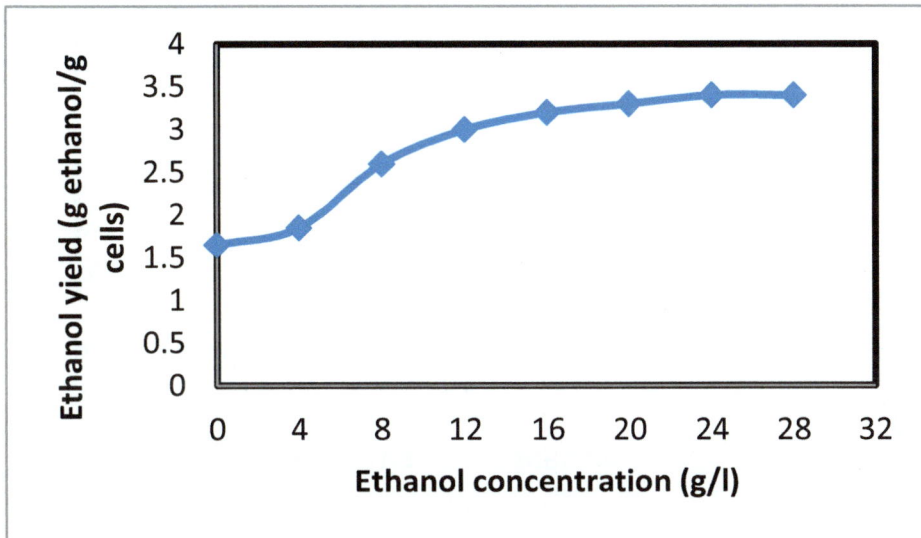

Fig. (4). The concentration of ethanol and yield of ethanol effects are observed through Hinshelwood and hyperbolic models.

The ethanol concentration and specific growth rate effects are well observed through the hyperbolic model is shown in Fig. (**3**). This was predicted by analyzing the specific growth rate with its maximum value (μmax).

The progressive decrease in the fermenting organism's specific growth rate with respect to the maximum rate was observed on the ethanol concentration increase. This linear behavior was observed through the Hinshelwood model which is the same as the previous research work [86, 87]. This shows the inhibition effect of ethanol on the fermenting organism-specific growth rate is primarily because of a decrease in the biomass yield shown in Fig. (**2**).

The ethanol specific production rate and the ethanol yield is a linear function shown in Fig. (**5**). The Hinshelwood model determined on specific ethanol production rate decrease gradually as the concentration of ethanol progressively increases. The model observed as a linear function between the concentration of ethanol and the specific productivity of ethanol. The same result was observed by Daugulis & Swaine [88] (Fig. **6**).

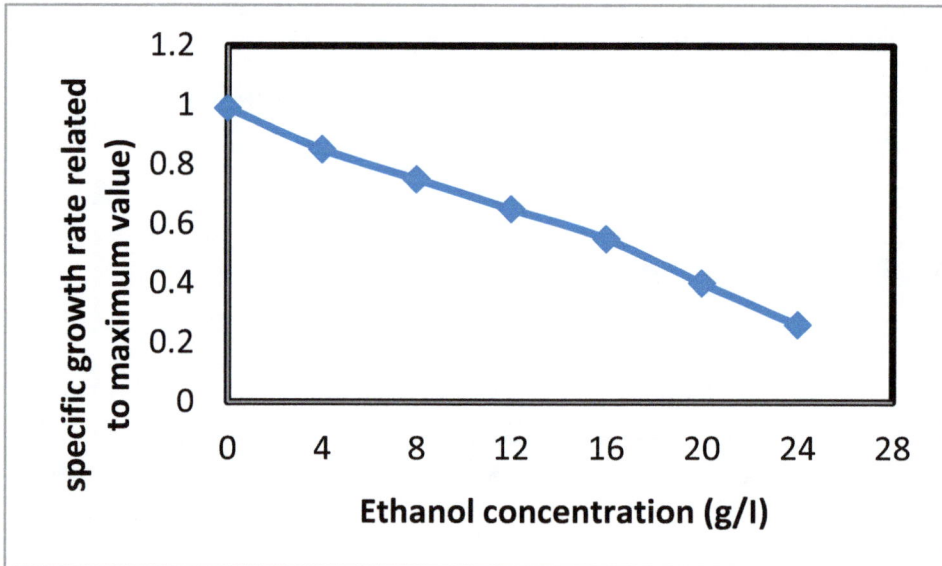

Fig. (5). The specific growth rate and ethanol concentration effects are observed through Hinshelwood and hyperbolic models.

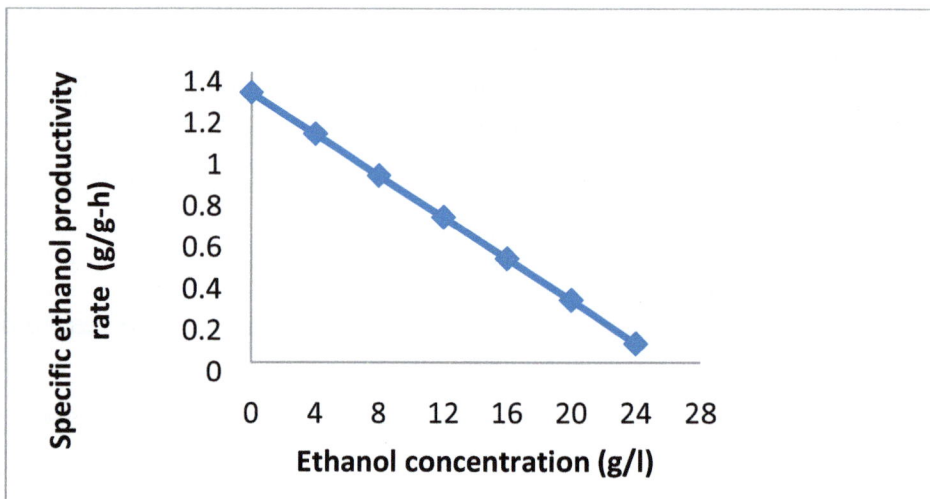

Fig. (6). The specific ethanol productivity rate that effects on concentration of ethanol as predicted through hyperbolic and Hinshelwood models.

CONCLUSION

The cause of ethanol inhibition from ethanol fermentation using fungi *Saccharomyces cerevisiae* was studied well.

- The validation of the Hinshelwood model on kinetics growth of cells was observed on inhibition of the product through dynamic steps on fermenting organisms on ethanol production.
- The linear relationships persist on the relative specific growth rate of the organism and the concentration of ethanol and also the productivity rate and concentration of ethanol.
- So, using continuous fermentation product removal can solve the ethanol inhibition problem by maintaining the ethanol concentration below the inhibitory level.

LIST OF ABBREVIATIONS

μ_{eff} effective specific growth

x biomass concentration

μ specific growth rate

t_d the time interval

Ks substrate saturation constant

S substrate concentration

μ_{max} the maximum specific growth rate

Xo biomass initial concentration

Xs biomass stationary phase

bct toxin concentration

CONSENT FOR PUBLICATION

Not applicable.

CONFLICT OF INTEREST

The author(s) confirms that there is no conflict of interest.

ACKNOWLEDGMENT

Authors are thankful to Mr. R. Praveen Kumar, Professor, Anna bioresearch foundation, Arunai engineering college and Dr. G. Baskar, Professor, Department of Biotechnology, St. Joseph's College of Engineering, for providing all technical assistance.

REFERENCES

[1]   Xia Y, Fang HHP, Zhang T. Recent studies on thermophilic anaerobic bioconversion of lignocellulosic biomass. RSC Advances 2013; 3: 15528-42.
        [http://dx.doi.org/10.1039/c3ra40866c]

[2]   Karimi K, Ed. Lignocellulose-based bioproducts. Cham, Switzerland: Springer 2015.
        [http://dx.doi.org/10.1007/978-3-319-14033-9]

[3]   Koppram R, Tomás-Pejó E, Xiros C, Olsson L. Lignocellulosic ethanol production at high-gravity: challenges and perspectives. Trends Biotechnol 2014; 32(1): 46-53.
        [http://dx.doi.org/10.1016/j.tibtech.2013.10.003] [PMID: 24231155]

[4]   Kosaric N, Ng DCM, Russell I, Stewart GS. Ethanol production by fermentation: An alternative liquid fuel. Adv Appl Microbiol 1980; 26: 147-227.
        [http://dx.doi.org/10.1016/S0065-2164(08)70334-4]

[5]   Kuhar S, Nair LM, Kuhad RC. Pretreatment of lignocellulosic material with fungi capable of higher lignin degradation and lower carbohydrate degradation improves substrate acid hydrolysis and the eventual conversion to ethanol. Can J Microbiol 2008; 54(4): 305-13.
        [http://dx.doi.org/10.1139/W08-003] [PMID: 18389003]

[6]   Boyer LJ, Vega JL, Klasson KT, Clausen EC, Gaddy JL. The effects of furfural on ethanol production by Saccharomyces cerevisiae in batch culture. Biomass Bioenergy 1992; 3(1): 41-8.
        [http://dx.doi.org/10.1016/0961-9534(92)90018-L]

[7]   Cao G, Ren N, Wang A, *et al.* Acid hydrolysis of corn stover for biohydrogen production using thermoanaerobacterium thermosaccharolyticum W16. Int J Hyd Ener 2009; 34: 7182-8.
        [http://dx.doi.org/10.1016/j.ijhydene.2009.07.009]

[8]   Kothari UD, Lee YY. Inhibition effects of dilute-acid prehydrolysate of corn stover on enzymatic hydrolysis of Solka Floc. Appl Biochem Biotechnol 2011; 165(5-6): 1391-405.
        [http://dx.doi.org/10.1007/s12010-011-9355-3] [PMID: 21909630]

[9]   Alriksson B, Cavka A, Jönsson LJ. Improving the fermentability of enzymatic hydrolysates of lignocellulose through chemical in-situ detoxification with reducing agents. Bioresour Technol 2011; 102(2): 1254-63.
        [http://dx.doi.org/10.1016/j.biortech.2010.08.037] [PMID: 20822900]

[10]   Klinke HB, Thomsen AB, Ahring BK. Inhibition of ethanol-producing yeast and bacteria by degradation products produced during pre-treatment of biomass. Appl Microbiol Biotechnol 2004; 66(1): 10-26.
        [http://dx.doi.org/10.1007/s00253-004-1642-2] [PMID: 15300416]

[11]   Clark T, Mackie KL. Fermentation inhibitors in wood hydrolysates derived from the softwood Pinus radiata. J Chem Biotechnol 1984; B34: 101-10.
        [http://dx.doi.org/10.1002/jctb.280340206]

[12]   Maiorella BL. Ethanol.Comprehensive Biotechnology. Oxford: Pergamon Press 1985; Vol. 3: pp. 861-914.

[13]   Gírio FM, Fonseca C, Carvalheiro F, Duarte LC, Marques S, Bogel-Łukasik R. Hemicelluloses for fuel ethanol: A review. Bioresour Technol 2010; 101(13): 4775-800.
        [http://dx.doi.org/10.1016/j.biortech.2010.01.088] [PMID: 20171088]

[14]   Morohoshi N. Chemical characterization of wood and its components.Wood and cellulosic chemistry. New York: Marcel Dekker, Inc. 1991; pp. 331-92.

[15]   Sadashivam S, Manikam A. Biochemical methods. New age international publishers 2006.

[16]   Mathapati PR, Ghasghase NV, Kulkarni MK. Study of *saccharomyces cerevisiae* 3282 for the production of tomato wine. International Journal of Chemical Sciences & Applications 2010; 1(1): 1-15.

[17]   Mosier N, Wyman C, Dale B, *et al.* Features of promising technologies for pretreatment of lignocellulosic biomass. Bioresour Technol 2005; 96(6): 673-86.
[http://dx.doi.org/10.1016/j.biortech.2004.06.025] [PMID: 15588770]

[18]   Chandel AK, Kapoor RK, Singh A, Kuhad RC. Detoxification of sugarcane bagasse hydrolysate improves ethanol production by Candida shehatae NCIM 3501. Bioresour Technol 2007; 98(10): 1947-50. a
[http://dx.doi.org/10.1016/j.biortech.2006.07.047] [PMID: 17011776]

[19]   Balat M, Balat H, Oz C. Progress in bioethanol processing. Pror Energy Combust Sci 2008; 34: 551-73.
[http://dx.doi.org/10.1016/j.pecs.2007.11.001]

[20]   Parajo JC, Dominguez H, Dominguez JM. Charcoal adsorption of wood hydrolysates for improving their fermentability: influence of the operational conditions. Bioresour Technol 1996; 157: 179-85.
[http://dx.doi.org/10.1016/0960-8524(96)00066-1]

[21]   Wilson JJ, Deschatelets L, Nishikawa NK. Comparative fermentability of enzymatic and acid hydrolysates of steam pretreated aspen wood hemicellulose by Pichia stipitis CBS 5776. Appl Microbiol Biotechnol 1989; 31: 592-6.
[http://dx.doi.org/10.1007/BF00270801]

[22]   Wickramasinghe SR, Grzenia DL. Adsorptive membranes, and resins for acetic acid removal from biomass hydrolysates. Desalination 2008; 234: 144-51.
[http://dx.doi.org/10.1016/j.desal.2007.09.080]

[23]   Zhuang J, Liu Y, Wu Z, Sun Y, *et al.* Hydrolysis of wheat straw hemicellulose and detoxification of the hydrolysate for xylitol production. BioResources 2009; 4: 674-86.

[24]   Converti A, Perego P, Dominguez JM. Xylitol production from hardwood hemicellulose hydrolysates by *P. tannophilus, D. hansenii and C. guilliermondii.* Appl Biotechnol Biochem 1999; 82: 141-51.
[http://dx.doi.org/10.1385/ABAB:82:2:141]

[25]   Palmqvist E, Hahn-Hagerdal B, Szengyel Z, Zacchi G, *et al.* Simultaneous detoxification and enzyme production of hemicelluloses hydrolysates obtained after steam pretreatment. Enzyme Microb Technol 1997; 20: 286-93.
[http://dx.doi.org/10.1016/S0141-0229(96)00130-5]

[26]   Cho DH, Lee YJ, Um Y, Sang BI, Kim YH. Detoxification of model phenolic compounds in lignocellulosic hydrolysates with peroxidase for butanol production from *Clostridium beijerinckii.* Appl Microbiol Biotechnol 2009; 83(6): 1035-43.
[http://dx.doi.org/10.1007/s00253-009-1925-8] [PMID: 19300996]

[27]   Wang G, Zhang S, Xu W, Qi W, *et al.* Efficient saccharification by pretreatment of bagasse pith with ionic liquid and acid solutions simultaneously. Energy Convers Manage 2015; 89: 120-6.
[http://dx.doi.org/10.1016/j.enconman.2014.09.029]

[28]   Kuhad RC, Gupta R, Khasa YP, Singh A, *et al.* Bioethanol production from pentose sugars: Current status and future prospects. Renew Sustain Energy Rev 2011; 15: 4950-62.
[http://dx.doi.org/10.1016/j.rser.2011.07.058]

[29]   Parawira W, Tekere M. Biotechnological strategies to overcome inhibitors in lignocellulose hydrolysates for ethanol production: review. Crit Rev Biotechnol 2011; 31(1): 20-31. [Review].
[http://dx.doi.org/10.3109/07388551003757816] [PMID: 20513164]

[30]   Zafar S, Owais M. Ethanol production from crude whey by *Kluyveromyces marxianus.* Biochem Eng J 2006; 27: 295-8.
[http://dx.doi.org/10.1016/j.bej.2005.05.009]

[31]   Talebnia F, Karakashev D, Angelidaki I. Production of bioethanol from wheat straw: An overview on pretreatment, hydrolysis and fermentation. Bioresour Technol 2010; 101(13): 4744-53.
[http://dx.doi.org/10.1016/j.biortech.2009.11.080] [PMID: 20031394]

[32] Nigam JN. Continuous ethanol production from pineapple cannery waste using immobilized yeast cells. J Biotechnol 2000; 80(2): 189-93.
[http://dx.doi.org/10.1016/S0168-1656(00)00246-7] [PMID: 10908799]

[33] Shruti A, Byadgi PB. Kalburgi. Production of Bioethanol from Waste Newspaper. Procedia Environ Sci 2016; 35: 555-62.
[http://dx.doi.org/10.1016/j.proenv.2016.07.040]

[34] Stepanov N, Efremenko E. Immobilised cells of Pachysolen tannophilus yeast for ethanol production from crude glycerol. N Biotechnol 2017; 34: 54-8.
[http://dx.doi.org/10.1016/j.nbt.2016.05.002] [PMID: 27184618]

[35] Wirawan F, Cheng CL, Kao WC, Lee DJ, *et al.* Cellulosic ethanol production performance with SSF and SHF processes using, immobilized Zymomonas mobilis. Appl Energy 2012; 100: 19-26.
[http://dx.doi.org/10.1016/j.apenergy.2012.04.032]

[36] Vučurović VM, Razmovski RN, Popov SD. Ethanol production using *Saccharomyces cerevisiae* cells immobilised on corn stem ground tissue. Proc Nat Sci Matica Srpska Novi Sad 2009; 116: 315-22.

[37] Sudarshan S. Lakhawat, Gajendra K. Aseri, and Vinod S.Gaur. Comparative study of ethanol production using yeast and fruits of vitis lanata roxb. Int J Adv Biotech Res 2011; 2(2): 269-77.

[38] Saha P, Baishnab AC, Alam F, Khan MR, *et al.* Production of bio-fuel (bio-ethanol) from biomass (Pteris) by fermentation process with yeast. Procedia Eng 2014; 90: 504-9.
[http://dx.doi.org/10.1016/j.proeng.2014.11.764]

[39] Patni N, Pillai SG, Dwivedi AH. Wheat as a Promising Substitute of Corn for Bioethanol Production. Procedia Eng 2013; 51: 355-62.
[http://dx.doi.org/10.1016/j.proeng.2013.01.049]

[40] Ito T, Nakashimada Y, Senba K, Matsui T, Nishio N. Hydrogen and ethanol production from glycerol-containing wastes discharged after biodiesel manufacturing process. J Biosci Bioeng 2005; 100(3): 260-5.
[http://dx.doi.org/10.1263/jbb.100.260] [PMID: 16243274]

[41] Behera S, Mohanty RC, Ray RC. Comparative study of bio-ethanol production from mahula (Madhuca latifolia L.) flowers by Saccharomyces cerevisiae and Zymomonas mobilis. Appl Energy 2010; 87: 2352-5.
[http://dx.doi.org/10.1016/j.apenergy.2009.11.018]

[42] Aimaretti N, Ybalo C. Valorization of carrot and yeast discards for the obtention of ethanol. Biomass Bioenergy 2012; 42: 18-23.
[http://dx.doi.org/10.1016/j.biombioe.2012.03.022]

[43] Nigam JN. Ethanol production from wheat straw hemicellulose hydrolysate by Pichia stipitis. J Biotechnol 2001; 87(1): 17-27.
[http://dx.doi.org/10.1016/S0168-1656(00)00385-0] [PMID: 11267696]

[44] O'Brien DJ, Roth LH, McAloon AJ. Ethanol production by continuous fermentation–pervaporation: a preliminary economic analysis. J Membr Sci 2000; 166: 105-11.
[http://dx.doi.org/10.1016/S0376-7388(99)00255-0]

[45] Cazetta ML, Celligoi MA, Buzato JB, Scarmino IS. Fermentation of molasses by Zymomonas mobilis: effects of temperature and sugar concentration on ethanol production. Bioresour Technol 2007; 98(15): 2824-8.
[http://dx.doi.org/10.1016/j.biortech.2006.08.026] [PMID: 17420121]

[46] Gumienna M, Szambelan K, Lasik M, Jeleń H, *et al.* Use of Saccharomyces cerevisiae and Zymomonas mobilis for bioethanol production from sugar beet pulp and raw juice 2013; 12(18): 2464-70.

[47] Goldschmidt F. From cellulose to ethanol: engineering microorganisms to produce biofuel. Institute of

Biogeochemistry and Pollutant Dynamics 2008; pp. 1-7.

[48]   Ahmed FM, Rahman SR, Gomes DJ. Saccharification of sugarcane bagasse by enzymatic treatment for bioethanol production. Malays J Microbiol 2012; 8(2): 97-103.

[49]   Prasad MP, Rekha S, Tamilarasan M, Subha KS. Production of bioethanol using various agricultural raw materials by two step enzymatic process. J Adv Biotechnol 2009; 9: 41-3.

[50]   Kumar SA, Pushpa A. Saccharification by fungi and ethanol production by Bacteria using lignocellulose materials. IRJP 2012; 3(5): 411-4.

[51]   Taherzadeh MJ, Karimi K. Enzyme-based hydrolysis process for ethanol from lignocellulosic materials: a review, enzyme-based ethanol. BioResources 2007; 2(4): 707-38.

[52]   Murugan CS, Rajendran S. Bioethanol production from Agave leave using *Saccharomyces cerevisiae* (MTCC 173) and Zymomonas mobilis (MTCC 2427). Int J Microbio Res 2013; 4(1): 23-6.

[53]   Geetha S, Kumar A, Deiveekasundaram M. Ethanol production from degrained sunflower head waste by Zymomonas mobilis and *Saccharomyces cerevisiae*. IJASR 2013; 3(4): 93-102.

[54]   Sun Y, Cheng J. Hydrolysis of lignocellulosic materials for ethanol production: a review. Bioresour Technol 2002; 83(1): 1-11.
[http://dx.doi.org/10.1016/S0960-8524(01)00212-7] [PMID: 12058826]

[55]   Zhao Xin-Qing, Zi Li-Han, Bai Feng-Wu, Lin Hai-Long, *et al.* Bioethanol from Lignocellulosic Biomass, Adv Biochem Engin/Biotechnol 2012; 128: 25-51.

[56]   Ghose TK. Measurement of cellulase activities. Pure Appl Chem 1987; 59: 257-68.

[57]   Miller GL. Use of dinitro salicylic acid reagent for determination of reducing sugar. Chem 1959; 31: 426-8.

[58]   Williams MB, Reese HD. Colorimetric determination of ethyl alcohol. Anal Chem 1950; 22(12): 1556-61.
[http://dx.doi.org/10.1021/ac60048a025]

[59]   Caputi A, Ueda M, Brown T. Spectrophotometric determination of ethanol in wine. Am J Enol Vitic 1968; 19: 160-5.

[60]   Kiss AA, David J, Suszwalak PC. Enhanced bioethanol dehydration by extractive and azeotropic distillation in dividing-wall columns. Separ Purif Tech 2012; 86: 70-8.
[http://dx.doi.org/10.1016/j.seppur.2011.10.022]

[61]   Luong JH. Kinetics of ethanol inhibition in alcohol fermentation. Biotechnol Bioeng 1985; 27(3): 280-5.
[http://dx.doi.org/10.1002/bit.260270311] [PMID: 18553670]

[62]   Lee KJ, Skotnicki ML, Tribe DE, Rogers PL. Kinetic studies on a highly productive strain of Zymomonas mobilis. Biotechnol Lett 1980; 2: 339-44.
[http://dx.doi.org/10.1007/BF00138666]

[63]   Yomano LP, York SW, Ingram LO. Isolation and characterization of ethanol-tolerant mutants of Escherichia coli KO11 for fuel ethanol production. J Ind Microbiol Biotechnol 1998; 20(2): 132-8.
[http://dx.doi.org/10.1038/sj.jim.2900496] [PMID: 9611822]

[64]   Jones RP, Greenfield PF. Ethanol and the fluidity of the yeast plasma membrane. Yeast 1987; 3(4): 223-32.
[http://dx.doi.org/10.1002/yea.320030403] [PMID: 3332975]

[65]   Kubota S, Takeo I, Kume K, *et al.* Effect of ethanol on cell growth of budding yeast: genes that are important for cell growth in the presence of ethanol. Biosci Biotechnol Biochem 2004; 68(4): 968-72.
[http://dx.doi.org/10.1271/bbb.68.968] [PMID: 15118337]

[66]   Malacrinò P, Tosi E, Caramia G, Prisco R, Zapparoli G. The vinification of partially dried grapes: a comparative fermentation study of Saccharomyces cerevisiae strains under high sugar stress. Lett Appl

Microbiol 2005; 40(6): 466-72.
[http://dx.doi.org/10.1111/j.1472-765X.2005.01713.x] [PMID: 15892744]

[67]   Hottiger T, Schmutz P, Wiemken A. Heat-induced accumulation and futile cycling of trehalose in *Saccharomyces cerevisiae*. JBacteriol 1987; 169(12): 5518-22.

[68]   Panek AC, Vânia JJ, Paschoalin MF, Panek D. Regulation of trehalose metabolism in Saccharomyces cerevisiae mutants during temperature shifts. Biochimie 1990; 72(1): 77-9.
[http://dx.doi.org/10.1016/0300-9084(90)90176-H] [PMID: 2160289]

[69]   Oliva.neto P. Estudo de diferentes fatores que influenciam o crescimento da população bacteriana contaminante da fermentação alcoólica por leveduras. Tese (Doutorado)-Faculdade de Engenharia de Alimentos- Unicamp- Campinas 1995; 112.

[70]   Egamberdiev NB, Jerusalimsky A. Continuous cultivation of microorganisms Czechoslovak academy of sciences, Prague. 1968.

[71]   Ghose TK, Tyagi RD. Rapid ethanol fermentation of cellulose hydrolysate. II. Product and substrate inhibition and optimization of fermentor design. Biotechnol Bioeng 1979; 21(8): 1401-20.
[http://dx.doi.org/10.1002/bit.260210808]

[72]   Hinshelwood CN. The chemical kinetics of the bacterial cell. London, UK: Clarendon Press Oxford 1946.

[73]   Amenaghawon NA, Okieimen CO, Ogbeide SE. Kinetic modelling of ethanol inhibition during alcohol fermentation of corn stover using *Saccharomyces cerevisiae*. Int J Engine Res 2012; 2(4): 798-803.

[74]   Hoppe G, Hansford G. Ethanol inhibition of continuous anaerobic yeast growth. Biotechnol Lett 1982; 41: 39-44.
[http://dx.doi.org/10.1007/BF00139280]

[75]   Lee JM. Computer simulation in ethanol fermentation.Biomass conversion processes for energy and fuels. New York: Plenum 1988.

[76]   Zafar S, Owais M, Saleemuddin M, Husain S. Batch kinetics and modelling of ethanolic fermentation of whey. Int J Food Sci Technol 2005; 40(6): 597-604.
[http://dx.doi.org/10.1111/j.1365-2621.2005.00957.x]

[77]   Longhi LGS, Luvizetto DJ, Ferreira LS, Rech R, Ayub MAZ, Secchi AR. A growth kinetic model of Kluyveromyces marxianus cultures on cheese whey as substrate. J Ind Microbiol Biotechnol 2004; 31(1): 35-40.
[http://dx.doi.org/10.1007/s10295-004-0110-4] [PMID: 14758555]

[78]   Lin Y, Tanaka S. Ethanol fermentation from biomass resources: current state and prospects. Appl Microbiol Biotechnol 2006; 69(6): 627-42.
[http://dx.doi.org/10.1007/s00253-005-0229-x] [PMID: 16331454]

[79]   Suja Malar RM, Thyagarajan T. Modelling of continuous stirred tank reactor using artificial intelligence techniques. Int J Simul Model 2009; 8(3): 145-55.
[http://dx.doi.org/10.2507/IJSIMM08(3)2.128]

[80]   Hinshelwood CN. The Chemical Kinetics of the Bacterial Cell. Oxford: Clarendon 1952; p. 105.

[81]   Sonnleitner B, Rothen SA, Kuriyama H. Dynamics of glucose consumption in yeast. Biotechnol Prog 1997; 13(1): 8-13.
[http://dx.doi.org/10.1021/bp960094+] [PMID: 9041706]

[82]   Baei MS, Mahmoudi M, Yunesi H. A kinetic model for citric acid production from apple pomac by Aspergillus niger. Afr J Biotechnol 2008; 7(19): 3487-9.

[83]   Ocloo FC, Ayernor GS. Production of alcohol from cassava flour hydrolysate. J Brew Distil 2010; 1(2): 15-21.

[84]   Warren RK, Hill GA, Macdonald DG. Improved bioreaction kinetics for the simulation of continuous ethanol fermentation by Saccharomyces cerevisiae. Biotechnol Prog 1990; 6(5): 319-25.
[http://dx.doi.org/10.1021/bp00005a002]

[85]   Taylor F, Kurantz MJ, Goldberg N, Craig JC. Kinetics of continuous fermentation and stripping of ethanol. Biotechnol Lett 1998; 20(1): 67-72.
[http://dx.doi.org/10.1023/A:1005339415979]

[86]   Holzberg I, Finn RK, Steinkraus KH. Steinkraus. A kinetic study of the alcoholic fermentation of grape juice. Biotechnol Bioeng 1967; 9(3): 413-27.
[http://dx.doi.org/10.1002/bit.260090312]

[87]   Aiba S, Shoda M, Nagatani M. Kinetics of product inhibition in alcohol fermentation. Biotechnol Bioeng 1968; 10(6): 845-64.
[http://dx.doi.org/10.1002/bit.260100610] [PMID: 10699849]

[88]   Daugulis AJ, Swaine DE. Examination of substrate and product inhibition kinetics on the production of ethanol by suspended and immobilized cell reactors. Biotechnol Bioeng 1987; 29(5): 639-45.
[http://dx.doi.org/10.1002/bit.260290513] [PMID: 18576495]

CHAPTER 9

Fermentation Strategies to Minimize Product Inhibition in Bioethanol Production

Luciana Porto de Souza Vandenberghe[*], **Nelson Libardi Junior**, **Cristine Rodrigues**, **Joyce Gueiros Wanderley Siqueira** and **Carlos Ricardo Soccol**

Biotechnology and Bioprocess Engineering Department, Federal University of Parana, Curitiba-PR, Brazil

Abstract: Bioethanol is the most used biofuel worldwide. Its use contributes to the reduction of fossil fuel consumption and environmental pollution. It is mainly produced from sucrose, which is available in alternative media. Yeast, mainly *Saccharomyces cerevisiae,* are the most employed microorganisms for ethanol production. These strains usually present high productivity, high ethanol tolerance, and the ability to ferment different sugars that are included in the composition of the highly utilized feedstock. Nevertheless, there are some barriers to yeast fermentation to overcome. They are linked to inhibitors of ethanol production, including high temperature, high ethanol concentration, and the ability to ferment pentose sugars. The efficiency and productivity of ethanol can be enhanced by the use of genetically modified yeast strains, including hybrid and recombinant. Other possibilities of limiting bioethanol processing inhibition are metabolic engineering of the medium and yeast cell immobilization. This chapter highlights some aspects that involve fermentation strategies to minimize bioethanol inhibition during its production.

Keywords: Biofuel, Bioethanol, Ethanol, Fermentation, Genetic modified, Hybrid, Inhibition, Metabolic engineering, Pentoses, Recombinant strains, *Saccharomyces cerevisiae*, Sucrose, Yeast.

INTRODUCTION

According to Rastogi and Shrivastava [1], bioethanol has an annual market of US$58 billion. Approximately half of the global sugar produced is used for ethanol production, with the USA and Brazil as the global leaders. These countries primarily use sugar-based crops like corn and sugarcane, respectively. In Brazil, biofuel accounts for 27.5% of the fuel market. Biofuels like ethanol have a strong global insertion. In terms of cellulosic biofuel production, approximately

[*] **Corresponding author Luciana Porto de Souza Vandenberghe:** Biotechnology and Bioprocess Engineering Department, Federal University of Parana, Curitiba-PR, Brazil; E-mail: lvandenberghe@ufpr.br

10% of the global residues would satisfy 50% of the biofuel demand by 2030. Nevertheless, there are some barriers in yeast fermentation to overcome that are linked to the inhibition of ethanol production: medium composition and processing conditions, including high temperature and product (bioethanol) concentrations (Fig. **1**). Another important point is the ability of strains to ferment pentose sugars, which are common components of lignocellulosic biomass hydrolysates.

Some actions for high ethanol efficiency and productivity can be achieved by the use of genetically modified yeast strains, including hybrid and recombinant microorganisms. Yeast cell immobilization is another important tool for better biofuel production. This chapter highlights recent information about strategies to minimize bioethanol inhibition during its production.

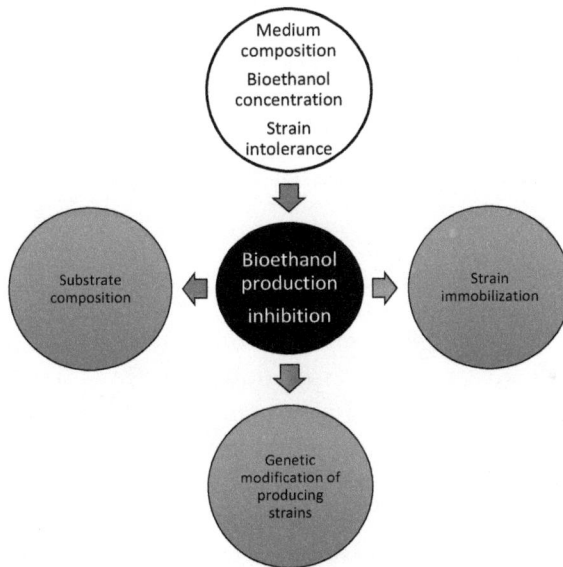

Fig. (1). The main causes of bioethanol production inhibition.

BIOETHANOL MICROBIAL STRAINS

Most microorganisms that produce ethanol by fermentation are mesophilic, with optimum metabolism between 30 and 37°C. Yeast has been used for centuries for alcoholic beverages (*e.g.*, beer and wine) production. Currently, *Saccharomyces* strains are widely used for biofuel production. Approximately 2,500 yeast species have been discovered, and this number is expected to reach 150,000, namely because these microorganisms can live in diverse habitats, including terrestrial, aerial and aquatic environments [2].

Saccharomyces cerevisiae and *Zymomonas mobilis* are the most well-known ethanol producers. These microorganisms can convert glucose, sucrose, and fructose to ethanol, but they are incapable of converting pentose sugars such as xylose. *S. cerevisiae* uses glycolysis to convert sugars, mainly glucose, into ethanol. Under anaerobic conditions, 1 mole of glucose generates 2 moles of ethanol, 2 moles of carbon dioxide, and 2 two adenosine triphosphate (ATP) molecules. *Z. mobilis*, a gram-negative bacterium, has been used for ethanol fermentation because it has a higher ethanol tolerance and glucose uptake and enhanced ethanol yield and productivity when compared to *S. cerevisiae*. *Z. mobilis* metabolizes glucose by the Entner-Doudoroff pathway to yield 1 mole of ATP [3].

Some species of the genera *Candida*, *Pichia*, *Pachysolen*, and *Schizosaccharomyces* can convert pentose sugars to ethanol. *Pichia stipitis* (NRRL-Y-7124), *S. cerevisiae* (RL-11), and *Kluyveromyces fragilis* (Kf1) are reportedly very good ethanol producers (from different types of sugars) [2].

The natural characteristics of each microorganism are beneficial for ethanol production. Thermotolerant yeast like *Kluyveromyces marxianus* can co-ferment hexoses and pentoses as well as tolerate temperatures from 42 to 45°C. The ability to ferment hexose sugars reduces the costs associated with the previous hydrolysis of feedstocks.

In general, yeast can sustain the main bottlenecks of the majority of ethanol fermentation processes, and thus they are the most suitable microorganisms for this purpose. The microorganism must have the simultaneous ability to produce high ethanol yields (> 90%) and tolerate concentrations up to 40 g/L, levels that allow productivities of 1g/L/h [2]. Yeast can grow in the simple and inexpensive culture medium, present resistance to inhibitors and contaminants, and tolerate a wide range of pH, mainly acidic conditions; all of these factors make them less susceptible to contamination [4]. Additionally, the property of floc formation is an advantage for the recovery and reuse of the biomass at the end of each fermentation batch. *Saccharomyces* strains are generally recognized as safe (GRAS), and thus they are still the most common microorganisms used for alcoholic beverages and other fermented food production as well as for biofuel production because they present some essential characteristics [3].

The co-culture of different strains has been also proposed as a method to overcome the problem of consumption of five and six-carbon sugars. For example, *S. cerevisiae* and *P. stipitis*, when cultivated together, consume glucose and xylose simultaneously, with a considerable augmentation in ethanol productivity. However, other problems, like the different ethanol tolerance of each strain, remain a barrier for this technology [4].

BIOETHANOL PRODUCTION FROM DIFFERENT FEEDSTOCKS

The main biomass-based sources for bioethanol production are sugar sources, such as sucrose and glucose, and starchy sources, which can be classified as first-generation or second-generation materials (including lignocellulosic sources). Sugar-based raw materials can be obtained from energy crops and fruit. Starchy crops for bioethanol production include cereals, tubers and roots, legumes, and green and immature fruit. Lignocellulosic biomass can be originated from energy crops, forest biomass, agricultural residues, aquatic plants, and municipal solid wastes [3, 5].

The main difference between the ethanol production processes is up-stream fermentation. Sugar-based materials are directly used for ethanol fermentation and obviate pretreatments such as milling, hydrolysis, and detoxification, all of which are required for starch and/or lignocellulose [5]. Starchy materials such as corn require processes like milling, liquefaction, and saccharification. For lignocellulosic materials, physical (milling and classification) and chemical and enzymatic pretreatments (acid and enzymatic hydrolysis) are required.

Sucrose Sources

The most frequently used feedstocks for ethanol production are sugar-based biomass, namely sugarcane, sugar beets, and sweet sorghum. They all have high sugar content and low conversion costs. Other feedstocks include fruit (grapes, dates, watermelon and apples), as well as sugar refinery wastes like cane and beet molasses [3, 5]. Sugar crops are advantageous because they present high sugar yields and low conversion costs and are seasonal crops.

Sugarcane is cultivated in tropical and subtropical regions. In Brazil, sugarcane juice is the main feedstock, accounting for 79% of the bioethanol production, because it contains favorable nutrients, minerals, and readily fermentable sugars that are ideal for bioethanol production [3]. The direct use of the medium, without any pretreatments, makes the process much less costly and competitive.

Sweet sorghum is a C4 plant that, contrary to sugarcane, can be cultivated in temperate and tropical areas; it has high levels of extractable sugars. Sugar beet is the major source of sugar in Europe and North America. It is typical of temperate climate, and the sugar is concentrated in the roots [3].

Molasses are generated in sugarcane and sugar beet refineries. They are characterized as dark, viscous, and rich in sugars, with variations that depend on the sugar extraction processes and the source material [3]. Molasses usually contain some inhibitors in their composition, including high concentrations of ions (sulfate, chlorate, and iron) and other components that may interfere in ethanol fermentation. Fruit that is discarded due to their low quality or unacceptable appearance can also be used as an alternative for ethanol production. However, the process yield is not comparable to that obtained with the use of the other sugar-based feedstock [3].

Starch Sources

The main characteristics that make starch sources interesting for ethanol production are their wide availability, storage stability for long periods (contrary to sucrose feedstocks), and high ethanol conversion. The main difference in comparison to sugar-based feedstock is that starch sources require a saccharification step. This process is performed through enzymatic hydrolysis for the conversion of starch into glucose, which is fermented by biocatalysts. High-performance amylolytic enzymes catalyze the hydrolysis reaction. Starch-based ethanol contributes to 60% of global ethanol production. Corn, sorghum grains, wheat, cassava, potatoes, and sweet potatoes are the most utilized starch ethanol sources [3, 6]. North America is the largest ethanol producer from corn, led by the United States of America—which produced 14.3 billion gallons in 2014 and exported roughly 825 million gallons of ethanol to 51 countries—followed by Asia, Europe, and South America.

Cassava has been described as a very promising carbohydrate source for ethanol production. Cassava is highly cultivated in developing countries, along with rice, wheat, corn, potatoes, and barley. It originated from South America and it is adapted to both tropical and subtropical climates, with tolerance to semi-arid conditions. The global cassava processing market compound annual growth rate (CAGR) was approximately 2.1% during 2010-2017; it reached a production volume of approximately 284.9 million tons in 2017 [7]. Sweet potato has been also described as a starch source for ethanol production; China is the dominant producer of cassava and sweet potato, representing 20% and 80%, respectively. In 2016, the global production of sweet potatoes reached 105.19 million metric tons [8].

Lignocellulose-based Feedstock

Lignocellulosic materials for ethanol production are divided according to their sources, such as energy crops, aquatic plants, forest materials, agricultural residues, and the organic portion of municipal solid wastes. The main components of lignocellulosic materials are cellulose (40-60% dry weight), hemicellulose (20-40%), and lignin (10-25%). The cellulose long chains of beta-glucose monomers are packed into microfibrils, which are firmly attached to each other by the hemicelluloses. These polymers are mainly composed of pentoses (*e.g.*, xylose and arabinose) and hexoses (including mannose, glucose, and galactose). Lignin is formed by phenolic compounds, including p-coumaryl, coniferyl, and sinapyl alcohols, and it is responsible for linking hemicellulose molecules by covalent bonds to provide the rigidity and resistance of the plant cell walls [1]. This crystallinity of the cellulose fibrils and the physical and chemical barriers of lignin underscore the need for pretreatments and hydrolysis, which are considered limiting steps for the bioethanol production from this kind of material.

Globally, the production of plant biomass is estimated to be 2.00×10^{11} tons per year, whereas 4-10% of this amount has the potential to be converted to biofuel [9]. Worldwide, corn, wheat, rice, and sugarcane are the greatest agricultural residue contributors.

The lignocellulosic ethanol production process comprises three main steps. Pretreatment (a) consists of mechanical and chemical processes that render the cellulose and hemicellulose more accessible and digestible for the subsequent steps. Hydrolysis (b) is performed by the use of chemicals and/or enzymes for the conversion of the polymers into sugars. Fermentation (c) is where the five and six-carbon sugars are consumed to produce ethanol, followed by separation and concentration of the ethanol by distillation-rectification-dehydration. The high pentose sugar content stimulates the use of strains that can convert five-carbon sugars. Among the three steps, the limiting one is hydrolysis (due to the yields of sugars and costs associated with the use of enzymes) [6].

Ethanol from cellulosic feedstock has been produced in a yield ranging from 0.21-0.27 g per g cellulosic biomass [4]. Each feedstock has its own advantages in terms of its potential for ethanol production. Various bioethanol production feedstocks and their ethanol yields are presented in Table **1**. Among the top three ethanol production feedstocks, corn represents the highest yield of ethanol volume per ton of feedstock and sugarcane the highest productivity per area.

Table 1. Bioethanol yield and potential yield with different feedstocks.

Feedstock	Ethanol Yield (L/ton)	Yield (L/ha)	References
Sugarcane	70-90	5,400-10,800	[10, 11]
Sugar beet	95-110	5,000-10,000	[10, 11]
Molasses	280	-	[3]
Corn	370-470	2,000-4,600	[10, 11]
Cassava	363-455	4,901	[3]

(Table 1) cont.....

Feedstock	Ethanol Yield (L/ton)	Yield (L/ha)	References
Sugarcane bagasse	280	-	[11]
Sorghum bagasse	250	1,796-6,591	[3]
Corn cob	510	-	[3]
Wheat straw	490	-	[3]
Corn stover	290	4,400	[11]
Sugar beet pulp	260	-	[11]

Ethanol yields are strongly dependent on the conversion efficiency, the nature of the biomass, process conditions, and the microorganisms that are employed in the fermentation process. A comparison of sugarcane, corn and sugar beet—the three main ethanol feedstocks in Brazil, USA, and Europe, respectively—was conducted by Goldenberg and Guardabassi [10]. They revealed that ethanol sugarcane production costs are 60% and 75% lower compared to corn and sugar beet, respectively. The authors presented the ethanolic yields of 6,470 L/ha (sugarcane), 4,180 L/ha (corn), and 5,500 L/ha (sugar beet), data that indicate the attractiveness of these feedstocks. The sustainability of the process was also investigated; the CO_2 emission reduction compared to gasoline was 84%, 30%, and 40% for sugarcane, corn and sugar beet, respectively.

The yields for ethanol production should be evaluated with different factors. For instance, ethanol production from corn is a more complex process compared to sucrose-based feedstocks. However, it has the highest ethanol yield. On the other hand, sugarcane has the highest productivity in terms of tons per ha. Comparing the two crops, corn ethanol production requires double the area to reach the ethanol yield of sugarcane. The cycles of harvesting for each kind of culture is another aspect that should be considered. The sugar conversion efficiency for each feedstock is also important; it ranges from 94.0 to 99.6% [11].

The inhibition effect derived from the pretreatment step before fermentation using lignocellulosic biomass comes from the lignocellulosic-derived by-products and pretreatment solvents. The main inhibitors are furans (furfural and hydroxymethylfurfural [HMF]), carboxylic acids (acetic, levulinic, and formic acids), phenolic compounds (syringaldehyde, vanillin, 4-hydrolxybanzaldehyde, coniferyl aldehyde, and syringic acid), ionic liquids (ILs), ethanol, and methanol [4, 12].

FEEDSTOCK PRETREATMENT AND ETHANOL PRODUCTION INHIBITION

In lignocellulosic biomass, lignin and hemicellulose protect cellulose structures from microbial and enzymatic attack. For bioethanol production, it is necessary to remove these structures so that cellulolytic enzymes can hydrolyze cellulose to glucose units for fermentation. This removal is mediated by a pretreatment process that can be physical, chemical, or biological [5, 12]. However, pretreatment processes, mainly thermochemical hydrolysis, usually release inhibitory components that decrease enzymatic action on cellulose digestion and glucose fermentation to ethanol. The higher the efficiency of hydrolysis, the greater the production of inhibitory compounds. The low production of undesired byproducts is one of the most important aspects to consider when choosing a pretreatment for the bioethanol production chain [3, 13, 14].

The inhibitory compounds can be categorized into groups: aliphatic acids, furan derivatives, and phenolic compounds; their formation is dependent on the type of biomass and the pretreatment conditions [3, 13, 14]. Besides, residual solvents used for treatments are also considered to be potential inhibitory components. All of them can be present in hydrolysates and biomass slurry. The biomass slurry demands copious amounts of water to wash it and remove all potential byproducts [12].

Thermochemical Pretreatments

The release of inhibitory compounds from lignocellulosic biomass pretreatment processes has been a concern for bioethanol production and it is being widely studied [12, 15]. The most common pretreatments are based on thermic and chemical catalysis, such as alkaline, dilute acid, steam explosion, ammonia-based catalysis, hot pressured water, oxidative reactions, and some other solvents.

Acid Pretreatments

Biomass acid pretreatments are usually performed with hydrochloric and sulfuric acid, as well as some organic acids and sulfur dioxide. This process is used to remove hemicellulose and it might release uronic acids, acetic acid, pentoses (mainly xylose), hexoses and furans, such as furfural, HMF, and 2-furoic acid. The formation of furfural and HMF is dependent on the conditions applied in the pretreatment; they are products of dehydration of pentoses and hexoses, respectively. Under highly severe conditions, furans can be degraded to levulinic and formic acids [12, 15, 16]. Furfural can decrease the activity of enzymes that are present in the catabolic cycles of fermentation, mainly pyruvate dehydrogenase and aldehyde dehydrogenase [17], and it can interfere with the absorption of cysteine and methionine [18]. Together with HMF, they can prolong the lag phase and consequently decrease cell growth and ethanol recovery. To reduce the cell damage caused by them, cells may redirect energy flows, an action that will decrease ATP and NADH levels. Additionally, furfural can amplify phenol and acetic acid toxicity and the number of reactive radicals, which will promote more damage to mitochondria, chromatin, vacuole membranes, and the actin cytoskeleton [12]. Acetic acid is the product of acetyl group degradation and can decrease the growth of specific yeast [12, 19].

The removal of hemicelluloses also releases compounds from its linkage to lignin, such as ferulic and ferulic acids. Some parts of lignin that are hydrolysable from acids are also a source of inhibitory components formation, including 4-hydroxybenzoic acid, 4-hydroxyenzaldehyde, vanillin, syringaldehyde, dihydroconiferyl alcohol, coniferyl aldehyde, syringaldehyde, and syringic acid [15, 17].

Extractive components can also inhibit bioethanol production; they are released upon acidic treatments. The composition of extractive components is highly dependent on the type of feedstock hydrolyzed. Some examples of inhibitory compounds that are formed are benzoic acid, benzyl alcohol, cinnamic acids, para and ortho-toluic acids, gallic acid, and soluble tannins. Certain benzoquinones were also recently found in some hydrolysates from biomass. Although these compounds are present and are potential inhibitors, they remain as trace elements at concentrations low enough to perhaps be assimilated by the yeast. Furthermore, the acidic process can release metal ions that affect enzymatic and microbial action, as they can be liberated from corrosion of equipment that is used in hydrolysis of from pH adjustment [15, 17]. It is important to consider the possibility of low concentrations of weak acids to improve glucose consumption and ethanol formation due to the enhancement of cellular division [20].

The removal of some of these undesired components can be mediated by evaporation of the volatile ones, chromatography techniques, activated charcoal, and extractive processes, among others [3]. For further utilization of the hydrolysate for ethanol production, it is indispensable to detoxify it because the released compounds may inhibit microbial growth [5, 16].

Steam Explosion

Like the dilute acid method, the steam explosion solubilizes hemicellulose and concentrates cellulose in the fiber. This procedure is based on applying overheated steam to the fiber followed by decompression. In this case, the process releases acetic, uronic, and levulinic acids. If an inorganic acid is employed to assist the catalysis, the mentioned byproducts in the previous section may also be present in the hydrolytic liquid fraction [5, 15]. The mechanism of inhibition is the same as described for acidic treatments.

Ammonia Fiber Expansion (AFEX)

The use of ammonia in lignocellulosic biomass pretreatment can be performed in several ways. One of them is the expansion of the fibers through a process similar to the steam explosion. However, it uses ammonia as the explosion agent. The inhibitor production is lower compared to other thermochemical pretreatment methods, with lower carboxylic acids relative to alkaline reactions and fewer furans than obtained from dilute acid treatments [1, 21]. Ammonia is also frequently used in recycled percolation (ARP) and soaking aqueous ammonia (SAA) [1].

Alkaline Pretreatment

The use of alkali as a catalyzer removes lignin from the biomass while leaving most hemicellulose and cellulose. It also decreases cellulose crystallinity and facilitates cellulase attacks. The most utilized catalyzers are sodium, calcium, and potassium hydroxides [3, 15, 22]. Alkaline hydrolysis generates a few inhibitory compounds, including carboxylic acids, acetic acid (from saponification of hemicelluloses acetyl groups), and phenolic components (from lignin degradation) [15].

Phenolics increase the lag phase, a phenomenon that reduces the production of alcohol and other fermentation products. In the same way, some of them can damage cells by crossing the membrane, increasing its fluidity, and cleaving internal cellular structures. Therefore, low molecular weight phenolic compounds become toxic to cell growth and glucose consumption. Additionally, they can increase the number of free radicals (reactive oxygen species) and thus cause DNA mutagenesis and cell death, besides structural damages. They can also deactivate some intracellular enzymes that are responsible for the metabolic pathways of some interesting products. Vanillin has been related to the repression of translation processes in yeast [12, 18]. The number, position and structure of side groups are crucial to inhibition abilities [18]. Some yeast and bacteria can convert phenolic compounds to less-damaging molecules, for example, the conversion of phenolic aldehydes to phenolic alcohols under specific conditions.

Hot Pressured Water

The use of hot liquid water for pretreatment presents desirable characteristics: It does not produce high amounts of inhibitory compounds. The process is performed by reaction of high pressured water (normally with temperatures between 170 and 230°C and pressures above 5 MPa) with biomass to remove hemicellulose. The absence of chemical catalyzers usually does not promote inhibitor production. This pretreatment is performed at neutral pH, but the process can release acids from the biomass, such as acetic acid from acetyl groups of hemicellulose. These compounds may favor inhibitor production as described in the acidic pretreatment section, although this action occurs at very low concentrations [1, 15, 16]. It is also possible to cleave some lignin linkages, depending on the severity of conditions applied in the pretreatment. This cleavage can mediate the release of phenolic compounds [12, 15].

Organosolv

Organosolv involves the use of organic solvents, like methanol, formic acid, ethanol, acetic acid, and others, to remove lignin. The solvents are usually mixed with an acidic catalyzer, such as hydrochloric or sulfuric acid [21]. Organosolv treatment generates three fractions of products: a dense liquid fraction of hemicellulose, a solid fraction of cellulose, and a solid fraction of lignin (considering that all solvent has been evaporated and recovered) [23]. As mentioned before, these solvents are themselves inhibitors, as they are present as residues in the biomass slurry. The solvent must be recovered to reduce costs and prevent their inhibitory action over cellulolytic enzymes and fermentative yeasts [24].

Oxidative Methods

Oxidative methods are used in the pulping industry, mainly with sulfite catalysis. These methods cause the release of gluconic and glucaric acids and phenolic compounds. The variety of the generated compounds is because sulfites cause a partial degradation of all structures in lignocellulosic biomass. If it is performed under alkaline conditions, phenolics can be oxidized to furfural, furoic acid, and carboxylic acids [5, 12, 15]. The use of sulfite as a pretreatment in the bioethanol producing chain is being studied as sulfite pretreatment to overcome recalcitrance of lignocellulose (SPORL). It has shown efficiency for increasing sugar recovery and decreasing inhibitory compound production [5].

ILs

Recently, ILs have received more attention due to their ability to reduce cellulose crystallinity and thus increase its digestibility. It is also considered to be an interesting approach given the different characteristics that biomass components show when solubilized in ILs and also due to the high selectivity of each liquid used in the extraction. The method is applied at high temperatures and atmospheric pressures [25, 26]. However, the solvent concentrations that are employed require large amounts of water to wash them away or at least dilute them [27]. Nevertheless, the solvents remain at low concentrations in the biomass slurry. Besides, they are toxic for microbes, due to the interaction between the cations from ionic liquids and the mitochondria membrane, which is polarized and produces reactive oxygen species that cause the damages described above. IL residues may also act synergistically with other inhibitors. These observations and conclusions were obtained through experiments performed with imidazolium IL [12, 28, 29]. The mechanism of inhibition is dependent on the solvent used in the extraction.

Biological Pretreatment

Biological pathways are considered to be a viable alternative treatment to deal with the production of inhibitory molecules. The use of fungi and enzymes have been studied as lignin and hemicellulose removal methods through the consumption and sequential enzymatic hydrolysis of cellulose, as already applied when chemical pretreatments are used.

Some fungi can degrading lignin, cellulose, and hemicelluloses and can be used for biomass pretreatment. Rot (white, brown, and soft) fungi are the most frequently applied microorganisms in these processes [3, 16]. Brown rot fungi degrade cellulose, whereas soft and white rot degrade cellulose and lignin. Among them, white rot is more suitable for lignocellulose degradation. They can produce enzymes that hydrolyze lignin structure (laccases, manganese peroxidase, and lignin peroxidase) as well as polysaccharides [21, 29]. Microbial treatment is an environmentally friendly process, and it does not produce compounds that inhibit enzymatic hydrolysis or glucose fermentation [29, 30] due to the mild conditions of microbial biomass consumption. This type of treatment presents some disadvantages: it takes a long time, so the rate of hydrolysis is relatively low. Another option concerning microbial degradation of lignocellulosic biomass is the use of co-cultures. Studies show that consortiums of fungi degrade biomass more efficiently than single cultures [29].

Another option is the combination of two types of pretreatments, biological and physical or chemical under mild conditions. In this combination, it is possible to obtain high efficiency for fermentable sugars recovery while not allowing the release of undesired byproducts [29].

Enzymatic Hydrolysis

Enzymatic conversion of cellulose and hemicelluloses to fermentable sugars is already a reality in lignocellulosic ethanol production; it occurs after biomass delignification or polysaccharides concentration. The enzymatic hydrolysis is applied under milder conditions than acid conversion and does not generate inhibitory compounds. This step is eased by a previous treatment that increases cellulose digestibility.

The use of microbial enzymes to convert the structures of lignocellulosic biomass at the same time has been studied was described by Dhiman *et al.* [31]. The enzymatic consortium used by them eliminated hazardous chemical compounds. This study suggests that this use can be explored in future research because it promotes simultaneous pretreatment and hydrolysis without the production of inhibitory or non-environmentally friendly compounds [29]. For faster processes, two-step acid hydrolysis can be applied, with a first step based on diluted acid to remove hemicelluloses and acetyl groups and a subsequent reaction with concentrated acid to break cellulose.

FACTORS THAT AFFECT BIOETHANOL PRODUCTION

Besides pretreatment efficiency and byproduct production, other factors influence bioethanol production from lignocellulosic biomass. The fermentation step is performed using the hydrolysates previously obtained in the enzymatic or acid process of pretreated biomass, by yeast or bacteria [3, 21]. Fermentation conditions are crucial for the microbial strain to grow and produce the desired biomolecule. Factors such as pH, temperature, hydrolysate composition, and sugar concentrations, product concentration, and oxygen availability, among others, limit or increase the performance of fermentation processes.

High Temperature

Microbial strains that are commonly used in bioethanol production are mesophilic and cannot survive at high temperatures. Therefore, research about the isolation and engineering of these strains are being developed to discover or adapt them to extreme conditions of fermentation processes [32, 33]. Besides, new thermotolerant strains are an interesting tool for potential application in simultaneous saccharification and fermentation processes. They must act optimally at higher temperatures where cellulolytic enzymes show better activities [34, 35].

As mesophiles, ethanologenic yeast and bacterial strains are damaged by high temperatures, specifically when they are exposed to temperatures above 37-40°C. *S. cerevisiae* has its optimum temperature near 30°C. As the temperature increases, ethanol yields are directly proportional to temperature increase [20, 34].

In thermic conditions reach an extreme point, ethanol recovery can be severely decreased by the high osmotic stress in cells. Consequently, inhibitory compounds, such as sorbitol and levan, are produced. For yeast, high temperatures are an important cause of stress. These microorganisms produce heat-shock proteins that deactivate cellular ribosomes. Additionally, high temperatures can cause denaturation of enzymes, other proteins, and ribosomes [34]. Besides, high temperatures can disrupt membranes.

High Bioethanol Concentration

Ethanol concentration is one factor that can inhibit yeast growth and viability during fermentation. *S. cerevisiae*, which is traditionally employed in ethanol production, cannot grow in the presence of high alcohol concentrations. This fact certainly limits the potential to achieve high productivity. In this case, *Z. mobilis* shows higher tolerance to ethanol concentrations over 16% in batch conditions.

Continued removal of the produced ethanol during the process can be performed to eliminate this inhibition. Some methods have been studied to extract part of the ethanol from the aqueous fermentation broth, including selective adsorption; evaporation under vacuum; removal by pervaporation; sparging of inert gas through the broth and liquid-liquid extraction, during fermentation, with the use of water as an immiscible solvent. These techniques diminish ethanol-mediated inhibition to realize high production levels [36].

Substrate Composition

The medium composition greatly affects bioethanol production, from the substrate load (sugar concentration) to the presence or absence of inhibitory compounds. Furthermore, temperature and sugar concentrations have a complex relationship with microbial growth and ethanol production. Increasing sugar up to a certain concentration enhances the fermentation rate, but beyond this point the substrate will inhibit growth due to osmotic stress on cells. Industrial processes generally use higher sugar concentrations. However, this condition increases the fermentation time [20, 34]. A *S. cerevisiae* strain used by Lin *et al.* [35] reached its maximum capacity of glucose conversion to ethanol using a concentration of 40 kg/m^3 glucose load; it is inhibited above 80 kg/m^3. It is possible to control substrate inhibition by using fed-batch fermentation, starting with a low concentration of sugars and feeding the medium during the process.

The presence of inhibitors from previous chemical pretreatment and enzymatic hydrolysis is another factor of medium composition for ethanol yield. The presence of pentoses in the medium can represent a problem because the majority of microorganisms, especially *Z. mobilis* and *S. cerevisiae*, are incapable of fermenting these sugars. Some yeast from the genera *Candida*, *Pichia*, *Pachysolen*, and *Schizosaccharomyces* assimilate pentose sugars [21]. However, some studies demonstrated that pentose-fermenting microorganisms are more susceptible to hemicellulose hydrolysate composition compared to hexose-fermenting ones [37].

The conditions applied during pretreatment of the lignocellulosic substrates produce different sub-products that can act as inhibitors in the production of ethanol, for example, HMF, other phenolic compounds, and organic acids. The amount and type of inhibitors that are formed depend primarily on the conditions used during pretreatment. Different techniques can be utilized to eliminate the action of these inhibitors, based on the removal or neutralizing that includes the use of ion exchange, active charcoal, evaporation, enzymatic action, or addition of $Ca(OH)_2$. The best method depends on its selectivity to remove the inhibitors, the cost, and the facility to use [3].

Other Factors

Some process characteristics, including pH and fermentation time, are also crucial in fermentation to ethanol. A short processing time reduces yield because it affects the microbial growth rate. On the other hand, long times cause inhibition by the product. A longer time is required when high loads of sugar are used as a substrate [20, 35].

One of the major problems for long-term fermentations is the possibility of bacterial contamination [35], which is directly affected by medium composition and pH. The latter factor also influences yeast growth and metabolism, which consequently impacts byproduct formation and ethanol production. The H^+ concentration may affect the permeability of some nutrients into the cell. For yeast, the ideal pH is approximately 4 to 5, with some of them favoring even more acidic pH, around 3 to 5. Lower pH influences the nutrient transport of cell membranes. The main factors that affect bioethanol fermentation and its solutions are summarized in Table **2**.

Table 2. Main factors that influence bioethanol production.

	Issue	Influence	Common Solutions
Medium composition	Initial sugar concentration	Osmotic stress on cells	Fed-batch processes; a lower concentration of sugars
	Presence of inhibitory compounds	Damage to cells	Detoxification of hydrolysates and biomass slurry
Temperature	High temperatures	Denaturation of proteins and ribosomes; cell stress	Use of thermotolerant microorganisms; genetic engineering to increase resistance
Product Accumulation		Toxicity to cells	Shorter process time; continuous fermentation.

(Table 2) cont.....

	Issue	**Influence**	**Common Solutions**
pH	Changes in H+ concentration	Changes in nutrient transport through the membrane	pH control during the process
	Contamination	Bacterial contamination at higher pH	

IMMOBILIZATION OF CELLS AS A SOLUTION TO REDUCE BIOETHANOL PRODUCTION INHIBITION

Ethanol fermentation is traditionally performed using free yeast cells in the fed-batch Melle-Boinot process. The repeated batches strategy has been employed by using self-flocculant strains. The immobilization of yeast cells in support materials permits the use of a continuous process for ethanol production. It allows the retention of the biocatalysts inside the reactor, a design that enhances yields and productivities and reduces bioreactor volumes and costs. However, there are some reports about the continuous fermentation process with immobilized cells, which are still scarce due to technical and economic problems [6, 38, 39].

The immobilization of yeast cells would increase cell stability, tolerance to highly concentrated substrates, ethanol yield, and volumetric productivity, reduce final product inhibition, contamination risks, and downstream costs, and protect against toxic substances and the possibility of cell recycling for repeated batches [3]. These characteristics are of special interest when performing fermentation of lignocellulosic ethanol.

Continuous immobilized yeast fermentation systems are conducted in five types of bioreactors, where the carrier is solid, the medium is the liquid, and feed gases are the gas phases. Packed-bed reactors, fluidized-bed reactors, air-lift reactors, stirred-tank reactors, and membrane reactors can be employed [38]. However, there is a limit on the use of such configurations due to the low cost and high productivity requirements for the bioethanol fermentation process.

Immobilization techniques are divided into four categories: a) attachment/ adsorption to solid surfaces (wood chips, brewer's spent grains, porous glass, loofah sponge, orange peels, corncob pieces, sorghum bagasse, sugarcane stalk, rice flour, cashew apple bagasse); b) entrapment within a porous matrix (agar, calcium alginate, k-carrageenan, polyacrylamide, alumina, PVA gel, cellulose, carboxymethyl cellulose, chitosan, gelatin); c) physical barriers (membrane filters and microcapsules); d) self-aggregation by flocculation [3, 4, 6, 40, 41].

Self-flocculation is cell aggregation by cell-to-cell adherence and formation of flocs. It is a simpler and cheaper technique when compared to the use of support materials [3]. This technique has been used in some Brazilian refineries with a flocculent *S. cerevisiae* strain. It permits the biomass recovery by its sedimentation in settlers and reduces the costs associated with centrifuges for cell separation in other related continuous processes. Self-flocculation is still a very simple and efficient immobilization when compared to the use of supporting materials, but it strongly depends on the flocculation characteristic of the strain [6]. The flocculant strains can protect each other from the hydrolysate inhibitors by forming flocs where the external layers of cells protect the inner layers of cells [6]. Many *S. cerevisiae* wild strains have the natural capacity to flocculate, a feature that could be isolated from the environment or even ethanol fermentation tanks.

Yeast immobilization permits the reusability of the biocatalysts from 3 to 15 cycles—until the cells are dead or yields are reduced—using supporting materials like sugarcane bagasse, lyophilized cellulose gel, and others [4]. Chen *et al.* [42] demonstrated that the immobilization of *S. cerevisiae* in a fibrous bed bioreactor increased ethanol productivity by 41.9% (6.48 g/L/h) during 22 cycles of repeated batches. Watanabe *et al.* [16] performed *S. cerevisiae* immobilization by entrapping the cells in resin beads, for a repeated batch simultaneous saccharification and fermentation process. The strains were maintained stable for five repeated cycles and produced 38 g/L of ethanol. Singh *et al.* [40] achieved the yields of 0.44 and 0.33 g of ethanol per g substrate using immobilized yeast cells in sugarcane bagasse and agar-agar for 10 and four repeated batches, respectively. Free and immobilized yeast cells produced in 0.35 and 0.38 g ethanol per g substrate for a repeated batch fermentation process using corn meal as hydrolysates as substrate and Ca-alginate beads as supporting material. The authors found that the immobilization resulted in higher ethanol tolerance and productivity, as well as lower substrate inhibition effects [41].

The immobilization of yeast cells for ethanol production has the potential to increase volumetric productivity and minimize the production costs by the reduction of the inhibition effects. The viability of immobilization techniques should be evaluated together with bioreactor and bioprocess design, the development of better supporting materials, and flocculent strains to overcome the limitations of this approach for full-scale bioethanol production [6].

GENETIC MODIFICATION OF BIOETHANOL-PRODUCING STRAINS

As presented above, there are some challenges to yeast during ethanol fermentation. Optimal temperatures stimulate yeast growth and metabolism, but higher temperatures (35–45°C) may retard the process or even inhibit growth [43, 44]. Yeast growth rate and metabolism are significantly affected by temperature. A factor that influences ethanol production is the inhibition of yeast growth [45, 46]. High alcohol levels in the medium lead to ethanol production inhibition [47]. Another problem with bioethanol fermentation is the inability of yeast to ferment pentose sugars. Ethanol production at the industrial scale would certainly be augmented with the use of yeast that can tolerate high ethanol titers in the medium [20, 48].

Several methods have been employed for the improvement of the traditional ethanol-producing strains. Traditional and modern techniques, including the conventional methodologies of mutagenesis and screening as well as evolutionary engineering, genome shuffling, and gTME strategies, have been used to increase yields and productivities and result in desired characteristics [6].

Hybrid yeast strains that can ferment pentose may be obtained by genetic engineering (Table **3**). They usually ferment pentose and hexose sugars to ethanol. Protoplast fusion was employed to obtain a *S. cerevisiae* and xylose-fermenting yeasts such as *Pachysolen tannophilus, Candida Shehatae*, and *P. stipitis* [49]. A genetically engineered *S. cerevisiae* and co-culture of two strains were developed to produce bioethanol from xylose with high yield [20]. Recombinant DNA technology was also used as a solution to upregulate stress tolerance genes to overcome inhibitory situations [50]. Xylose reductase and xylitol dehydrogenase genes from *Scheffersomyces stipitis* were introduced into *S. cerevisiae* to develop a strain that could ferment xylose [20].

Engineered yeast strains can convert cellulose to ethanol more efficiently compared to unmodified yeast strains. Some yeast can catabolize hexoses, such as glucose, into ethanol, without proceeding to the final oxidation product, which is CO_2. "Crabtree-positive" yeast, such as *S. cerevisiae*, accumulate ethanol under oxygenated conditions. Other yeast, such as *Candia Albicans,* which is Crabtree negative, catabolize sugars into CO_2 in the presence of oxygen. Six-carbon carbohydrates repress the oxidative respiration pathway in Crabtree-positive yeast where energy for growth is generated *via* glycolysis. Alcohol dehydrogenase plays an important role in fermentative metabolism (ADH1; EC 1.1.1.1). *S. cerevisiae* contains two genes that encode ADH. ADH1 is expressed constitutively and is responsible for the catalysis of the reduction of acetaldehyde to ethanol during the fermentation of glucose and also the reverse reaction, but with lower efficiency. The expression of ADH2 is induced by reduced concentrations of intracellular glucose; its substrate is ethanol [51]. ADH1 and ADH2 were sequenced, and transcriptome analysis revealed the structure and DNA binding elements [52]. Furthermore, the ADH gene was subjected to advanced synthetic biology studies, which have focused on re-engineering for greater substrate specificity and improvement of catalytic activity as well as engineering the yeast genome with protein-coding genes. These advancements are expected to improve tolerance to ethanol and promote the ability of the selected yeast strains to catalyze a wide range of carbon sources. Novel genes that encode ADHs are being studied using metagenomic approaches, with the generation of a representative number of variants [20].

The implementation of high cell density fermentation, especially for continuous and repeated batch operational conditions, pushed the development of yeast cells with the self-flocculation capacity. This research produced the rapid sediment from the fermenting broth. Genetic engineering efforts have been made to transfer the flocculation capacity from other strains to *S. cerevisiae*. In China, a pilot plant operation for ethanol production was validated using the flocculent *Saccharomyces* strain SPSC01, developed by the fusion of a flocculent *Schizosaccharomyces pombe* with *S. cerevisiae* K2 [6, 53].

Table 3. Engineered strains for bioethanol production.

Strain	Improvement	Medium	Ethanol Yield	References
Candida shehatae NCL 3501	Co-ferments xylose and glucose	Rice straw	0.45 g/g by autohydrolysis 0.5 g/g by immobilized cells	[54]
Clostridium thermocellum DSM 1313	Higher ethanol production	Not specified	0.8 g/L at 0.5 g/L cellobiose	[55]
Clostridium thermocellum YD01	Higher ethanol production	Not specified	1.33 mol ethanol/mol glucose	[56]
Clostridium thermocellum YD02	Higher ethanol production	Not specified	1.28 mol ethanol/mol glucose	[56]
Escherichia coli KO11	Ferments xylose and glucose	Sugarcane bagasse	31.5 g/L (91.5%)	[57]
Escherichia coli FBR5	Ferments xylose	Xylose	0.5 g/g xylose	[58]
Escherichia coli FBR5	Ferments xylose and arabinose	Rice hull	2.25% (w/v)	[59]
Pichia stipitis A	Adapted to high concentrations of hydrolysate	Wheat straw	0.41 gp/gs	[60]
Pichia stipitis NRRL Y-7124	Adapted to high concentrations of hydrolysate	Wheat straw	0.35 gp/gs	[60]
Pichia stipitis BCC15191	Ferments xylose and glucose	Sugarcane bagasse	8.4 g/L after 24 h fermentation	[61]

(Table 3) cont.....

Strain	Improvement	Medium	Ethanol Yield	References
Saccharomyces cerevisiae D5a	Improved ethanol yield	Rice hull	0.58% (w/v) or 100% theoretical yield	[59]
Saccharomyces cerevisiae 590. E1	Ferments glucose and cellobiose	Whatman paper	1.09% from 2% glucose 1.16% from 2% cellobiose	[62]
Saccharomyces cerevisiae 590. E1	Ferments cellulose without additional enzymatic hydrolysis	Corn stover	63% theoretical ethanol	[62]
Saccharomyces cerevisiae RWB 217	Ferments xylose and glucose	2% glucose + 2% xylose	0.43 g/g of sugars	[63]
Saccharomyces cerevisiae RWB 218	Ferments xylose and glucose	2% glucose + 2% xylose	0.4 g/g of sugars	[63]
Zymomonas mobilis ZM4 (pZB5)	Ferments both xylose and glucose	Stillage	11 g/L with supplementation of 10 g/L glucose 28 g/L with supplementation of 5 g/L yeast extract and 40 g/L glucose	[64]
Zymomonas mobilis AX 101	Ferments glucose, xylose, and arabinose	Various agriculture wastes	3.54 g/L/h (with no acetic acid) 1.17 g/L/h (with acetic acid)	[65]
Thermoanaerobacterium saccharolyticum ALK2	Improved ethanol yield; ferments glucose, xylose, mannose, and arabinose	Not specified	37 g/L	[66]
Thermoanaerobacterium mathranii BG1L1	Improved ethanol yield	Wheat straw	0.39-0.42 g/g sugars	[67]

Modified from Aditiya *et al.* [21].

CONCLUSION

Yeast is a commonly employed microorganism in bioethanol production; they can use different feedstocks as a substrate. However, some feedstocks require physical and chemical pretreatment, which generates components, including pentose sugars, which are not assimilated by ethanol-producing yeasts. Besides, inhibitory components are liberated and drastically affect bioethanol synthesis. As a solution, some interesting strategies are being employed that may ameliorate some of these problems. One of them is the choice of performant yeast strains that have been selected by their ability to produce ethanol from different types of feedstocks. Another would be the use of continuous fermentation methods with cell immobilization that leads to high productivities, avoiding cell inhibition by the presence of high concentrations of ethanol in the medium. In this case, yeast cell immobilization is performed by an adsorption method with the use of calcium alginate as a yeast carrier. Cell immobilization has shown some advantages in ethanol production: high cell density, easy separation from the medium, high substrate conversion, less inhibition, short reaction time, and cell recycling. Finally, yeast genetic modification may provide advantages for bioethanol production, such as gains in productivity and desired characteristics.

CONSENT FOR PUBLICATION

Not applicable.

CONFLICT OF INTEREST

The author(s) confirms that there is no conflict of interest.

ACKNOWLEDGEMENTS

Author(s) want to thank CAPES and CNPq, Brazil, for financial support.

REFERENCES

[1]   Rastogi M, Shrivastava S. Recent advances in second-generation bioethanol production: An insight to pretreatment, saccharification and fermentation processes. Renew Sustain Energy Rev 2017; 80: 330-40.
[http://dx.doi.org/10.1016/j.rser.2017.05.225]

[2]   Phaiboonsilpa N, Chysirichote T, Champreda V, Laosiripojana N. Fermentation of xylose, arabinose, glucose, their mixtures and sugarcane bagasse hydrolyzate by yeast *Pichia stipitis* for etanol production. Energy Reports 2020; 6: 710-3.
[http://dx.doi.org/10.1016/j.egyr.2019.11.142]

[3]   Zabed H, Sahu JN, Suely A, Boyce AN, *et al.* Bioethanol production from renewable sources: Current perspectives and technological progress. Renew Sustain Energy Rev 2017; 71: 475-501.
[http://dx.doi.org/10.1016/j.rser.2016.12.076]

[4]   Tesfaw A, Assefa F. Current Trends in Bioethanol Production by *Saccharomyces cerevisiae*: Substrate, Inhibitor Reduction, Growth Variables, Coculture, and Immobilization. Int Sch Res Notices 2014; 2014532852
[http://dx.doi.org/10.1155/2014/532852] [PMID: 27379305]

[5]   Vohra M, Manwar J, Manmode R, Padgilwar S, *et al.* Bioethanol production: Feedstock and current technologies. J Environ Chem Eng 2014; 2: 573-84.
[http://dx.doi.org/10.1016/j.jece.2013.10.013]

[6]   Mussatto SI, Dragone G, Guimarães PMR, *et al.* Technological trends, global market, and challenges of bio-ethanol production. Biotechnol Adv 2010; 28(6): 817-30.
[http://dx.doi.org/10.1016/j.biotechadv.2010.07.001] [PMID: 20630488]

[7]   Available from: https://www.prnewswire.com/news-releases/global-cassava-processing-market-r-port-2018-2023-industry-trends-share-size-growth-opportunity-and-forecasts-300639999.html [cited June 25th, 2018].

[8]   Statistica. Available from: https://www.statista.com/statistics/812343/global-sweet-potato-production/ [cited June 25th, 2018].

[9]   Saini JK, Saini R, Tewari L. Lignocellulosic agriculture wastes as biomass feedstocks for second-generation bioethanol production: concepts and recent developments. 3 Biotech 2015; 5: 337-53.

[10]   Goldenberg J, Guardabassi P. The potential for first-generation ethanol production from sugarcane. Biofuels Bioprod Biorefin 2010; 4: 17-24.
        [http://dx.doi.org/10.1002/bbb.186]

[11]   Manochio C, Andrade BR, Rodriguez RP, Moraes BS. Ethanol from biomass: A comparative overview. Renew Sustain Energy Rev 2017; 80: 743-55.
        [http://dx.doi.org/10.1016/j.rser.2017.05.063]

[12]   Wang S, Sun X, Yuan Q. Strategies for enhancing microbial tolerance to inhibitors for biofuel production: A review. Bioresour Technol 2018; 258: 302-9.
        [http://dx.doi.org/10.1016/j.biortech.2018.03.064] [PMID: 29567023]

[13]   Mitchell VD, Taylor CM, Bauer S. Comprehensive analysis of monomeric phenolics in diluted acid plants hydrolysates. Bioenergy Resources 2014; 7: 654-69.
        [http://dx.doi.org/10.1007/s12155-013-9392-6]

[14]   Cavka A, Jönsson LJ. Detoxification of lignocellulosic hydrolysates using sodium borohydride. Bioresour Technol 2013; 136: 368-76.
        [http://dx.doi.org/10.1016/j.biortech.2013.03.014] [PMID: 23567704]

[15]   Jönsson LJ, Martín C. Pretreatment of lignocellulose: Formation of inhibitory by-products and strategies for minimizing their effects. Bioresour Technol 2016; 199: 103-12.
        [http://dx.doi.org/10.1016/j.biortech.2015.10.009] [PMID: 26482946]

[16]   Sarkar N, Gosh SK, Bannerjee S, Aikat K. Bioethanol production from agricultural wastes. Renew Energy 2012; 37: 19-27.
        [http://dx.doi.org/10.1016/j.renene.2011.06.045]

[17]   Taherzadeh MJ, Karimi L. Acid-based hydrolysis processes for ethanol from lignocellulosic materials: a review. BioResources 2007; 2: 472-99.

[18]   van der Pol EC, Bakker RR, Baets P, Eggink G. By-products resulting from lignocellulose pretreatment and their inhibitory effect on fermentations for (bio)chemicals and fuels. Appl Microbiol Biotechnol 2014; 98(23): 9579-93.
        [http://dx.doi.org/10.1007/s00253-014-6158-9] [PMID: 25370992]

[19]   Pampulha ME, Loureiro-Dias MC. Energetics of the effect of acetic acid on growth of *Saccharomyces cerevisiae*. FEMS Microbiol Lett 2000; 184(1): 69-72.
        [http://dx.doi.org/10.1111/j.1574-6968.2000.tb08992.x] [PMID: 10689168]

[20]   Mohd Azhar SH, Abdulla R, Jambo SA, *et al.* Yeasts in sustainable bioethanol production: A review. Biochem Biophys Rep 2017; 10: 52-61.
        [http://dx.doi.org/10.1016/j.bbrep.2017.03.003] [PMID: 29114570]

[21]   Aditiya HB, Mahlia TMI, Chong WT, Nur H, *et al.* Second generation ethanol production: a critical review. Renew Sustain Energy Rev 2016; 66: 631-53.
        [http://dx.doi.org/10.1016/j.rser.2016.07.015]

[22]   Kim JS, Lee YY, Kim TH. A review on alkaline pretreatment technology for bioconversion of lignocellulosic biomass. Bioresour Technol 2016; 199: 42-8.
        [http://dx.doi.org/10.1016/j.biortech.2015.08.085] [PMID: 26341010]

[23]   Zhang K, Pei Z, Wang D. Organic solvent pretreatment of lignocellulosic biomass for biofuels and biochemicals: A review. Bioresour Technol 2016; 199: 21-33.
        [http://dx.doi.org/10.1016/j.biortech.2015.08.102] [PMID: 26343573]

[24]   Zhao X, Cheng K, Liu D. Organosolv pretreatment of lignocellulosic biomass for enzymatic

hydrolysis. Appl Microbiol Biotechnol 2009; 82(5): 815-27.
[http://dx.doi.org/10.1007/s00253-009-1883-1] [PMID: 19214499]

[25]   Lienqueo ME, Ravanal MC, Pezoa-Conte R, *et al.* Second generation bioethanol from *eucalyptus globulus labill* and *nothofagus pumilio*: Ionic liquid pretreatment boosts the yields. Ind Crops Prod 2016; 80: 148-55.
[http://dx.doi.org/10.1016/j.indcrop.2015.11.039]

[26]   Silveira MHL, Morais ARC, da Costa Lopes AM, *et al.* Current pretreatment technologies for the development of cellulosic ethanol and biorefineries. ChemSusChem 2015; 8(20): 3366-90.
[http://dx.doi.org/10.1002/cssc.201500282] [PMID: 26365899]

[27]   Li C, Tanjore D, He W, *et al.* Scale-up and evaluation of high solid ionic liquid pretreatment and enzymatic hydrolysis of switchgrass. Biotechnol Biofuels 2013; 6(1): 154.
[http://dx.doi.org/10.1186/1754-6834-6-154] [PMID: 24160440]

[28]   Dickinson Q, Bottoms S, Hinchman L, *et al.* Mechanism of imidazolium ionic liquids toxicity in *Saccharomyces cerevisiae* and rational engineering of a tolerant, xylose-fermenting strain. Microb Cell Fact 2016; 15: 17.
[http://dx.doi.org/10.1186/s12934-016-0417-7] [PMID: 26790958]

[29]   Sindhu R, Binod P, Pandey A. Biological pretreatment of lignocellulosic biomass--An overview. Bioresour Technol 2016; 199: 76-82.
[http://dx.doi.org/10.1016/j.biortech.2015.08.030] [PMID: 26320388]

[30]   Potumarthi R, Baadhe RR, Nayak P, Jetty A. Simultaneous pretreatment and sacchariffication of rice husk by Phanerochete chrysosporium for improved production of reducing sugars. Bioresour Technol 2013; 128: 113-7.
[http://dx.doi.org/10.1016/j.biortech.2012.10.030] [PMID: 23196230]

[31]   Dhiman SS, Haw JR, Kalyani D, Kalia VC, Kang YC, Lee JK. Simultaneous pretreatment and saccharification: green technology for enhanced sugar yields from biomass using a fungal consortium. Bioresour Technol 2015; 179: 50-7.
[http://dx.doi.org/10.1016/j.biortech.2014.11.059] [PMID: 25514402]

[32]   Techaparin A, Thanonkeo P, Klanrit P. High-temperature ethanol production using thermotolerant yeast newly isolated from Greater Mekong Subregion. Braz J Microbiol 2017; 48(3): 461-75.
[http://dx.doi.org/10.1016/j.bjm.2017.01.006] [PMID: 28365094]

[33]   Balat M. Production of bioethanol from lignocellulosic materials *via* the biochemical pathway: a review. Energy Convers Manage 2011; 52: 858-75.
[http://dx.doi.org/10.1016/j.enconman.2010.08.013]

[34]   Choudhary J, Singh S, Nain L. Thermotolerant fermenting yeasts for simultaneous saccharification and fermentation of lignocellulosic biomass. Electron J Biotechnol 2016; 21: 82-92.
[http://dx.doi.org/10.1016/j.ejbt.2016.02.007]

[35]   Lin Y, Zhang W, Li C, *et al.* Factors affecting ethanol fermentation using *Saccharomyces cerevisiae* BY4742. Biomass Bioenergy 2012; 47: 395-401.
[http://dx.doi.org/10.1016/j.biombioe.2012.09.019]

[36]   Ajit A, Sulaiman AZ, Chisti Y. Production of bioethanol by *Zymomonas mobilis* in high-gravity extractive fermentations. Food Bioprod Process 2017; 102: 123-35.
[http://dx.doi.org/10.1016/j.fbp.2016.12.006]

[37]   Klinke HB, Thomsen AB, Ahring BK. Inhibition of ethanol-producing yeast and bacteria by degradation products produced during pre-treatment of biomass. Appl Microbiol Biotechnol 2004; 66(1): 10-26.
[http://dx.doi.org/10.1007/s00253-004-1642-2] [PMID: 15300416]

[38]   Verbelen PJ, De Schutter DP, Delvaux F, Verstrepen KJ, Delvaux FR. Immobilized yeast cell systems for continuous fermentation applications. Biotechnol Lett 2006; 28(19): 1515-25.

[http://dx.doi.org/10.1007/s10529-006-9132-5] [PMID: 16937245]

[39]   Brányik T, Vicente AA, Dostálek P, Teixeira JA. Continuous beer fermentation using immobilized yeast cell bioreactor systems. Biotechnol Prog 2005; 21(3): 653-63.
[http://dx.doi.org/10.1021/bp050012u] [PMID: 15932239]

[40]   Singh A, Sharma P, Saran AK, Singh N, *et al.* Comparative study on ethanol production from pretreated sugarcane bagasse using immobilized *Saccharomyces cerevisiae* on various matrices. Renew Energy 2013; 50: 488-93.
[http://dx.doi.org/10.1016/j.renene.2012.07.003]

[41]   Nikolić S, Mojović L, Pejin D, Rakin M, Vukašinović M. Production of bioethanol from cornmeal hydrolyzates by free and immobilized cells of *Saccharomyces cerevisiae* var. *ellipsoideus.* Biomass Bioenergy 2010; 34: 1449-56.
[http://dx.doi.org/10.1016/j.biombioe.2010.04.008]

[42]   Chen Y, Liu Q, Zhou T, *et al.* Ethanol production by repeated batch and continuous fermentations by *Saccharomyces cerevisiae* immobilized in a fibrous bed bioreactor. J Microbiol Biotechnol 2013; 23(4): 511-7.
[http://dx.doi.org/10.4014/jmb.1209.09066] [PMID: 23568205]

[43]   Watanabe I, Miyata N, Ando A, Shiroma R, Tokuyasu K, Nakamura T. Ethanol production by repeated-batch simultaneous saccharification and fermentation (SSF) of alkali-treated rice straw using immobilized *Saccharomyces cerevisiae* cells. Bioresour Technol 2012; 123: 695-8.
[http://dx.doi.org/10.1016/j.biortech.2012.07.052] [PMID: 22939189]

[44]   Tofighi A, Mazaheri Assadi M, Asadirad MHA, Zare Karizi S. Bio-ethanol production by a novel autochthonous thermotolerant yeast isolated from wastewater. J Environ Health Sci Eng 2014; 12: 107.
[http://dx.doi.org/10.1186/2052-336X-12-107] [PMID: 25937930]

[45]   Alexandre H, Charpentier C. Biochemical aspects of stuck and sluggish fermentation in grape must. J Ind Microbiol Biotechnol 1998; 20: 20-7.
[http://dx.doi.org/10.1038/sj.jim.2900442]

[46]   Attfield PV. Stress tolerance: the key to effective strains of industrial baker's yeast. Nat Biotechnol 1997; 15(13): 1351-7.
[http://dx.doi.org/10.1038/nbt1297-1351] [PMID: 9415886]

[47]   Fiedurek J, Skowronek M, Gromada A. Selection and adaptation of Saccharomyces cerevisae to increased ethanol tolerance and production. Pol J Microbiol 2011; 60(1): 51-8.
[http://dx.doi.org/10.33073/pjm-2011-007] [PMID: 21630574]

[48]   Fonseca GG, Heinzle E, Wittmann C, Gombert AK. The yeast *Kluyveromyces marxianus* and its biotechnological potential. Appl Microbiol Biotechnol 2008; 79(3): 339-54.
[http://dx.doi.org/10.1007/s00253-008-1458-6] [PMID: 18427804]

[49]   Kumari R, Pramanik K. Bioethanol production from Ipomoea carnea biomass using a potential hybrid yeast strain. Appl Biochem Biotechnol 2013; 171(3): 771-85.
[http://dx.doi.org/10.1007/s12010-013-0398-5] [PMID: 23892623]

[50]   Doğan A, Demirci S, Aytekin AO, Şahin F. Improvements of tolerance to stress conditions by genetic engineering in *Saccharomyces cerevisiae* during ethanol production. Appl Biochem Biotechnol 2014; 174(1): 28-42.
[http://dx.doi.org/10.1007/s12010-014-1006-z] [PMID: 24908051]

[51]   Raj SB, Ramaswamy S, Plapp BV. Yeast alcohol dehydrogenase structure and catalysis. Biochemistry 2014; 53(36): 5791-803.
[http://dx.doi.org/10.1021/bi5006442] [PMID: 25157460]

[52]   Alper H, Moxley J, Nevoigt E, Fink GR, Stephanopoulos G. Engineering yeast transcription machinery for improved ethanol tolerance and production. Science 2006; 314(5805): 1565-8.

[http://dx.doi.org/10.1126/science.1131969] [PMID: 17158319]

[53] Zhao XQ, Bai FW. Yeast flocculation: New story in fuel ethanol production. Biotechnol Adv 2009; 27(6): 849-56.
[http://dx.doi.org/10.1016/j.biotechadv.2009.06.006] [PMID: 19577627]

[54] Abbi M, Kuhad RC, Singh A. Fermentation of xylose and rice straw hydrolysate to ethanol by *Candida shehatae* NCL-3501. J Ind Microbiol 1996; 17(1): 20-3.
[http://dx.doi.org/10.1007/BF01570143] [PMID: 8987687]

[55] Tripathi SA, Olson DG, Argyros DA, *et al.* Development of pyrF-based genetic system for targeted gene deletion in Clostridium thermocellum and creation of a pta mutant. Appl Environ Microbiol 2010; 76(19): 6591-9.
[http://dx.doi.org/10.1128/AEM.01484-10] [PMID: 20693441]

[56] Deng Y, Olson DG, Zhou J, Herring CD, Joe Shaw A, Lynd LR. Redirecting carbon flux through exogenous pyruvate kinase to achieve high ethanol yields in *Clostridium thermocellum*. Metab Eng 2013; 15: 151-8.
[http://dx.doi.org/10.1016/j.ymben.2012.11.006] [PMID: 23202749]

[57] Takahashi C, Lima K, Takahashi D, Alterthum F. Fermentation of sugar cane bagasse hemicellulosic hydrolysate and sugar mixtures to ethanol by recombinant *Escherichia coli* KO11. World J Microb Biot 2000; p. 16.

[58] Qureshi N, Dien BS, Nichols NN, Saha BC, Cotta MA. Genetically engineered *Escherichia coli* for ethanol production from xylose: substrate and product inhibition and kinetic parameters. Food Bioprod Process 2006; 84: 114-22.
[http://dx.doi.org/10.1205/fbp.05038]

[59] Nichols NN, Hector RE, Saha BC, Frazer SE, Kennedy GJ. Biological abatement of inhibitors in rice hull hydrolyzate and fermentation to ethanol using conventional and engineered microbes. Biomass Bioenergy 2014; 67: 79-88.
[http://dx.doi.org/10.1016/j.biombioe.2014.04.026]

[60] Nigam JN. Ethanol production from wheat straw hemicellulose hydrolysate by *Pichia stipitis*. J Biotechnol 2001; 87(1): 17-27.
[http://dx.doi.org/10.1016/S0168-1656(00)00385-0] [PMID: 11267696]

[61] Buaban B, Inoue H, Yano S, *et al.* Bioethanol production from ball milled bagasse using an on-site produced fungal enzyme cocktail and xylose-fermenting *Pichia stipitis*. J Biosci Bioeng 2010; 110(1): 18-25.
[http://dx.doi.org/10.1016/j.jbiosc.2009.12.003] [PMID: 20541110]

[62] Xin Qing Z, Qian L, Lei Yu H, *et al.* Exploration of a natural reservoir of flocculating genes from various *Saccharomyces cerevisiae* strains and improved ethanol fermentation using stable genetically engineered flocculating yeast strains. Process Biochem 2012; 47: 1612-9.
[http://dx.doi.org/10.1016/j.procbio.2011.06.009]

[63] Kuyper M, Toirkens MJ, Diderich JA, Winkler AA, van Dijken JP, Pronk JT. Evolutionary engineering of mixed-sugar utilization by a xylose-fermenting *Saccharomyces cerevisi*ae strain. FEMS Yeast Res 2005; 5(10): 925-34.
[http://dx.doi.org/10.1016/j.femsyr.2005.04.004] [PMID: 15949975]

[64] Davis L, Jeon Y-J, Svenson C, Roger P, Pearce J, Peiris P. Evaluation of wheat stillage for ethanol production by recombinant *Zymomonas mobilis*. Biomass Bioenergy 2005; 29: 49-59.
[http://dx.doi.org/10.1016/j.biombioe.2005.02.006]

[65] Lawford H, Rousseau J. Performance testing of Zymomonas mobilis metabolically engineered for cofermentation of glucose, xylose, and arabinose. Biotechnology for Fuels and Chemicals. 1rst ed.. Humana Press 2002; pp. 429-8.

[66] Shaw AJ, Podkaminer KK, Desai SG, *et al.* Metabolic engineering of a thermophilic bacterium to

produce ethanol at high yield. Proc Natl Acad Sci USA 2008; 105(37): 13769-74.
[http://dx.doi.org/10.1073/pnas.0801266105] [PMID: 18779592]

[67]   Georgieva TI, Mikkelsen MJ, Ahring BK. Ethanol production from wet-exploded wheat straw hydrolysate by thermophilic anaerobic bacterium *Thermoanaerobacter* BG1L1 in a continuous immobilized reactor. Appl Biochem Biotechnol 2008; 145(1-3): 99-110.
[http://dx.doi.org/10.1007/s12010-007-8014-1] [PMID: 18425616]

CHAPTER 10

Toxicity and Structural Activity Relationship of Persistent Organic Pollutants

Ankur Khare[1,2], Pradip Jadhao[1,2], Sonam Paliya[1,2] and Kanchan Kumari[1,2,*]

Environmental Impact and Sustainability Division, CSIR-NEERI, Nagpur, India

Academy of Scientific and Innovative Research (AcSIR), Ghaziabad-201002, India

Abstract: Persistent Organic Pollutants (POPs) are organic compounds of mainly anthropogenic origin posing a huge threat to human health and the ecosystem. Though the production and intended use of major POPs are banned by Stockholm Convention, still some POPs are being used in most developing countries around the globe. Although, environmental levels of some POPs, such as Polychlorinated biphenyls (PCBs) have declined, newly added POPs in the list of conventions, such as Polybrominated Diphenyl Ethers (PBDEs), Perfluorooctanesulfonate (PFOS) have emerged as new challenges. Exposure to POPs has been associated with a wide spectrum of health effects, including developmental, carcinogenic immunologic, reproductive, and neurotoxic effects. It is of major concern that the neurotoxic effects of some POPs have been observed in humans at low environmental concentrations. This chapter focuses on various POPs like PCBs, PBDEs, and PFOS as a representative chemical class of POPs and discusses the possible modes (s) of action for the neurotoxic effects with an emphasis on comparing dose-response and structure-activity relationships (SAR) with other structurally related chemicals. There are sufficient epidemiological and experimental studies carried out in different parts around the globe showing that PBDEs and PFOS exposure is associated with motor and cognitive deficits in humans and animal models. Several potential mechanisms were presumed for PBDEs and PFOs induced neurotoxic effects and alteration in neurotransmitter systems. Among them, the intracellular signaling processes and hormonal imbalance impacting the activity of thyroid hormone were reported as predominant. All these potential mechanisms are discussed in detail in the chapter. In addition to this, SAR will be highlighted for examining the toxicity of other relevant and structurally similar POPs to assess if they have a common mode(s) of action. Potency factors for several other POPs will also be described focusing on their effects on intracellular signaling processes and enzymatic activity and cell signaling pathways. This chapter is a comprehensive review, describes the alteration of enzymatic pathways and their associated toxicity at the biochemical level in different models for environmentally relevant POPs.

* **Corresponding author Kanchan Kumari:** Environmental Impact and Sustainability Division, CSIR-NEERI, Nagpur, Academy of Scientific and Innovative Research (AcSIR), Ghaziabad-201002, India; E-mail: onlinekanchan1@gmail.com

Keywords: Neurotoxicity, Persistent Organic Pollutants, Polychlorinated Biphenyls (PCBs), Polybrominated Diphenyl Ethers (PBDEs), Perfluorooctane Sulfonate(PFOS), Structure-Activity Relationships (SAR), Thyroid hormones.

INTRODUCTION

Persistent Organic Pollutants (POPs) are a group of persistent and extremely toxic chemicals in the environment even in trace amounts. They degrade very slowly. They are lipophilic (fat-loving), they have a tendency of bioaccumulation in adipose tissue of organisms and can travel from lower to higher trophic levels in the food chain. They have the ability of long-range transport and can be detected in the environment and biota of regions far away from its source of production. Therefore, they are a subject of regional, national and global concern. The United Nations Environment Programme (UNEP) Stockholm Convention (SC) on POPs was adopted in May 2001 and entered into force on17 May 2004. POPs are commonly known as Dirty Dozen which comprises of Polychlorinated biphenyls (PCBs), aldrin, dieldrin, endrin, polychlorinated dibenzofurans (PCDF), polychlorinated dibenzo-p-dioxins (PCDD), dichlorodiphenyltrichloroethane (DDT), hexachlorobenzene, mirex, toxaphene, chlordane, and heptachlor (Table **1**). Apart from the dirty dozen, some more chemicals were further added in the list of POPs called as new POPs. To protect human health and environment from POPs, article 16 of the SC has recommended the conference of the Parties, to check the effectiveness of the Convention. Apart from that, a Global Monitoring Plan (GMP) was also established by the SC to provide a framework for the necessary data collection on POPs from all regions. The main objective of this GMP was to identify the concentration of POPs over a time period. Level of POPs like PCBs, PBDEs have been accounted to be highest in the species at the top of the trophic level (polar bear, killer whales, eagles, and human beings). Apart from that, the concentration of POPs in every individual is accounted to be more than that of their ancestors [1 - 4]. Degradation of POPs occurs very slowly in the environment. They remain in the environment for a longer duration even if the provenance of POPs is instantly abolished. Scientific studies have revealed that a majority of the human population carries significant amounts of POPs in body fat, causing various health effects like endocrine and immune system disruption, cancer, reproductive and developmental problems, abnormal behavior and neurological disorders [5, 6]. POPs, upon their release in the environment, get transported by different transport media like air and water currents and reach places far from their origin point. This complete journey is a complex process consisting of numbers of "hops"; each hop is a sequence of three-stage which includes evaporation, atmospheric transportation followed by condensation at lower temperatures. Scientifically this complete phenomenon is termed as "grasshopper effect" (Fig. **1**). Similar to this process, POPs get transported to long distances and get widely distributed in a short period. Vast water reservoirs like oceans, glaciers, icebergs, and huge mountains are known to be the ultimate fate of such chemicals.

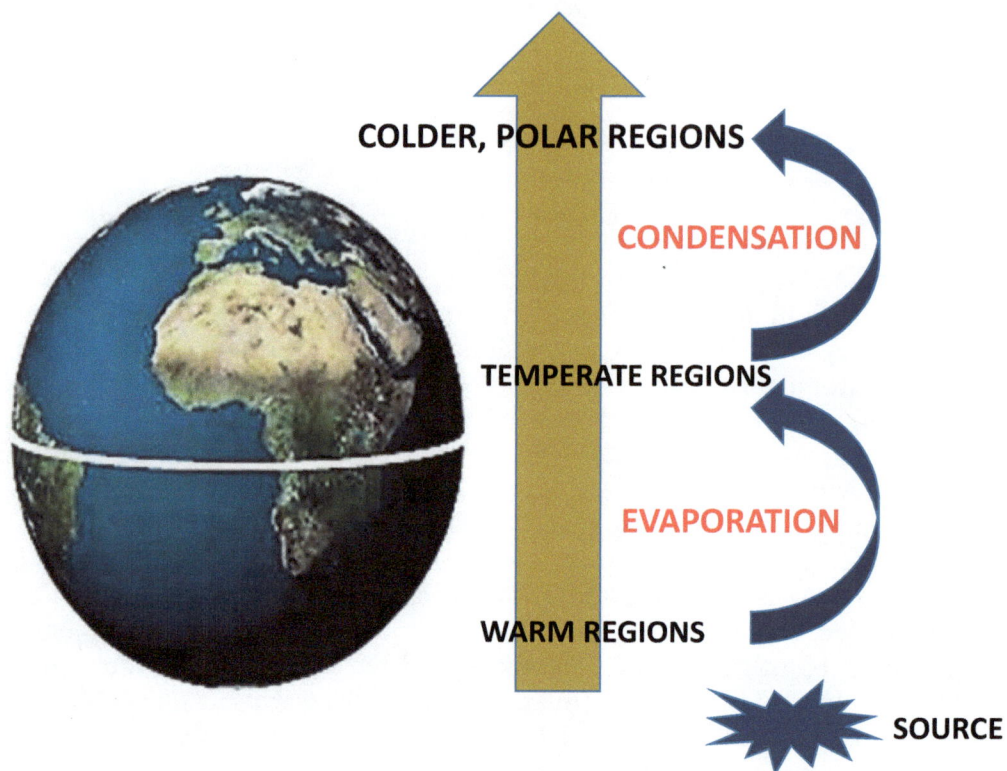

Fig. (1). Grasshopper Effect.

Table 1. Status of Twelve POPs in Stockholm Convention.

S.No.	Name	Listed Under	Category
1.	Aldrin	Annex A	Pesticide
2.	Chlordane	Annex A	Pesticide
3.	DDT	Annex B	Pesticide
4.	Dieldrin	Annex A	Pesticide
5.	Endrin	Annex A	Pesticide
6.	Heptachlor	Annex A	Pesticide
7.	Hexachlorobenzene	Annex A and Annex C	Pesticide/ Industrial Chemical
8.	Mirex	Annex A	Pesticide
9.	Toxaphene	Annex A	Pesticide

(Table 1) cont.....

S.No.	Name	Listed Under	Category
10.	Polychlorinated Biphenyls (PCBS)	Annex A with specific exemptions and under Annex C	Industrial Chemical
11.	Dioxin (PCDD)	Annex C	By-product
12.	Furans (PCDF)	Annex C	By-product

DIFFERENT POPS (PCBS, PBDES, AND PFOS) KNOWN AND ALTERATION IN ENZYMATIC ACTIVITIES

Although the intended use of POPs has been banned in many of the developed and developing countries, still traces of some chemicals are being detected in many parts around the world, leading to environmental degradation. Likewise, PCBs, whose environmental concentration has fallen to a large extent, still it was detected in the food chain due to a high degree of persistence. Similar is the case with some emerging POPs like PBDEs and PFOS. This chapter focuses on the toxicological behavior of these groups of chemicals, emphasizing their properties, toxicity mechanism, and Structural Activity Relationship (SAR) with their biological counterparts. A general overview of the compounds dealt in detail in this chapter, and their status in the Stockholm Convention is given in Table **2**.

Table 2. Status and Use of PCBs, PBDEs, and PFOS.

S.No	Chemical	Use	Listed in Stockholm Convention
1.	PCBs	Transformer oil	Annex A, with specific exemption in Annex C
2.	PBDEs	Electronic goods, Plastic Products	Annex A
3.	PFOS	Teflon coatings in non-stick cookware, Carpet Manufacturing	Annex B

Polychlorinated Biphenyls (PCBs)

Polychlorinated Biphenyls (PCBs) are man-made organochlorine compounds having 209 congeners. Due to the thermal stability and fire-resistant properties, it finds a significant application in the manufacture of transformer oils and also has several other industrial applications [7]. The Environmental Protection Agency (EPA) has restricted the production and use of PCBs in 1979 due to concern over the environment and human health [8]. PCBs are highly persistent in the environment; in fact, traces of PCBs are still being detected where its production was carried out more than a decade back [9]. PCBs are chemically synthesised by the attachment of one or more chlorine atoms at the Xs (Fig. **2**) to a pair of Benzene rings. At present, a leading source of PCBs exposure is the environmental recycling of PCBs which were previously released into the environment.

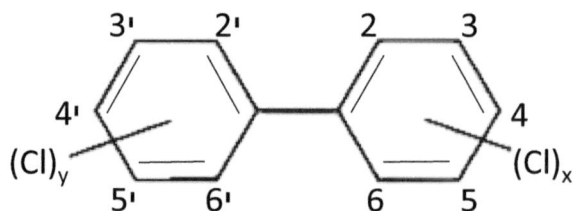

Fig. (2). General Structure of PCBs.

PCBs have ecotoxicological significance as environmental contaminants and listed under Annex A with specific exemptions under Annex C by the Stockholm Convention. PCBs bio-accumulate into the fat tissues due to its lipophilic properties. The route of exposure of PCB is mainly by the gastrointestinal tract after exposure *via* diet. PCBs impair steroidal production [10] and also have adverse effects like renal failure, hepatic, and reproductive disorders [11].

Polybrominated Diphenyl Ether (PBDEs)

PBDEs are most widely used as flame retardants having a wide range of applications like manufacturing of heat resistant tubing, electrical wires, and other household materials. The structure of PBDEs contains a diphenyl ring with bromine atoms. There are 209 different congeners of PBDEs known depending upon the position and number of bromine atoms. Mainly PBDEs have three commercial mixture *i.e.*, Penta-, Octa- and Deca-BDE (Fig. **3**)

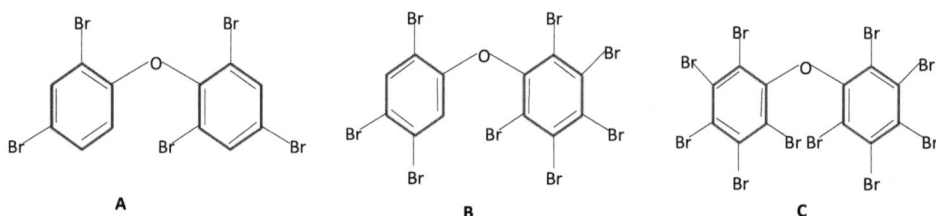

Fig. (3). (**A**) Penta BDE; (**B**) Octa BDE; (**C**) Deca BDE.

PBDEs are persistent in the environment leading to their accumulation in animal tissues, blood as well as in different life forms. The International Agency for Research on Cancer (IARC) also reported sufficient evidence of carcinogenicity in experimental animals due to PBDEs exposure. PBDEs are relatively similar to PCBs which are well known endocrine disruptors. All PBDEs are technical products, including decaBDE, which are known for showing thyroid disrupting properties. Alteration in the activity of several vitamins and thyroid hormones linked with liver dysfunction is also associated with PBDEs exposure [12, 13].

The liver serves as the primary site for the metabolism of xenobiotic compounds. PBDEs are known to biotransform in animals and form hydroxylated metabolites. The route of formation of hydroxylated metabolites in fishes is different from those of humans and rodents. This complete metabolism is mediated by the oxidative cytochrome P-450 pathway, which needs to be explored in detail. This peculiar metabolic pathway where oxidative cytochrome P450–mediated pathways dominate to produce hydroxylated PBDEs is distinguished from that of rodent and human metabolism.

Perfluorooctane Sulfonate (PFOS) and Perfluorooctanoic Acid (PFOA)

Perfluorooctane Sulfonate (PFOS) is a group of compounds comprising of eight to ten carbon atoms with the sulfonate group (Fig. **4**). Perfluorooctanoic acids (PFOA) are simple organic acids formed synthetically due to the degradation of precursor compounds like fluorotelomeric alcohols. The chemicals have surface-active properties due to which they are used in a variety of industrial applications. These compounds find their applications in stain-resistant coatings for carpets and fabrics, paints, fire-fighting foams, insecticides formulations, floor polishes, oil well surfactants, *etc*. The compounds of PFOS are generally formed by the degradation of higher compounds or high molecular weight polymers of PFOS that can be simple salts like lithium, potassium, ammonium or polymers containing PFOS. Due to the polarity of Per Fluorinated Compounds (PFCs) and its congeners, they get easily dissolved in water and soil particles and find their way to sea and oceans and also contaminate the marine life. Many scientific studies in the past have accounted for an increase in the concentration of PFCs in the environmental biota because of widespread occurrence and bioaccumulative potential in animal and human tissues [14].

Per Fluorinated Compounds (PFCs) exert direct or indirect effects on the biological systems of the earth, including plants, animals, and humans. Recent studies have shown that exposure of living organisms to environmental pollution results in increased risks of diseases and the development of severe health issues. These chemicals affect the biological systems such as reproductive systems, nervous systems, the circulatory system of humans and animals resulting in the development of several diseases and deformities such as arthritis, neurotoxicity, thyroid dysfunction and carcinogenesis.

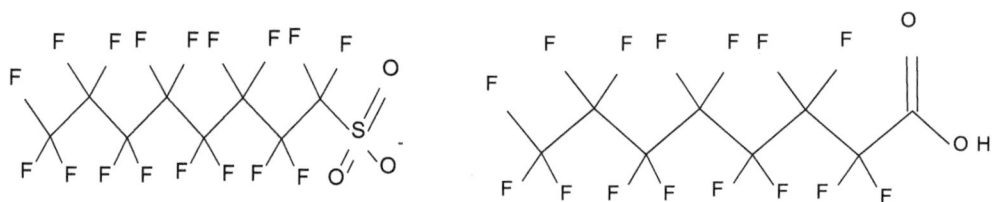

Fig. (4). General structure of PFOS and PFOA.

A large number of studies have been carried out so far on the fate and exposure of POPs on animal models for getting an accurate idea about the exposure route and impact of POPs on Biological systems. The general population gets exposed to POPs *via* different pathways that include dietary exposure, occupational exposure, ambient exposure and product exposure. Table **3** summarizes some of the studies carried on different animals exposed to POPs for assessing different health effects.

Table 3. Toxicity of different POPs in animal models.

S. No.	Chemical	Animal/ Species	Duration/ Route of Exposure	Effect in Biological System	References
1.	PCB (commercial mixtures)	Rat	Aroclor 1221 (0, 0.1, 1, and 10 mg/kg). In utero exposed female offspring (F1) and (F2)	In both generations, the litter sex ratio was skewed toward females.	[15]
2.	PCB (Aroclor 1260)	Female Goat	Orally dosed with PCB 153 in corn oil from day 60 of gestation until delivery.	Altered bone composition in offspring	[16]
3.	PCB (Aroclor 1260)	Female Rat	Pre- and/or postnatal exposure to PCB 77. Pregnant rats were treated with oil or PCB dissolved in oil (2 mg/kg b.w.) on gestation days 6–18	None of the treatments affected female sexual behaviour	[17]
4.	PCB (congener mixture)	Rats	PCB 126 + PCB 153; PCB 126 alone; TCDD+PCB 126 By Gavage/2 years	Increased incidence of gingival squamous cell hyperplasia and carcinoma	[18]
5.	PBDEs	Mouse	14 days	Delayed ontogeny in males and females, with reduced serum T4 levels in Pnd 21 males	[19]

(Table 3) cont.....

S. No.	Chemical	Animal/ Species	Duration/ Route of Exposure	Effect in Biological System	References
6.	PBDEs	Mouse	Once	Impairment in habituation at two and four months and decrement in the activity	[20]
7.	PBDEs	Rats	3-days	Reduced levels of T4 in maternal serum	[21]
8.	PFOS	Rats	One hour	Nasal discharge, stained urogenital region	[22, 23]
9.	PFOS	Rats	90-day	Increased relative and absolute liver weight	[22, 23]
10.	PFOS	Rats	14 weeks	Liver and kidney weight	[22, 23]
11.	PFOA	Rats	28 days	Muscular weakness	[22, 23]
12.	PFOA	Monkeys	90 day	Death with signs of anorexia, swollen face and eyes	[23]

NEUROTOXIC EFFECTS OF POPS

Exposure of POPs has a broad spectrum of health hazards, including carcinogenicity, immunological cytotoxicity, neurotoxicity, developmental and reproductive effects. Concern about the POPs is that they have neurotoxic and developmental effects at the lower environmental level. There are several epidemiological and experimental findings exhibiting the association of motor and cognitive deficits with the POPs concentration in human and animal systems. There are several potential modes of actions hypothesize for depicting the neurotoxic effects of POPs, but an alteration in the thyroid hormone system, neurotransmitter system, and imbalanced intracellular signaling are the predominant modes for neurotoxicity of POPs.

A wide range of POPs has been proven lipophilic and neurotoxic. The neurotoxic effect of these organic pollutants is due to their non-hydrogen bonding ability, biodegradation resistant by the action of proteolysis in organisms. The human brain has 50% dry weight made up of lipids whereas the other body organs have only 6 to 20% of fats. Hence this lipid environment of the brain creates an appropriate platform for interaction of these neurotoxic compounds as they have lipophilic and non-hydrogen-bondable nature.

The brain and spinal cord control all the functions of the body such as memory, speech, learning, hearing and muscular movements. The peripheral nervous system connects the Central Nervous System (CNS) to sensory organs, blood vessels, glands, and muscles. The vulnerability of the peripheral nervous system is higher than the CNS for neurotoxin because it exists outside. The neurotoxic

POPs interact with the CNS by the lipophilic environment of critical sites *viz.* glial cells and myelin, neurotransmitter system, neurons and blood vessels feeding the central nervous system. The PCB and Aroclor 1242 are found to be transported to the stomach within 60 minutes and further distribution in the brain, kidney, and lungs within 24 hours [24]. The blood-brain barrier partially protects the CNS from toxins as they have a layer of cells that only allow the hydrogen bondable compounds to pass through in nerve tissues and counter the non-hydrogen bondable molecules to pass. A study on neurotoxic effects on chlordecone confirmed that they affect the nervous system at the developmental stage. Scientific studies have revealed that memory impairment and fine body tremors were observed in rodents upon exposure to chlordecone [25, 26]. Chlordecone is known to cross the placental barrier leading to neurological disorders in humans [27].

Possible Mode of Action(s)

Although there are several epidemiological and experimental results for motor and cognitive deficits in human and animals due to developmental exposure of POPs but, the cellular and molecular mechanisms responsible for this impairment are not yet known [28 - 30]. Some tools are hypothesizing for the developmental neurotoxic mode of action of POPs, but the predominant among them are following: (i) Effects of Pops on Neurotransmitters System, (ii) Dose-Response Effects of Pops on Intracellular Signalling

Effects on Neurotransmitters System

It has been hypothesized that changes in levels of neurotransmitters such as dopamine (DA) and catecholamine are associated with the observed neurobehavioral changes. Disturbing levels of neurotransmitters such as acetylcholine, serotonin, and norepinephrine also of cyclic guanosine monophosphate have been shown in neurochemical studies of DDT exposure [31]. In the case of chlordecone, some individuals complained about tremors and memory loss after several years of cessation of exposure [32]. The exact mechanism behind the chlordecone neurotoxicity is not precisely known, but it is believed that inhibition of ATPases (Na^+-K^+ and Mg_2^+ATPases) and consequent inhibition of catecholamine uptake is responsible [33] for the toxicity.

Dose-Response Effects on Intracellular Signalling

Intracellular signaling or signal transduction is a process to transfer extracellular signals into the cytosol and nucleus of the cell. These signal molecules are essential for nervous system function and also play a vital role in the development of the nervous system [34, 35]. Any interference in the mechanism of signal transduction has the potential to disturb neuronal function and their development, which will further convert into the behavioral changes. By interacting with specific cell receptors, growth factors, hormones, and neurotransmitters, they transfer the signals or messages from one cell to another and act as a first messenger. This signal transduction will further activate or inhibit specific enzymes and open the ion channels. This intracellular signaling and pathways get activated by particular receptors and known to have an active cross-talk link between the receptor signals.

In recent years, investigations on several neurotoxic compounds showed that signal transduction pathways are potential targets for these neurotoxicants [36 - 38]. These neurotoxic substances have the potential to interfere in the normal functioning of the nervous system and structure at any step(beginning from the extracellular stimulus to transcriptional level). Chemicals can cause an imbalance in the nervous system function either directly or by altering the morphology of the nervous system through the disruption of a potential mechanism called intracellular signaling [39]. Kodavanti and Tilson, 2000, found that neurotoxicants may affect brain function and morphology at a critical phase of development by altering the intracellular signaling. As evidence [40] DDT inhibits Ca_2^+-ATPase (an ecto-ATPase, located outside of the cell membrane) which acts to maintain high external calcium concentrations. DDT alters this function of Ca_2^+-ATPase hence lower calcium concentration in the external environment contributing to membrane instability and repetitive firing.

Calcium is a crucial factor for cellular functioning, particularly for the signal transduction in the nervous system where the movement of calcium ions across the membrane and maintenance of its gradient is fundamental for nerve excitation. POPs like organochlorines and pyrethroids affect the above process by altering the calcium transport activity of ATPases. There are two major calcium-stimulated ATPases in nerves *viz.* Ca-Mg ATPase and Na-Ca ATPase. The increment in free calcium at the presynaptic terminal promotes the release of neurotransmitters; such type of mechanism documented after the exposure of cyclodiene and pyrethroid insecticides where Ca-Mg ATPase may be a causative factor [40].

Effect on Different Cellular Pathways and Molecules Involved

Endocrine or hormone disrupters are the external agents and or chemicals that interfere with the synthesis, secretion, transport, binding, action, or elimination of hormones. Hormones are protein molecules, which regulate reproduction, development, and behavior in the human body [41]. Endocrine Disruptor (ED) molecules additionally may affect the reproductive system and also alter the function of the central and peripheral nervous system [42]. Several organochlorine pesticides, PCBs, PBDEs, PCDDs/DFs which are listed as POPs by Stockholm Convention, act as Endocrine Disruptors (EDs). There are several congeners of PCBs and PCDDs/DFs which have their specific toxicity and mode of action.

The toxic response of commercial PCB mixtures has a broad spectrum which depends on several aspects including chlorine content, duration of exposure, dose and species, *etc.* Various laboratory experiments with exposure of commercial PCB mixtures on animals, fish, and wildlife have been studied [43] and documented the immunotoxicity, developmental toxicity, biochemical effects as well as carcinogenicity.

Chlordecone, along with metabolites of DDT like methoxychlor, is known to have estrogenic and antiandrogenic properties and have the activity as ERa agonists or ERb antagonists [44]. Similar to other organochlorine compounds like dieldrin, endosulfan, toxaphene are known to have weak estrogenic activity. Exposure of such compounds to animals may cause hormonally mediated adverse health hazards, for example when p,p'-DDE given to fetal, pubertal and adult male rats, a clear antiandrogenic activity was documented [45].

Effect on Aryl Hydrocarbon Receptor Pathway

Chemical compounds like 2, 3, 7, 8-tetrachlorodibenzo-p-dioxin (TCDD), a halogenated aromatic compound, have the toxicity factor of 1, which confirms its maximum toxicity. Toxicity of TCDD is used to determine the relative (*i.e.*, toxic equivalents) toxicity of individual halogenated aromatic compounds. Twelve congeners of Dioxin-like PCBs and 17 congeners of PCDD/DFs have Toxicity Equivalent Factors (TEFs) lower than TCDD. Congeners of PCBs who have the most substitution of a chlorine atom at non-ortho positions are known to be most toxic congeners. These highly toxic PCB congeners exhibit Ah-receptor mediated toxicity similar to TCDD and also induces a variety of microsomal enzymes in the liver. It has been widely documented in cell culture (mammalian) and different lab-based studies that Aryl Hydrocarbon Hydroxylase (AHH) induction associated with cytochrome P-450 due to dose-dependent exposure of TCDD causes toxicity which can be reflected as elevation in the activity of AHH and EROD (markers of *CYP1A* activity).

Toxic effects on Ah-receptor results in a broader range of biological imbalances including alteration in metabolic pathways atrophy in thyroid system, bodyweight loss, hepatotoxicity, developmental and reproductive effects, neoplasia, impaired immune response, chloracne, and related skin lesions.

TCDD can induce genes involved in cell proliferation, xenobiotic-metabolizing enzymes, as well as apoptosis, and inflammation [46]. The primary mode of TCDD toxicity is binding with Aryl hydrocarbon Receptor (AhR), resulting in nuclear translocation of AhR followed by heterodimerization with Aryl hydrocarbon Receptor Nuclear Translocator (ARNT), and linking with xenobiotic responsive elements present in promoters of target genes [47, 48]. Other dioxins, individual PCBs such as PCB-16, and furans have a similar mechanism of action as TCDD with different potential [49].

The action of dioxins on cells specifically by the ligand-activated induction of transcription through the cytosolic Aryl hydrocarbon Receptor (AhR) has been concluded in several studies (Fig. **5**). Aryl hydrocarbon receptor occurs on different cell *viz.* lymphocytes in the thymus, liver, and lungs. Upon binding of dioxins with AhR, this complex faces conformational changes and move towards the nucleus and binds with specific regulatory sequences and the NF-kb induction of transcription takes place [50, 51].

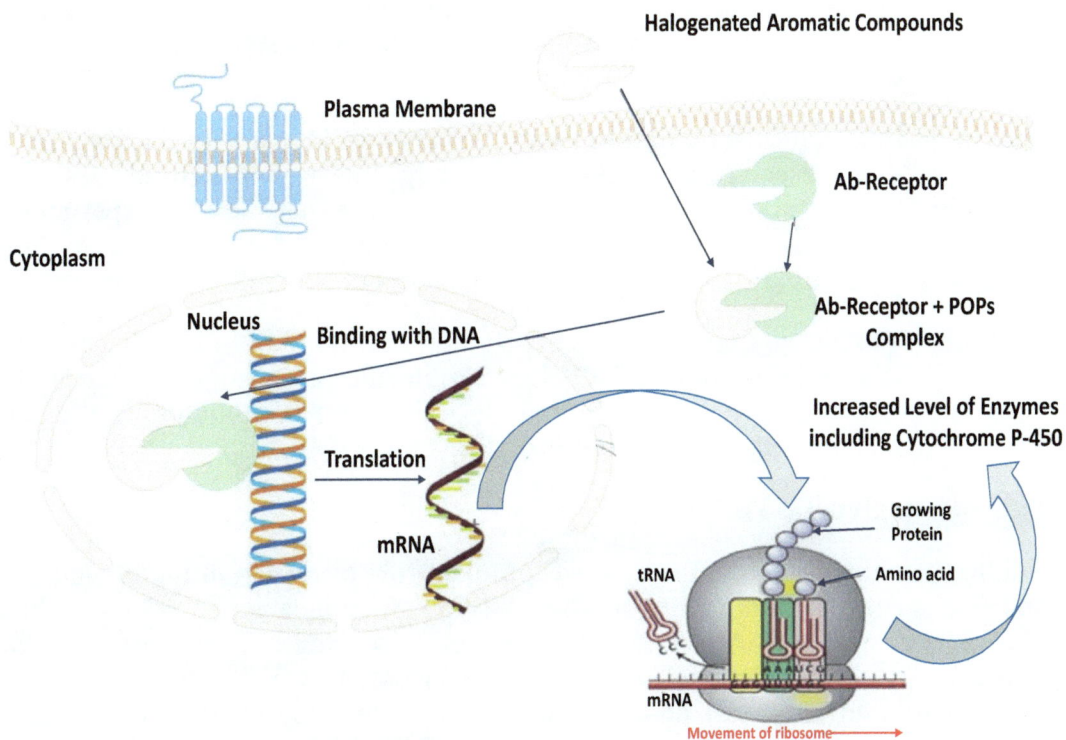

Fig. (5). Possible mechanism of toxic action of POPs by altering Aryl hydrocarbon Receptor pathway.

Effect on the Immune System

POPs enter the body of the host *via* different mechanisms like inhalation through the air, ingestion by the food chain. Upon their entry into the biological system, they pass through the mucosal barrier and interact with the immune system. Upon their interaction with the immune system, a series of chemical processes are initiated within the system (Activation of B and T cells). Several studies have proven that environmental pollutants like PCBs, DDT, PBDEs induce apoptosis in hepatic cells. Apart from hepatic apoptosis these pollutants significantly contribute to apoptosis of immune cells. The apoptosis of T cells is collectively termed as Activation-Induced Cell Death (AICD) [52]. Camacho *et al.*, 2001, showing that in superantigen-primed T cells, Fas-dependent activation induces cell death. In another study, TCDD was documented as an inducer of suppression of CD4+T cells and cell death, which may be mediated by Fas/FasL interactions [53]. The effect of TCDD in mice on T cell-derived cytokine production was observed in a study. According to Ito *et al.*, 2002, and Fujimaki *et al.*, 2002 [54, 55], an initial increment in IL-2 caused by TCDD was observed, but this increment was suppressed on day 4, and the production of Th-2 cell-derived cytokines IL-4, IL-5, and IL-6 were also documented to be significantly decreased as compared with control mice.

Alteration in B cell differentiation by exposure of TCDD is also documented in several types of research, marked decrease in IgM secretion and the number of antibody-forming plasma cells was observed in experiments. Experimental findings also suggest that in TCDD mediated inhibition of B-cell differentiation, cyclin-dependent kinase inhibitors may be an essential intracellular target.

Increment (40%) in double-positive cells in murine fetal thymus cells and overexpression of CD44 and MHC class I molecule on thymocytes and higher positive selection rate (phenomenon may increase the autoimmunity in the host) have been documented to be induced by dioxins [56].

Pathway of Genetic Toxicity

Studies on genetic effects found a predominant role of dioxin in the induction of specific genes including cytochrome P-450, oncogenes, Interleukin1-beta, plasminogen activator inhibitor, and cytokines. According to [57] Crawford *et al.*, 2003, TCDD found attached to increased expression of NF-kb, c-jun, and p27 (kipl) genes. In another study [58] some genes *viz.* insulin-like growth factor binding protein-6 and Interleukin-5R alpha subunit genes have been found up-regulated although the genes *viz.* CD14 showing a down-regulated action. PCBs are also documented for the induction of genes that affects cell differentiation, proliferation, and alteration resulting in disruption of cell function and imbalanced homeostasis in a broader sense.

Effect on Thyroid Hormone System

There is a wide range of man-made and naturally occurring chemicals which interact with the endocrine system of human, and wildlife and disrupt hormone homeostasis, metabolism, and different cellular and physiological processes [59]. The main target of this endocrine-disrupting toxic group is the thyroid hormone system.

An adequate Hypothalamus-Pituitary-Thyroid (HPT) axis is vital for the appropriate function of the cardiovascular system and neurological system [60]. Low-level exposure with POPs like DDT, PBDEs, PCBs, and their primary metabolites have been found to disrupting this essential HPT-axis [61 - 63].

Effect of PCB on Thyroid Hormone System

Polychlorinated Biphenyl (PCB) are a class of chemical compounds that are no longer synthesized due to their hazardous impacts, but they are still present in the environment due to their implications and uses in a variety of products before banning [64, 65]. PCB disrupts the normal functioning of the thyroid hormone system by binding with thyroid hormone transport proteins like transthyretin [66]. PCB congener and the target tissue disruption mechanism of thyroid receptor by PCB may be agonistic or antagonistic. The mRNA expressions of thyroid hormones also get effected by PCBs in the pituitary gland and liver if their metabolites are hydroxylated by the metabolic enzyme cytochrome *P450 1A1* which acts as thyroid receptor (TR) agonist [67]. PCBs upon binding with Thyroid Receptor (TR) may also antagonize triiodothyronine (T3) and inhibit the TR-mediated gene activation. *In-vivo* studies in rats reported in a reduction in the level of total and free serum thyroxine levels followed by progenitor cell proliferation and cell death [68]. In human PCBs are commonly found to interfere TR receptor signaling mainly *via* TRβ complex and also an alteration in the maturation of white matter during early development and differentiation of oligodendrocytes [69].

Effect of Polybrominated Diphenyl Ethers on Thyroid Hormone System

PBDEs are a class of bioaccumulative and recalcitrant halogenated compound which has emerged as a leading environmental pollutant. These chemicals are intended to decrease the ignition rate and thus finds its uses in various industrial sectors. Brominated flame retardants are used as a coating on electrical wires, manufacture of polyurethane foam and many electronic types of equipment. However, the uses and implications of the majority of PBDEs congeners are banned, still, due to their high persistence and degradation resisting properties, a considerable population is exposed to it. Due to the similar structure of PBDE and thyroid hormone molecules, PBDE can bind with thyroid hormone receptor and inhibits binding of T3 to TRs which results in suppression of T3 action at environmentally relevant doses [70, 71].

Other possible mechanisms of PBDE action on the thyroid hormone system are, competitive binding with serum thyroid transporters (*e.g.*, thyroid-binding globulin and transthyretin), inhibition of deiodinase activity and upregulation of clearance enzymes (*e.g.*, glucuronidases) [72, 73]. In a study, a decrease in T4 activity and downregulation of expression of TR genes was found upon exposure of decaBDE in Zebrafish [74]. A similar negative effect of PBDEs was found [75, 76, 77] on the thyroid hormone system in several studies, but the association and impact of PBDEs vary with congeners. Another study [78] found a positive association between PBDE concentration (BDE-47, BDE-99, and BDE-100) during pregnancy and the level of T3 in cord blood. In pregnant women, both higher and lower concentrations of T3 were reported associated with PBDEs [79, 80]. Interconversion of T3 and T4 metabolites takes place in the biological system, PBDEs molecules have a stronger affinity to bind at enzyme active site due to Structure-Activity Relationship (SAR) with T4 molecule, leading to an alteration in the enzymatic system. Several epidemiological studies suggesting the effect of PBDE exposure on the neurophysiological motor and cognitive function have also been reported [78, 81, 82]. Thyroid disruption, impacting the early brain development in neonates is one the predominant mechanism associated with PBDEs [81, 82].

Effect of Perfluoroalkyl Substances (PFASs) on the Thyroid hormone system

Perfluoroalkyl substances are organic chemicals that are extensively used in cookware, furniture, and textiles [83, 84]. Although major US companies phased out perfluorooctane sulfonate (PFOs) and perfluorooctanoic acid (PFOA) since 2002, PFOS and PFOA have been detected in a study carried out by National Health and Nutrition Examination Survey (NHANES) in 2013-14 [85]. The study confirms the concentration of PFAs in children even after phasing out of these chemicals. PFAs upregulate the deiodinase in the thyroid gland by the action of interference in the binding of thyroid hormone to transthyretin [86, 87]. Thyroid disruption by PFASs also alters other processes of the endocrine system and influence the sex-specific development.

MAJOR PATHWAY OF POPS METABOLISM AND ENZYMATIC ALTERATIONS

Effect on Cytochrome P (CYP) Enzyme

CYP enzymes are monooxygenase enzymes that are found in all life forms. Particularly in the mammalian body, it is primarily found in the Endoplasmic Reticulum (ER) of the liver cells. Human Cytochrome P enzymes are membrane-associated proteins that use *haem* (iron) for the oxidation of molecules to convert them into water-soluble forms. CYP enzymes help in the synthesis of endogenous compounds like steroids, a phenomenon known as Steroidogenesis. There are several subsets of enzymes in the CYP family-like *CYP11A1, CYP11B1, CYP17A1,* and *CYP19A.* CYP enzyme plays a vital role in the preservation of life. It also plays a critical role in protecting against possible harmful effects of various endogenous compounds ingested during diet throughout a lifetime. The detoxification of molecular oxygen to water in tissues is the primary role of CYP [88]. CYP enzyme plays a vital role in the metabolism of Xenobiotic compounds, catalysing the initial reaction for the breakdown of PCBs and other halogenated hydrocarbons.

PCBs, undergo biotransformation *via* CYP enzymes forming hydroxy metabolites followed by conjugation mediated by the phase II enzymes to Glucoronide, sulfate, glutathione conjugate for consequent defecation. Some of the polar metabolites are formed during the process, which causes biochemicals changes in tissue (Fig. **6**). Different phenolic metabolites like monohydroxy (phenolic), dihydroxy, poly-hydroxylated, sulphur containing and methyl ether are major by-products of PCBs metabolites. Oxygenation of PCBs may occur directly by the addition of hydroxylated metabolite (OH-PCB), or by epoxide formation. The epoxide produced gets rearranged, producing an OH-PCB and /or react with glutathione to yield a mercapturic acid or a methyl sulfone conjugate. This Methyl sulfone-PCBs produced gets covalently attached to proteins in tissues causing induction of cytochrome P450 enzymes or exert endocrine-related effects [89].

Steroidogenesis is a complex enzymatic process involving different forms of CYP. Most of the PCBs that gets metabolized in the body are excreted *via* urine and bile. PCBs are easily oxygenated by *CYP450* in which *CYP1A* and *CYP2B* are the two key isoenzymes known which act as a catalyst in this conversion reaction. To get excreted from the body, these lipophilic compounds need to be detoxicated first by converting to more water-soluble forms.

Hormones   PCBs other xenobiotic

CYP Enzymes ⇅ → Polar metabolites of PCBs

PCBs ─────────────────────────→

O$_2$, NADPH, CYP reductase

Toxic effects   Deactivation   Bioactivation

Elimination   Biochemical effects   Toxic effects

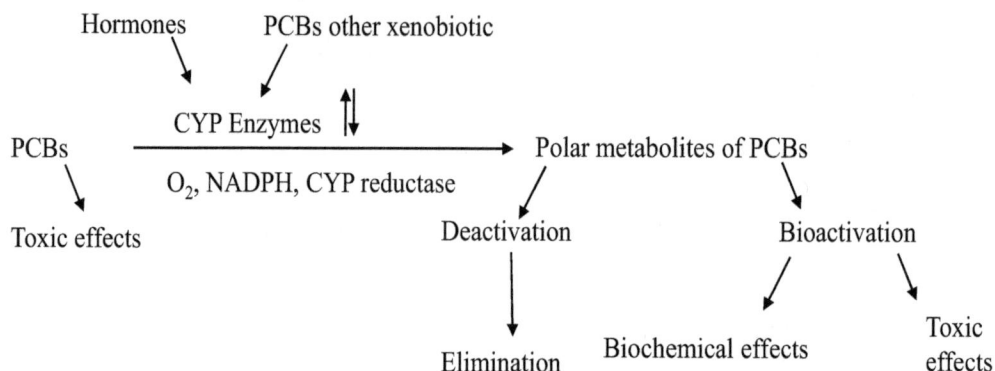

Fig. (6). CYP dependent biotransformation of PCBs.

A previous study carried out by Barber *et al.*, 2006, reported the upregulation of the *CYP1A2* gene in human MCF breast cancer cells due to exposure of BDE-99, similar mechanism *viz* formation of OH metabolites was responsible for the upregulation of CYP genes [90]. Another study conducted by Lundgren *et al.*, 2007 [91] reported an increase in both the CYP activity and Pentoxyresorufin- O-depentylase (PROD) in mice exposed to BDE-99. A higher level of *CYP2B* and PROD was accounted for in exposed mice. A similar kind of study conducted to depict the impact of PFCs on CYP and associated isoenzymes like *CYP1A2, CYP2A6, CYP2B6* resulted in alteration in CYP mechanism due to PFOS exposure [92].

Effect on EROD Activity

Estimation of ethoxyresorufin-O-deethylase (EROD) activity is the best biomarker for embryonic mortality due to Xenobiotic exposure. Estimation of EROD activity is a biomarker of exposure of certain planar halogenated and similar organic compounds. EROD is a sensitive indicator in fishes and other animals, especially in *in-vivo* conditions and provides direct evidence of induction of cytochrome P450-dependent monooxygenases (the *CYP1A* subfamily specifically) caused by manmade chemicals [93]. The activity of EROD is generally ranked with the activity of isoenzyme *CYP1A1*. In rodents, this is mainly catalysed by CYP form rather than *CYP1A1* [81]. This activity upon the administration of PCB increased to 147 times in rodents than normal while in hamster it rose to only three times. The same study accounted for the increase in EROD activity in bulls to five times upon chronic exposure to PCB [94].

Effect of PCBs on Xenobiotic Biotransformation Enzymes

Biotransformation is the metabolic conversion of endogenous and xenobiotic chemicals to more water-soluble compounds. Xenobiotic biotransformation is accomplished by a limited number of enzymes with broad substrate specificities. This family of enzymes in the liver converts steroid hormones into water-soluble metabolites to get excreted [95]. Blom and Forlin in 1997 investigated exposure of PCB on the induction of biotransformation enzymes in which they concluded that the methyl sulfonyl PCB metabolites are the final product of PCBs metabolism. These metabolites once produced bind with the renal system, particularly kidney with high affinity than other organs [96]. However, whether these metabolites cause toxicity linked to PCBs to a significant extent in the kidney is still needs to be explored.

STRUCTURE ACTIVITY RELATIONSHIP (SAR)

Qualitative and quantitative modeling of a chemical structure relating to its biological activity is collectively known as Structure Activity Relationship (SAR). This modeling approach is applied for predicting and characterizing the toxicity of a chemical structure in the biological environment. A structure activity relationship (SAR) relates the attribute of a chemical structure with biological activity or effect due to that chemical. It can be both a quantitative and or qualitative consideration. The basic principle behind SAR is the chemical structure of compound reflecting its chemical as well as physical attributes along with other properties like reactivity, stability, and interaction with other compounds within a biological system, which in terms also regulates its chemical, biological and toxicological properties. The underlying mechanism behind SAR is an attempt to understand and reveal how different properties relevant to activities of a specific compound are encrypted within and revealed by its chemical structure.

Why SAR?

The biological characteristics of emerging chemical compounds are often inferred from previous similar existing chemicals whose impact is already known. A computer-based modeling approach for screening of an emerging toxicant and its effects on the biological system for qualitative biological activity (SAR) as well as quantitative biological potency (QSAR) are being used for setting many diverse problems. Further to that, acceleration in knowledge for understanding the toxicological endpoints of emerging chemical pollutants, SAR plays a key role in such chemical investigations.

SAR dwells at the intersection the following three components:

1- Chemistry (Chemical structure, properties)
2- Biology (biological significance, role in biological system)
3- Statistics (method of interpretation)

This focused intersection of three components has formulated the science behind the SAR (Fig. **7**).

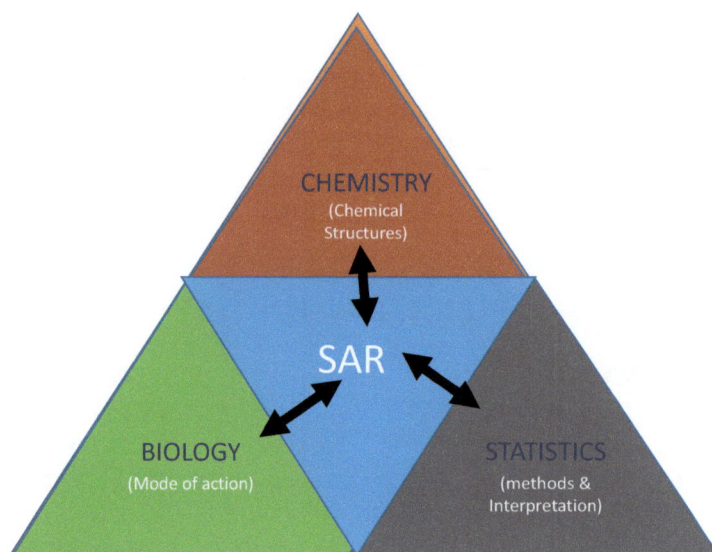

Fig. (7). The Science of SAR.

SARs have long been used in various chemical and pharmaceutical industries, for designing of different chemicals with commercially desirable properties. The compounds with desired therapeutic and pharmacologic activities are being designed using SAR. For the protection of environmental health, SAR is currently widely used for the prediction of ecological risk and human health effects. SAR mechanism has earlier been used for recognition and close resemblances for "coplanar PCBs, TCDD, and other toxicants" [97]. This, in turn, has led to the establishment of Toxic Equivalency Factor (TEFs) and Toxicity Equivalents (TEQ) for several toxicants like dioxins and other POPs. SAR is even being used to help pharmaceutical industries for designing safer chemicals used for commercial use because of their desirable properties.

Elements of SAR and SAR Models

SAR works on several elements when used for the classification of any toxic substances and or its similar chemical compound and its biological fate. Some key elements of SAR are:

i- Background data for the biological toxicity of the chemical compound
ii- Structural properties (Stereochemistry, the position of bonds, angles)
iii- Appropriate methods for chemical analysis (qualitative as well as quantitative)

The most important element for a SAR model to work is the background information of any toxicant, which in turn accounts for its associated physical and chemical properties (Fig. **8**); the best example for this is the Octanol/ water partition coefficient that determines the lipophilic character of chemicals. Structural properties like 3D, 2D and or position of atoms and bond angles are essential elements for the representation of the molecular structure. At last, a well-established and appropriate method for analysis (qualitative as well as quantitative) is required for establishing the activity and the extent up to which the toxicants will pose a threat to the biological system. The probable chemical and biological mechanism at every step should be considered to derive a suitable SAR model that can strongly strive and fit in the scientific domain.

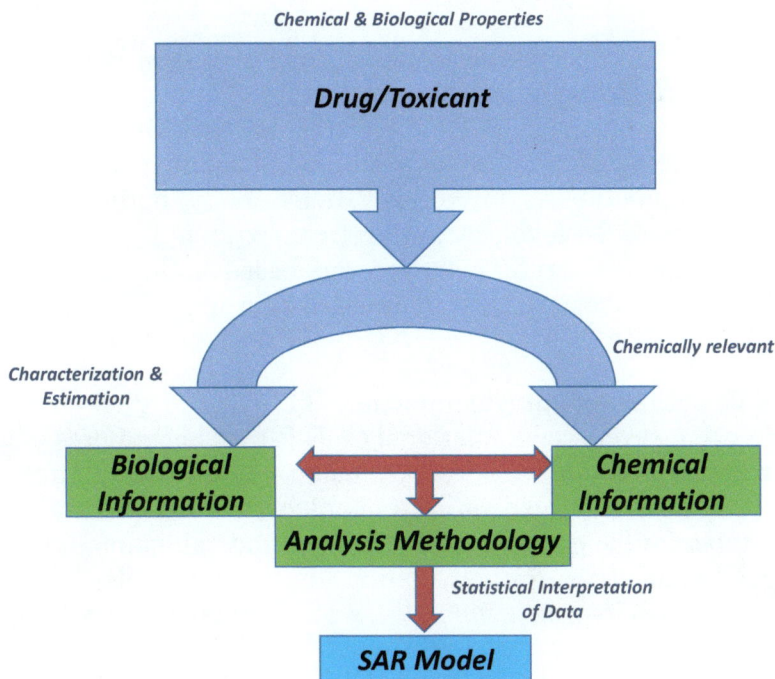

Fig. (8). Elements of SAR.

SAR and Associated Toxicity Mechanism

POPs especially PCBs and PBDEs belong to the group of polyhalogenated aromatic hydrocarbons having a strong array of biological toxicity. PCBs, PBDEs and other POPs like TCDD are known to produce biological reactions *via* an enzymatic receptor-mediated response (binding with cytosolic aryl hydrocarbon (Ah) receptor) followed by genetic induction which has been discussed in the earlier part of the chapter. The structural similarity between biological molecules and POPs enables them to bind at the enzyme active site creating malfunctioning in the biological system. A good example of this is a structural similarity between a T4 and polyhalogenated aromatic compounds. PCBs, PBDEs, and other POPs closely resemble the structure of Thyroid hormones (T3 and T4), and thus, the potential of SAR has increased on predicting the thyroid disrupting properties of such compounds (Fig. **9**). The basic mechanism behind these common properties is non-ortho, meta-para positions substitution on biphenyl rings that promote the coplanarity among the structures of PCBs, TCDD and other toxicants having planer similarities. SAR has been widely used for predicting the activity of various chemical and their associated congeners. Many *in vivo* and *in-vitro* studies have reported that POPs like PCBs, PBDEs, TCDD induce stress by binding with the Ah receptor in different organisms which has revealed a common toxicity mechanism of these pollutants [97].

Few studies have only been investigated for the effects of PFOSA and PBDE on TH homeostasis reporting a tendency for increased free plasma T4 in response to different PBDEs Congeners (BDE-28, BDE-15, and BDE-183) in employees working in an electronic recycling unit [93, 94]. In another study carried out in the general U.S. population exposed to PBDEs, body burdens were positively associated with their T4 level, including free T4, total T4, and urinary T4. This was one of the largest studies carried out based upon the linkage of PBDE exposure to TH homeostasis and demonstrated a negative association between PBDEs and T3 and TSH [95].

The use of Quantitative biological potency (QSAR) for research in toxicology is being done using several computational tools [98]. This requires a huge database for storing the information like chemical name, chemical properties, their toxicity mechanism, software for generation of molecular descriptors along with other simulation tools for the generation of equations and validation of QSAR. Some of the computational tools for calculating toxicity variables along with their applications for toxicity prediction in different compounds are shown in Fig. (**10**).

The selection of right mathematical tools, approach, and right molecular descriptors plays a crucial role in achieving toxicity endpoints and or relevantly high-quality modeling using SAR models [99].

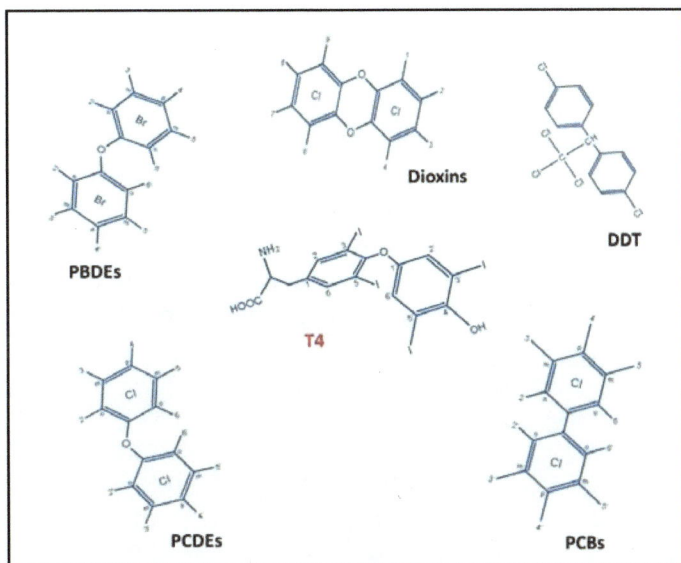

Fig. (9). Structural similarity between a T4 and polyhalogenated aromatic compounds.

Fig. (10). SAR Software tools and their applications for toxicity prediction.

The application of SAR now a day has reduced the manifestation time, test costs, and animal usage. SAR has found variable linkages with different scientific variables problems like biological, chemical as well as statistical. It has evolved as a multidisciplinary field resolving a wide range of issues like human health-related disorders, toxicological studies, pollution prevention, and risk assessment. SAR has also been used as a tool for studying Endocrine disrupting properties and environmental fate of highly toxic chemicals like POPs. As the scientific research community is entering into a new era, SAR automatically draws its validity and linkages with various mechanisms *via* databases and scientific understandings *i.e.*, interacting with the chemicals at different life forms and life-giving processes at even molecular level. For animal bioassays, SARs is being used for property measurements and animal testing, because of their potential to reduce the need for more efficient screening of chemicals, having a wide range of toxicity endpoints. Thus, SAR can be collectively called as conjugation of tools and techniques (physical, chemical and biological) that will further advance the scientific communities for the betterment of society in terms of improved ecological and human health.

Thus, SAR can be a useful tool against environmental health protection, with strategic applications toward identifying the greatest chemical hazards.

CONCLUSION

There are sufficient epidemiological and experimental studies carried in different parts around the globe showing that PCB, PBDEs, and PFOS exposure associated with motor and cognitive deficits in humans and animal models. Although several potential modes (s) of actions were reported for PCB, PBDEs and PFOS neurotoxic effects and by alteration in neurotransmitter systems, amending the intracellular signaling processes and hormonal imbalance impacting activity of thyroid hormone are the predominant ones. Most of the toxic effects are exerted by exploiting the enzymatic surface receptor-mediated binding. This property can be used as a tool for screening of toxic effects of chemicals such as POPs with structural similarity. Thus, SAR as conjugation of a tool and techniques (physical, chemical, and biological) will further advance the scientific communities for the betterment of society in terms of improved ecological and human health. SAR is an interdisciplinary area that finds solutions to a broad range of problems and toxicity endpoints. Toxicological sciences now a day's entering in such a time frame where SAR in conjugation with biochemistry, organic chemistry, and physical sciences will further boost our knowledge and scientific comprehension on vital life-giving procedures and can fabricate benefits for a healthy society and better outcomes.

CONSENT FOR PUBLICATION

Not applicable.

CONFLICT OF INTEREST

The author(s) confirms that there is no conflict of interest.

ACKNOWLEDGEMENTS

The authors are thankful to Director, CSIR-NEERI and Head, EIS Division for their continuous support and encouragement. All the authors are also thankful to Ms. Anshika Singh and Mr. Deepak Marathe, AcSIR Ph.D. students for their crucial indirect support.

REFERENCES

[1] Humphrey HEB. Evaluation of changes in the levels of polychlorinated biphenyls (PCB) in human tissue Final Report on US FDA contract. Lansing: Michigan Department of Public Health 1976.

[2] Humphrey HEB. Human exposure to persistent aquatic contaminants: a PCB case study.Toxic Contamination in Large Lakes. Chelsea, MI: Lewis Publishing 1988; Vol. 1: pp. 237-8. a

[3] Humphrey HEB. Chemical contaminants in the Great Lakes: the human health perspective.Toxic Contaminants and Ecosystem Health: A Great Lakes Focus. John Wiley 1988; pp. 153-64. b

[4] Humphrey HEB. Population studies of Great Lakes residents exposed to environmental chemicals.Cancer Growth and Progression Vol 5, Comparative Aspects of Tumor Development. Massachusetts: Kluwer, Norwell 1989.

[5] Pauwels A, Covaci A, Weyler J, *et al.* Comparison of persistent organic pollutant residues in serum and adipose tissue in a female population in Belgium, 1996-1998. Arch Environ Contam Toxicol 2000; 39(2): 265-70.
[http://dx.doi.org/10.1007/s002440010104] [PMID: 10871430]

[6] Katsoyiannis A, Samara C. Persistent organic pollutants (POPs) in the conventional activated sludge treatment process: fate and mass balance. Environ Res 2005; 97(3): 245-57.
[http://dx.doi.org/10.1016/j.envres.2004.09.001] [PMID: 15589233]

[7] Borja J, Taleon DM, Auresenia J, Gallardo S. Polychlorinated biphenyls and their biodegradation. Process Biochem 1999–2013; 40
[http://dx.doi.org/10.1016/ j. procbio.2004; 08.006]

[8] Ross G. The public health implications of polychlorinated biphenyls (PCBs) in the environment. Ecotoxicol Environ Saf 2004; 59: 275-91.
[http://dx.doi.org/10.1016/j. ecoenv.2004; 06.003]

[9] Stella T, Covino S, Burianová E, *et al.* Chemical and microbiological characterization of an aged PCB-contaminated soil. Sci Total Environ 2015; 533: 177-86.
[http://dx.doi.org/10.1016/j.scitotenv.2015.06.019] [PMID: 26156136]

[10] Freeman HC, Uthe JF, Sangalong G. The use of hormone metabolism in studies in assessing the sublethal effects of marine pollution. Rapp P-V Reun- Cons Int Explor Mer 1980; 179: 16-22.

[11] Faroon O, Ruiz P. Polychlorinated biphenyls: New evidence from the last decade. Toxicol Ind Health 2016; 32(11): 1825-47.
[http://dx.doi.org/10.1177/0748233715587849] [PMID: 26056131]

[12] Ellis-Hutchings RG, Cherr GN, Hanna LA, Keen CL. Polybrominated diphenyl ether (PBDE)-induced alterations in vitamin A and thyroid hormone concentrations in the rat during lactation and early postnatal development. Toxicol Appl Pharmacol 2006; 215(2): 135-45.
[http://dx.doi.org/10.1016/j.taap.2006.02.008] [PMID: 16580039]

[13] Zhou T, Ross DG, DeVito MJ, Crofton KM. Effects of short-term in vivo exposure to polybrominated diphenyl ethers on thyroid hormones and hepatic enzyme activities in weanling rats. Toxicol Sci 2001; 61(1): 76-82.
[http://dx.doi.org/10.1093/toxsci/61.1.76] [PMID: 11294977]

[14] OECD (Organisation for Economic Co-operation and Development). Hazard assessment of perfluorooctane sulfonate (PFOS) and its salts. ENV/JM/RD (2002)17/FINAL. Joint Meeting of the Chemicals Committee and the Working Party on Chemicals, Pesticides, and Biotechnology, Environment Directorate, Organisation for Economic Co-operation and Development (Paris) 2002. Available at http://www.oecd.org/dataoecd/23/18/2382880.pdf

[15] Steinberg RM, Walker DM, Juenger TE, Woller MJ, Gore AC. Effects of perinatal polychlorinated biphenyls on adult female rat reproduction: development, reproductive physiology, and second generational effects. Biol Reprod 2008; 78(6): 1091-101.
[http://dx.doi.org/10.1095/biolreprod.107.067249] [PMID: 18305224]

[16] Lundberg R, Lyche JL, Ropstad E, et al. Perinatal exposure to PCB 153, but not PCB 126, alters bone tissue composition in female goat offspring. Toxicology 2006; 228(1): 33-40.
[http://dx.doi.org/10.1016/j.tox.2006.08.016] [PMID: 17007988]

[17] Cummings JA, Nunez AA, Clemens LG. A cross-fostering analysis of the effects of PCB 77 on the maternal behavior of rats. Physiol Behav 2005; 85(2): 83-91.
[http://dx.doi.org/10.1016/j.physbeh.2005.04.001] [PMID: 15878184]

[18] Yoshizawa K, Walker NJ, Jokinen MP, et al. Gingival carcinogenicity in female Harlan Sprague-Dawley rats following two-year oral treatment with 2,3,7,8-tetrachlorodibenzo-p-dioxin and dioxin-like compounds. Toxicol Sci 2005; 83(1): 64-77.
[http://dx.doi.org/10.1093/toxsci/kfi016] [PMID: 15509667]

[19] Rice DC, Reeve EA, Herlihy A, Zoeller RT, Thompson WD, Markowski VP. Developmental delays and locomotor activity in the C57BL6/J mouse following neonatal exposure to the fully-brominated PBDE, decabromodiphenyl ether. Neurotoxicol Teratol 2007; 29(4): 511-20.
[http://dx.doi.org/10.1016/j.ntt.2007.03.061] [PMID: 17482428]

[20] Johansson N, Viberg H, Fredriksson A, Eriksson P. Neonatal exposure to deca-brominated diphenyl ether (PBDE 209) causes dose-response changes in spontaneous behaviour and cholinergic susceptibility in adult mice. Neurotoxicology 2008; 1;29(6): 911-.

[21] Cai Y, Zhang W, Hu J, Sheng G, Chen D, Fu J. Characterization of maternal transfer of decabromodiphenyl ether (BDE-209) administered to pregnant Sprague-Dawley rats. Reprod Toxicol 2011; 31(1): 106-10.
[http://dx.doi.org/10.1016/j.reprotox.2010.08.005] [PMID: 20851178]

[22] ENV/JM/RD(2002)17/Final, OECD. 2002.

[23] The Toxicology of the Perfluoroalkyl Acids – Perfluorooctane Sulfonate (PFOS) and Perfluorooctanoic acid (PFOA). TOX/2005/07 2005.

[24] Neskovic N, Vojinovic V, Vuksa M. Abstract Vid-3 Fifth International Congress of Pesticide Chemistry. Kyoto. 1982.

[25] Mactutus CF, Tilson HA. Neonatal chlordecone exposure impairs early learning and retention of active avoidance in the rat. Neurobehav Toxicol Teratol 1984; 6(1): 75-83.
[PMID: 6201755]

[26] Mactutus CF, Tilson HA. Evaluation of long-term consequences in behavioral and/or neural function following neonatal chlordecone exposure. Teratology 1985; 31(2): 177-86.

[http://dx.doi.org/10.1002/tera.1420310202] [PMID: 2581329]

[27]   Dallaire R, Muckle G, Rouget F, *et al.* Cognitive, visual, and motor development of 7-month-old Guadeloupean infants exposed to chlordecone. Environ Res 2012; 118: 79-85.
[http://dx.doi.org/10.1016/j.envres.2012.07.006] [PMID: 22910562]

[28]   Kodavanti PRS, Tilson HA. Structure-activity relationships of potentially neurotoxic PCB congeners in the rat. Neurotoxicology 1997; 18(2): 425-41.
[PMID: 9291492]

[29]   Grova N, Schroeder H, Olivier JL, Turner JD. Epigenetic and Neurological Impairments Associated with Early Life Exposure to Persistent Organic Pollutants. Int J Genomics 2019; 20192085496
[http://dx.doi.org/10.1155/2019/2085496] [PMID: 30733955]

[30]   Tilson HA, Kodavanti PRS. Neurochemical effects of polychlorinated biphenyls: an overview and identification of research needs. Neurotoxicology 1997; 18(3): 727-43.
[PMID: 9339820]

[31]   Woolley DE. Neurotoxicity of DDT and possible mechanisms of action.Mechanisms of Action of Neurotoxic Substances. New York: Raven Press 1982; pp. 95-141.

[32]   Taylor JR. Neurological manifestations in humans exposed to chlordecone and follow-up results. Neurotoxicology 1982; 3(2): 9-16.
[PMID: 6186968]

[33]   Desaiah D. Biochemical mechanisms of chlordecone neurotoxicity: a review. Neurotoxicology 1982; 3(2): 103-10.
[PMID: 6186954]

[34]   Murphy S, McCabe N, Morrow C, Pearce B. Phorbol ester stimulates proliferation of astrocytes in primary culture. Brain Res 1987; 428(1): 133-5.
[http://dx.doi.org/10.1016/0165-3806(87)90091-5] [PMID: 3545395]

[35]   Girard PR, Kuo JF. Protein kinase C and its 80-kilodalton substrate protein in neuroblastoma cell neurite outgrowth. J Neurochem 1990; 54(1): 300-6.
[http://dx.doi.org/10.1111/j.1471-4159.1990.tb13315.x] [PMID: 2293618]

[36]   Costa LG. Ontogeny of second messenger systems.Handbook of Developmental Neurotoxicology. San Diego, CA: Academic Press 1998; pp. 275-84.
[http://dx.doi.org/10.1016/B978-012648860-9/50019-4]

[37]   Nihei MK, McGlothan JL, Toscano CD, Guilarte TR. Low level Pb(2+) exposure affects hippocampal protein kinase C gamma gene and protein expression in rats. Neurosci Lett 2001; 298(3): 212-6.
[http://dx.doi.org/10.1016/S0304-3940(00)01741-9] [PMID: 11165444]

[38]   Limke TL, Otero-Montañez JK, Atchison WD. Evidence for interactions between intracellular calcium stores during methylmercury-induced intracellular calcium dysregulation in rat cerebellar granule neurons. J Pharmacol Exp Ther 2003; 304(3): 949-58.
[http://dx.doi.org/10.1124/jpet.102.042457] [PMID: 12604669]

[39]   Kodavanti PRS, Tilson HA. Neurochemical effects of environmental chemicals: in vitro and in vivo correlations on second messenger pathways. Ann N Y Acad Sci 2000; 919: 97-105.
[http://dx.doi.org/10.1111/j.1749-6632.2000.tb06872.x] [PMID: 11083102]

[40]   Matsumura F, Ghiasuddin SM. Characteristics of DDT-sensitive Ca-ATPase in the axonic membrane.Neurotoxicology of Insecticides and Pheromones. New York: Plenum Press 1979; pp. 245-57.
[http://dx.doi.org/10.1007/978-1-4684-0970-3_12]

[41]   Stoke TE, Kavlock RJ. Pesticides as endocrine disrupting chemicals.Hayes' Handbook of Pesticide Toxicology. San Diego: Academic Press 2010; pp. 551-69.

[42]   Weiss B. The intersection of neurotoxicology and endocrine disruption. Neurotoxicology 2012; 33(6):

1410-9.
[http://dx.doi.org/10.1016/j.neuro.2012.05.014] [PMID: 22659293]

[43]   Giesy JP, Kannan K. Dioxin-like and non-dioxin-like toxic effects of polychlorinated biphenyls (PCBs): implications for risk assessment. Crit Rev Toxicol 1998; 28(6): 511-69.
[http://dx.doi.org/10.1080/10408449891344263] [PMID: 9861526]

[44]   Guzelian PS. Comparative toxicology of chlordecone (Kepone) in humans and experimental animals. Annu Rev Pharmacol Toxicol 1982; 22: 89-113.
[http://dx.doi.org/10.1146/annurev.pa.22.040182.000513] [PMID: 6177278]

[45]   Kelce WR, Stone CR, Laws SC, Gray LE, Kemppainen JA, Wilson EM. Persistent DDT metabolite p,p'-DDE is a potent androgen receptor antagonist. Nature 1995; 375(6532): 581-5.
[http://dx.doi.org/10.1038/375581a0] [PMID: 7791873]

[46]   Haarmann-Stemmann T, Bothe H, Abel J. Growth factors, cytokines and their receptors as downstream targets of arylhydrocarbon receptor (AhR) signaling pathways. Biochem Pharmacol 2009; 77(4): 508-20.
[http://dx.doi.org/10.1016/j.bcp.2008.09.013] [PMID: 18848820]

[47]   Barouki R, Coumoul X, Fernandez-Salguero PM. The aryl hydrocarbon receptor, more than a xenobiotic-interacting protein. FEBS Lett 2007; 581(19): 3608-15.
[http://dx.doi.org/10.1016/j.febslet.2007.03.046] [PMID: 17412325]

[48]   Puga A, Ma C, Marlowe JL. The aryl hydrocarbon receptor cross-talks with multiple signal transduction pathways. Biochem Pharmacol 2009; 77(4): 713-22.
[http://dx.doi.org/10.1016/j.bcp.2008.08.031] [PMID: 18817753]

[49]   Denison MS, Pandini A, Nagy SR, Baldwin EP, Bonati L. Ligand binding and activation of the Ah receptor. Chem Biol Interact 2002; 141(1-2): 3-24.
[http://dx.doi.org/10.1016/S0009-2797(02)00063-7] [PMID: 12213382]

[50]   Baccarelli A, Mocarelli P, Patterson DG Jr, *et al.* Immunologic effects of dioxin: new results from Seveso and comparison with other studies. Environ Health Perspect 2002; 110(12): 1169-73.
[http://dx.doi.org/10.1289/ehp.021101169] [PMID: 12460794]

[51]   Smialowicz RJ. The rat as a model in developmental immunotoxicology. Hum Exp Toxicol 2002; 21: 513-9.
[http://dx.doi.org/10.1191/0960327102ht290oa]

[52]   Camacho IA, Hassuneh MR, Nagarkatti M, Nagarkatti PS. Enhanced activation-induced cell death as a mechanism of 2,3,7,8-tetrachlorodibenzo-p-dioxin (TCDD)-induced immunotoxicity in peripheral T cells. Toxicology 2001; 165(1): 51-63.
[http://dx.doi.org/10.1016/S0300-483X(01)00391-2] [PMID: 11551431]

[53]   Dearstyne EA, Kerkvliet NI. Mechanism of 2,3,7,8-tetrachlorodibenzo-p-dioxin (TCDD)-induced decrease in anti-CD3-activated CD4(+) T cells: the roles of apoptosis, Fas, and TNF. Toxicology 2002; 170(1-2): 139-51.
[http://dx.doi.org/10.1016/S0300-483X(01)00542-X] [PMID: 11750091]

[54]   Ito T, Inouye K, Fujimaki H, Tohyama C, Nohara K. Mechanism of TCDD-induced suppression of antibody production: effect on T cell-derived cytokine production in the primary immune reaction of mice. Toxicol Sci 2002; 70(1): 46-54.
[http://dx.doi.org/10.1093/toxsci/70.1.46] [PMID: 12388834]

[55]   Fujimaki H, Nohara K, Kobayashi T, *et al.* Effect of a single oral dose of 2,3,7,8-tetrachlorodibenzo-p-dioxin on immune function in male NC/Nga mice. Toxicol Sci 2002; 66(1): 117-24.
[http://dx.doi.org/10.1093/toxsci/66.1.117] [PMID: 11861978]

[56]   Kerkvliet NI, Shepherd DM, Baecher-Steppan L. T lymphocytes are direct, aryl hydrocarbon receptor (AhR)-dependent targets of 2,3,7,8-tetrachlorodibenzo-p-dioxin (TCDD): AhR expression in both CD4+ and CD8+ T cells is necessary for full suppression of a cytotoxic T lymphocyte response by

TCDD. Toxicol Appl Pharmacol 2002; 185(2): 146-52.
[http://dx.doi.org/10.1006/taap.2002.9537] [PMID: 12490139]

[57] Crawford RB, Sulentic CE, Yoo BS, Kaminski NE. 2,3,7,8-Tetrachlorodibenzo-p-dioxin (TCDD)
      alters the regulation and posttranslational modification of p27kip1 in lipopolysaccharide-activated B
      cells. Toxicol Sci 2003; 75(2): 333-42.
      [http://dx.doi.org/10.1093/toxsci/kfg199] [PMID: 12883080]

[58] Park JH, Lee SW, Kim IT, Shin BS, *et al.* TCDD- up-regulation of IGFBP-6 and IL-5R alpha subunit
      genes in vivo and in vitro. Mol Cells 2001; 12: 372-9.

[59] Damstra T, Barlow S, Bergman A, Kavlock R, Van Der Kraak G. Global Assessment of the State-o-
      -Science of Endocrine Disruptors. International Programme on Chemical Safety 2002.

[60] Liappas I, Drago A, Zisakis AK, Malitas PN, *et al.* Thyroid hormone and affective disorders. Clinical
      Neuropsychiatry 2009; 6: 103-11.

[61] Boas M, Feldt-Rasmussen U, Main KM. Thyroid effects of endocrine disrupting chemicals. Mol Cell
      Endocrinol 2012; 355(2): 240-8. [PubMed: 21939731].
      [http://dx.doi.org/10.1016/j.mce.2011.09.005] [PMID: 21939731]

[62] Hagmar L. Polychlorinated biphenyls and thyroid status in humans: a review. Thyroid 2003; 13(11):
      1021-8. [PubMed: 14651786].
      [http://dx.doi.org/10.1089/105072503770867192] [PMID: 14651786]

[63] Salay E, Garabrant D. Polychlorinated biphenyls and thyroid hormones in adults: a systematic review
      appraisal of epidemiological studies. Chemosphere 2009; 74(11): 1413-9. [PubMed: 19108870].
      [http://dx.doi.org/10.1016/j.chemosphere.2008.11.031] [PMID: 19108870]

[64] Lauby-Secretan B, Loomis D, Grosse Y, *et al.* Carcinogenicity of polychlorinated biphenyls and
      polybrominated biphenyls. Lancet Oncol 2013; 14(4): 287-8.
      [http://dx.doi.org/10.1016/S1470-2045(13)70104-9] [PMID: 23499544]

[65] Xue J, Liu SV, Zartarian VG, Geller AM, Schultz BD. Analysis of NHANES measured blood PCBs in
      the general US population and application of SHEDS model to identify key exposure factors. J Expo
      Sci Environ Epidemiol 2014; 24(6): 615-21.
      [http://dx.doi.org/10.1038/jes.2013.91] [PMID: 24424407]

[66] Duntas LH, Stathatos N. Toxic chemicals and thyroid function: hard facts and lateral thinking. Rev
      Endocr Metab Disord 2015; 16(4): 311-8.
      [http://dx.doi.org/10.1007/s11154-016-9331-x] [PMID: 26801661]

[67] Giera S, Bansal R, Ortiz-Toro TM, Taub DG, Zoeller RT. Individual polychlorinated biphenyl (PCB)
      congeners produce tissue- and gene-specific effects on thyroid hormone signaling during development.
      Endocrinology 2011; 152(7): 2909-19.
      [http://dx.doi.org/10.1210/en.2010-1490] [PMID: 21540284]

[68] Naveau E, Pinson A, Gérard A, *et al.* Alteration of rat fetal cerebral cortex development after prenatal
      exposure to polychlorinated biphenyls. PLoS One 2014; 9(3)e91903
      [http://dx.doi.org/10.1371/journal.pone.0091903] [PMID: 24642964]

[69] Fritsche E, Cline JE, Nguyen NH, Scanlan TS, Abel J. Polychlorinated biphenyls disturb
      differentiation of normal human neural progenitor cells: clue for involvement of thyroid hormone
      receptors. Environ Health Perspect 2005; 113(7): 871-6.
      [http://dx.doi.org/10.1289/ehp.7793] [PMID: 16002375]

[70] Kitamura S, Kato T, Iida M, Jinno N, Suzuki T, Ohta S, *et al.* Antithyroid hormonal activity of
      tetrabromobisphenol A, a flame retardant, and related compounds: affinity to the mammalian thyroid
      hormone receptor, and effect on tadpole metamorphosis. Life Sci 2005; 76(14): 1589-601.

[71] Fini JB, Le Mével S, Turque N, *et al.* An *in vivo* multiwell-based fluorescent screen for monitoring
      vertebrate thyroid hormone disruption. Environ Sci Technol 2007; 41(16): 5908-14.
      [http://dx.doi.org/10.1021/es0704129] [PMID: 17874805]

[72]   Szabo DT, Richardson VM, Ross DG, Diliberto JJ, Kodavanti PR, Birnbaum LS. Effects of perinatal PBDE exposure on hepatic phase I, phase II, phase III, and deiodinase 1 gene expression involved in thyroid hormone metabolism in male rat pups. Toxicol Sci 2009; 107(1): 27-39.
[http://dx.doi.org/10.1093/toxsci/kfn230] [PMID: 18978342]

[73]   Hoffman K, Sosa JA, Stapleton HM. Do flame retardant chemicals increase the risk for thyroid dysregulation and cancer? Curr Opin Oncol 2017; 29(1): 7-13.
[http://dx.doi.org/10.1097/CCO.0000000000000335] [PMID: 27755165]

[74]   Han Z, Li Y, Zhang S, *et al.* Prenatal transfer of decabromodiphenyl ether (BDE-209) results in disruption of the thyroid system and developmental toxicity in zebrafish offspring. Aquat Toxicol 2017; 190: 46-52.
[http://dx.doi.org/10.1016/j.aquatox.2017.06.020] [PMID: 28686898]

[75]   Zhou T, Ross DG, DeVito MJ, Crofton KM. Effects of short-term *in vivo* exposure to polybrominated diphenyl ethers on thyroid hormones and hepatic enzyme activities in weanling rats. Toxicol Sci 2001; 61(1): 76-82.
[http://dx.doi.org/10.1093/toxsci/61.1.76] [PMID: 11294977]

[76]   Lee E, Kim TH, Choi JS, *et al.* Evaluation of liver and thyroid toxicity in Sprague-Dawley rats after exposure to polybrominated diphenyl ether BDE-209. J Toxicol Sci 2010; 35(4): 535-45.
[http://dx.doi.org/10.2131/jts.35.535] [PMID: 20686340]

[77]   Marteinson SC, Palace V, Letcher RJ, Fernie KJ. Disruption of thyroxine and sex hormones by 1,2dibromo (1,2dibromoethyl) cyclohexane (DBEDBCH) in American kestrels (Falco sparverius) and associations with reproductive and behavioral changes. Environ Res 2017; 154: 389-97.

[78]   Roze E, Meijer L, Bakker A, Van Braeckel KN, Sauer PJ, Bos AF. Prenatal exposure to organohalogens, including brominated flame retardants, influences motor, cognitive, and behavioral performance at school age. Environ Health Perspect 2009; 117(12): 1953-8.
[http://dx.doi.org/10.1289/ehp.0901015] [PMID: 20049217]

[79]   Stapleton HM, Eagle S, Anthopolos R, Wolkin A, Miranda ML. Associations between polybrominated diphenyl ether (PBDE) flame retardants, phenolic metabolites, and thyroid hormones during pregnancy. Environ Health Perspect 2011; 119(10): 1454-9.
[http://dx.doi.org/10.1289/ehp.1003235] [PMID: 21715241]

[80]   Zota AR, Park JS, Wang Y, Petreas M, Zoeller RT, Woodruff TJ. Polybrominated diphenyl ethers, hydroxylated polybrominated diphenyl ethers, and measures of thyroid function in second trimester pregnant women in California. Environ Sci Technol 2011; 45(18): 7896-905.
[http://dx.doi.org/10.1021/es200422b] [PMID: 21830753]

[81]   Herbstman JB, Sjödin A, Kurzon M, *et al.* Prenatal exposure to PBDEs and neurodevelopment. Environ Health Perspect 2010; 118(5): 712-9.
[http://dx.doi.org/10.1289/ehp.0901340] [PMID: 20056561]

[82]   Chen MH, Ha EH, Liao HF, *et al.* Perfluorinated compound levels in cord blood and neurodevelopment at 2 years of age. Epidemiology 2013; 24(6): 800-8.
[http://dx.doi.org/10.1097/EDE.0b013e3182a6dd46] [PMID: 24036611]

[83]   Olsen GW, Burris JM, Ehresman DJ, *et al.* Half-life of serum elimination of perfluorooctanesulfonate, perfluorohexanesulfonate, and perfluorooctanoate in retired fluorochemical production workers. Environ Health Perspect 2007; 115(9): 1298-305.
[http://dx.doi.org/10.1289/ehp.10009] [PMID: 17805419]

[84]   Lindstrom AB, Strynar MJ, Libelo EL. Polyfluorinated compounds: past, present, and future. Environ Sci Technol 2011; 45(19): 7954-61.
[http://dx.doi.org/10.1021/es2011622] [PMID: 21866930]

[85]   Ye X, Kato K, Wong LY, Jia T, *et al.* Per and polyfluoroalkyl substances in sera from children 3 to 11 years of age participating in the National Health and Nutrition Examination Survey 2013–2014

[86]   Yu WG, Liu W, Jin YH. Effects of perfluorooctane sulfonate on rat thyroid hormone biosynthesis and metabolism. Environ Toxicol Chem 2009; 28(5): 990-6.
[http://dx.doi.org/10.1897/08-345.1] [PMID: 19045937]

[87]   Weiss JM, Andersson PL, Lamoree MH, Leonards PE, van Leeuwen SP, Hamers T. Competitive binding of poly- and perfluorinated compounds to the thyroid hormone transport protein transthyretin. Toxicol Sci 2009; 109(2): 206-16.
[http://dx.doi.org/10.1093/toxsci/kfp055] [PMID: 19293372]

[88]   Lewis DFV, Hlavica P. Interactions between redox partners in various cytochrome P450 systems: functional and structural aspects. Biochim Biophys Acta 2000; 1460(2-3): 353-74. [PubMed].
[http://dx.doi.org/10.1016/S0005-2728(00)00202-4] [PMID: 11106776]

[89]   Letcher RJ, Klasson-Wehler E, Bergman A. Methyl sulfone and hydroxylated metabolites of polychlorinated biphenyls.The Handbook of Environmental Chemistry, Vol 3 Part K, New Types of Persistent Halogenated Compounds (Paasivirta J, ed). Berlin: Springer-Verlag 2000; 3: pp. 315-5.
[http://dx.doi.org/10.1007/3-540-48915-0_11]

[90]   Barber JL, Walsh MJ, Hewitt R, Jones KC, Martin FL. Low-dose treatment with polybrominated diphenyl ethers (PBDEs) induce altered characteristics in MCF-7 cells. Mutagenesis 2006; 21(5): 351-60.
[http://dx.doi.org/10.1093/mutage/gel038] [PMID: 16980705]

[91]   Lundgren M, Darnerud PO, Molin Y, Lilienthal H, Blomberg J, Ilbäck NG. Viral infection and PBDE exposure interact on CYP gene expression and enzyme activities in the mouse liver. Toxicology 2007; 242(1-3): 100-8.
[http://dx.doi.org/10.1016/j.tox.2007.09.014] [PMID: 17964055]

[92]   Narimatsu S, Nakanishi R, Hanioka N, Saito K, Kataoka H. Characterization of inhibitory effects of perfluorooctane sulfonate on human hepatic cytochrome P450 isoenzymes: focusing on CYP2A6. Chem Biol Interact 2011; 194(2-3): 120-6.
[http://dx.doi.org/10.1016/j.cbi.2011.09.002] [PMID: 21964418]

[93]   Machala M, Neča J, Drábek P, *et al.* Effects of chronic exposure to PCBs on cytochrome P450 systems and steroidogenesis in liver and testis of bulls (Bos taurus). Comp Biochem Physiol A Mol Integr Physiol 1998; 120(1): 65-70.
[http://dx.doi.org/10.1016/S1095-6433(98)10011-9] [PMID: 9773499]

[94]   Miller WL. Molecular biology of steroid hormone biosynthesis. Endocr Rev 1988; 9: 29553 18.

[95]   Blom S, Forlin L. Effect of PCBs on Xenobiotic Biotransformation enzymes in the liver and 21-hydroxylation in the head kidney of Rainbow trout. Aquat Toxicol 1997; 39: 215-30.
[http://dx.doi.org/10.1016/S0166-445X(97)00035-0]

[96]   Safe SH. Polychlorinated biphenyls (PCBs): environmental impact, biochemical and toxic responses, and implications for risk assessment. Crit Rev Toxicol 1994; 24(2): 87-149.
[http://dx.doi.org/10.3109/10408449409049308] [PMID: 8037844]

[97]   Mckinney JD, Singh P. Structure activity relationships in halogenated biphenyls; unifying hypothesis for structural specificity. Chem Biol 1981; Interact. 33: 271-83.

[98]   Pirhadi S, Sunseri J, Koes DR. Open source molecular modeling. J Mol Graph Model 2016; 69: 127-43.
[http://dx.doi.org/10.1016/j.jmgm.2016.07.008] [PMID: 27631126]

[99]   Satpathy R. Quantitative structure–activity relationship methods for the prediction of the toxicity of pollutants. Environ Chem Lett 2019; 17(1): 123-8.
[http://dx.doi.org/10.1007/s10311-018-0780-1]

Enzyme Inhibitors for Breast Cancer Therapy

Hariharan Jayaraman[1], Praveen Kumar Posa Krishnamoorthy[1], Lakshmi Suresh[1], Mahalakshmi Varadan[1], Aparna Madan[1] and Balu Ranganathan[2,3,*]

[1] *Department of Biotechnology, Sri Venkateswara College of Engineering, Sriperumbudur Tk. 602 117, Kancheepuram District, Tamil Nadu, India*

[2] *Palms Connect Sdn Bhd, Shah Alam 40460, Selangor Darul Ehsan, Malaysia*

[3] *Palms Connect LLC, Showcase Lane, Sandy, UT 84094, USA*

Abstract: Globally millions of women die of cancer, the most common cancer occurrence sites for women being breast, cervical and ovarian. Engineered enzyme inhibitors are a component of the drug regimen that has taken a drive into pharmaceutical international corporations' product portfolios which is by targeted delivery, moderating disease-free survival and leading to procrastination of death. Since 2002 the enzyme inhibitor anastrozole (Arimidex) is used as the first drug of choice for breast cancer which is available in the commercial market. Currently, there are several other FDA approved enzyme inhibitors like sulfonanilide analogs available in the pharmaceutical shelf decreasing aromatase expression and regulating enzyme activity for the treatment and cure of breast cancer.

Keywords: Anastrozole, Breast cancer, Celecoxib, Enzyme inhibitors, HSD17B, TLK.

INTRODUCTION

Breast cancer is one of the deadliest diseases occurring in women globally. Next to lung cancer, breast cancer ranks second accounting for 11.6% death. Whereas it ranks No.1 among the females with a high mortality rate subsequently followed by lung cancer and colorectal cancer [1]. Cancer is an uncontrolled proliferation of cells which is a result of changes in the genes such as oncogenes and tumor suppressor genes. Cancer is still a mystery due to the complex changes that occur in the cellular environment due to the genetic and non-genetic influences. The primary causative agents of cancer are much less understood until the 1980s. However, knowledge on the role of certain factors contributing to the etiology of breast cancer received attention in the 1980s. Risk factors associated with breast

---

* **Corresponding author Balu Ranganathan:** Palms Connect LLC, Showcase Lane, Sandy, UT 84094, USA; E-mail: ranga@palmsconnect.com

G. Baskar, K. Sathish Kumar & K. Tamilarasan (Eds.)

cancer include age, exposure to radiation, diet, body weight, family history and consumption of exogenous hormones as oral contraceptives [2]. Recent pieces of evidence suggest that single target based drug therapy is a limitation to efficiently treat cancer. Hence the pharmaceutical companies are currently exploring multiple target based drug therapy for their efficacy in controlling the disease as a more potent drug regimen. Currently, pharmaceutical companies are producing drugs for chemotherapy (small molecules), hormonal therapy (therapeutic hormones) and targeted therapy (therapeutic proteins) of breast cancer. Top 10 pharmaceutical companies which are involved in the production of oncology products in the year 2018 includes Roche, Novartis, Celgene, Johnson and Johnson, Bristol-Myers Squibb, Pfizer, Merck, AstraZeneca, Lilly and Abbvie [3]. Some of the breast cancer drugs manufactured by these companies are mentioned in the below table (Table. **1**) with their trade names and the type of therapy in which they are used.

The existing treatment strategies for breast cancer include unique or combination of chemotherapy, hormone therapy and targeted therapy. While chemotherapy is nonselective and hormonal therapy is partially specific to the estrogen producing and responding to cell types, targeted therapy specifically targets cells that overexpress HER2. Earlier the drugs which are used to treat cancer-targeted only single target, whereas now new drugs such as sorafenib and sunitinib are proven

Table 1. Breast cancer drugs.

S.No	Company	Brand Name	Generic Name	Therapy Type	Mechanism of Action
1.	Roche	Avastin	Bevacizumab	Targeted therapy	Blocks VEGF
2.	Roche	Herceptin	Transtuzumab	Targeted therapy	Block HER-2
3.	Pfizer	Ibrance	Palbociclib	Chemotherapy	CDK 4/6 Inhibitor
4.	Roche	Perjeta	Pertuzumab	Targeted therapy	Block HER-2
5.	Novartis	Afinitor	Everolimus	Chemotherapy	mTOR inhibitor
6.	Astra Zeneca	Nolvadex	Tamoxifen	Hormone therapy	Blocks Estrogen Receptors (SERM)
7.	Astra Zeneca	Faslodex	Fulvestrant	Hormone therapy	Blocks Estrogen Receptors (SERD)
8.	Arimidex	Arimidex	Anastrozole	Hormone therapy	Stops Estrogen synthesis (AIs)
9.	Novartis	Femara	Letrozole	Hormone therapy	Stops Estrogen synthesis (AIs)
10.	Pfizer	Aromasin	Exemestane	Hormone therapy	Stops Estrogen synthesis (AIs)

to have efficacy towards multiple targets such as PDGFR and VEGFR [4] making a milestone in the cancer treatment progression. A treatment regimen recommended to the breast cancer patient upon identification of the breast cancer subtype and the grade of the tumor. Because it is observed that breast cancer drugs respond differently to different molecular subtypes of cancer. For example, the drugs paclitaxel and doxorubicin showed a different gene expression profile in basal-like and erbB2+ subtypes of breast cancer than luminal and normal-like cancers [5].

Mammary Gland Anatomical Construction

The female breast is an extremely interlocking and tortuous organ involving lactation, emotional excitation and social blossom. It undergoes interchanged modifications in the life span of a female than any other part of the human body – from birth, upon puberty, during the periodical menstrual ovarian cycle, pregnancy leading to lactation and till menopause. The various parts of breast being:

- Lobule (tiny bulb-like structure) produces milk
- Lobe (15-20 sections of lobules)
- Duct, the tube through which milk travels through from lobes
- Nipple, milk is secreted through from the duct
- Areola

There is a complete absence of muscles and bones in the breast. Fat, connective tissue and ligaments act as the filler in between lobes and ducts which provides the shape, structure and aesthetics to the breasts. Basically carcinoma of the breast is seen in the lobules, ducts and rarely in the nipple.

Breast Carcinogenesis

Cancer in the Duct

Ductal carcinoma in situ (DCIS) is the earliest form of breast cancer where cancer is at the cellular level in the duct. Cancer cells are only present and not cancerous tissue. Cancerous cells are observed in the duct. It is being localized, contained to a local region and which has not spread across to the normal breast tissues (Fig. 1).

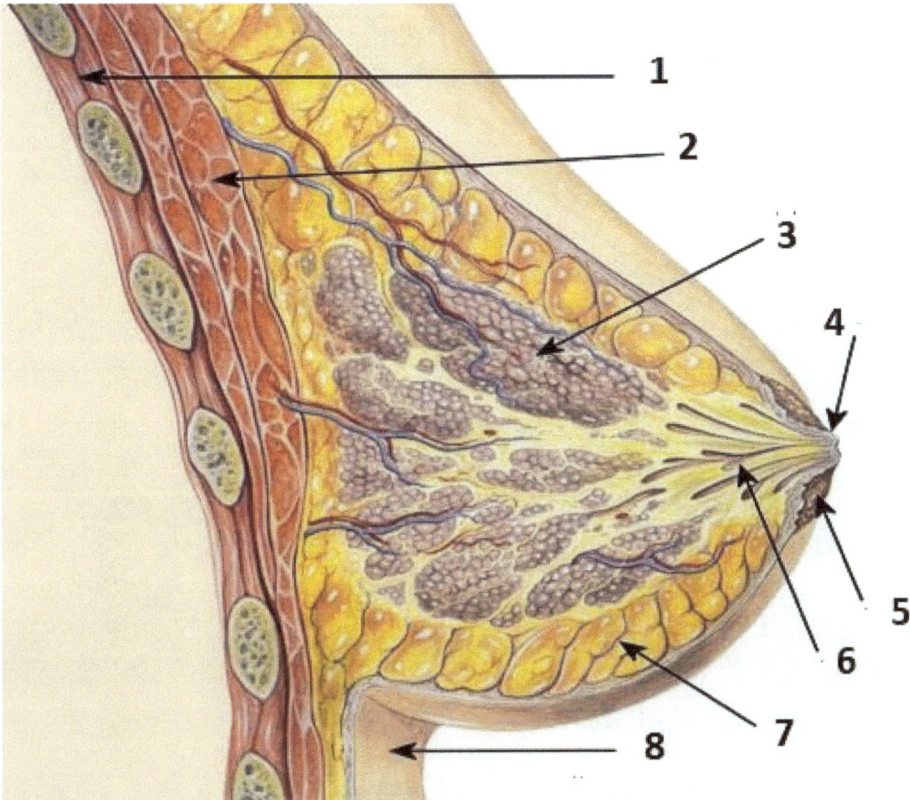

Fig. (1). Normal anatomy of the breast. 1. Chest wall 2. Pectoralis muscles 3. Lobules 4. Nipple 5. Areola 6. Milk duct 7. Fatty tissue 8. Skin.

(**Picture Credit**: By Original author: Patrick J. Lynch. Reworked by Morgoth666 to add numbered legend arrows. - Patrick J. Lynch, medical illustrator, CC BY 3.0, https://commons.wikimedia.org/w/index.php?curid=2676813)

Invasive ductal carcinoma (IDC): cancer invaded deeper inside the breast which has started and sharing the origin from ducts that have spread beyond and outside the place of origin (ducts). This is the most common type of breast cancer (approximately 60% of all breast cancers)

Cancer in the Lobules

Lobular carcinoma in situ: Cancer cells are observed in the lobules which are the milk-secreting component of the female breast (Fig. **2**).

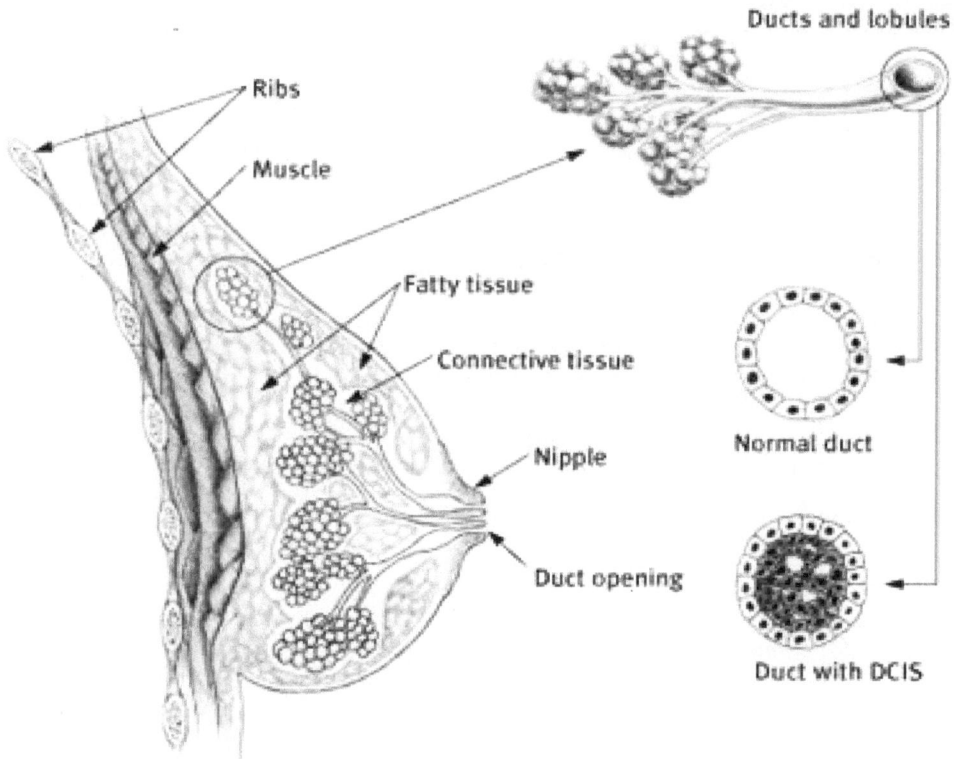

Fig. (2). Ductal carcinoma in situ (**Picture Credit**: Cancer council, Victoria, Australia).

Invasive lobular carcinoma: Originates in the lobules and spreads to the breast tissues which are present at the vicinity of origin and form a malignant tumorous mass, signs, and symptoms include a lump in the breast, swelling or skin thickening around the region of the lump, and change in breast profile and morphology. This subset comprises about 10-15% of invasive breast cancers, affects mostly women at their menopause.

Cancer in the Nipple

Paget's disease of the breast – an uncommon type of breast cancer that usually shows as a red, scaly rash involving the nipple.

Molecular subtypes of breast cancer: This type of breast cancer category is based on receptor activities.

HER2-positive: Human epidermal growth factor receptor-2 (HER2) -positive breast cancer is a molecular subtype of breast cancer that tests positive for this HER2 protein receptor, This protein promotes the growth of breast cancer cells.

Luminal A: This is the most common subtype. Estrogen receptor (ER) and progesterone receptor (PR) plays an active role in this subtype. The clinical samples give positive results for ER and PR testing.

Luminal B: In this subtype category, the clinical samples are ER-positive, PR negative and HER2 positive.

Triple-negative breast cancer: In this subtype category, the clinical samples are negative for all the three receptors, namely, ER, PR and HER2. The subtype usually originates from the ducts.

Enzymes Predominantly in the Breast

Aromatase

The main function of aromatase is to synthesize estrogen. Aromatase protein is expressed mainly in the following regions. ovaries in the premenopausal stage, the placenta of pregnant women and adipose fibroblast cells of women in the postmenopausal stage. An increase in aromatase expression is observed in breast adipose tissues of breast cancer patients [6]. The major function of this enzyme is the biocatalytic action of the terminal rate-limiting reaction for the biosynthesis of estrogen. The human form of the aromatase enzyme is located on chromosome 15 and is a constituent of the cytochrome P450 family as well as the result of the CYP19A1 gene. Aromatase contains a microsomal enzymatic complex which is present in a heterodimer structure. A steroid-binding site and a heme group are present on the catalytic segment of cytochrome P450 aromatase. The active site of aromatase is present within the enzyme and contains multiple closely packed hydrophobic residues. The enzyme is a 55-kDa protein of 503 amino acids.

In the homosapiens, the CYP19A1 gene situated at chromosome 15 translates to a cytochrome P450 constituent which is compiled of cytochrome P450 aromatase and omnipresent NADPH cytochrome P450 reductase. Specificity to androgen is primarily due to the structure in the crystallized form of the enzyme. Aromatase activity is high in the quadrant containing the tumor, where the primary repertoire is from the outer surface (peripheral) membrane conversion whereupon suppression leads to a substantial change in the magnitude of estrogens. Hence, aromatase inhibitors serve as a potential treatment for breast cancer [7].

Aromatase inhibitors (AIs) are categorized into steroidal and non-steroidal.

Steroidal inhibitors known as type 1 are similar in structure to the aromatase substrate, androstenedione. Due to their similarity, they adjoin with the substrate-binding site. Irreversible inactiveness results due to the conversion to a reactive intermediary involving covalent bonding. This includes formestane and exemestane. Non-covalent binding by type 2, non-steroidal inhibitors to the heme molecule of the enzyme (aromatase) suppresses the binding of androgens. Fadrozole, Vorozole and Anastrazole act as competitive inhibitors for androgen which is of reversible mechanism.

Cyclooxygenases

Cyclooxygenase (prostaglandin-endoperoxide synthase) is a heme-containing glycoprotein that belongs to a family of enzymes called COX. This enzyme plays a key role in metabolizing arachidonic acid to produce prostaglandins (PGs) and thromboxanes. There are two isoforms of COX: COX-1 and COX-2 which are 60% identical in its amino acid sequence [8 - 11].

Location

COX-1 is generally expressed at a high level in all tissues and cells in animal species and it is mapped to chromosome 9q32-9q33 in humans. The size of the gene is 22kb with 11 exons. The size of mRNA coding COX-1 is 2.8kb. COX-2 is generally inducible and its gene is located in chromosome 1q25.2-q25.3. Its gene and mRNA size: 8.3kb and 4.6kb respectively.

Mechanism

Membrane arachidonic acid, released by the action of phospholipase A_2, is subjected to the catalytic activity by COX to produce PGG_2 which is then reduced to PGH_2 by peroxidase activity of COX. This PGH_2 is highly reactive and produces cyclic endoperoxides which then converted to various PG_s by the action of tissue-specific isomerase (Fig. **3**).

COX-1 is generally referred to as "constitutive isoenzymes" because its gene is similar to "housekeeping genes" [12]. It is generally expressed in all tissues and cells. But COX-2 expression is inducible by various factors/compounds like inflammatory cytokines, growth factors and tumor promoters. So, COX-2 is known as "inducible isoenzymes". The induction of COX-2 is done by the upregulation of genes for transcription. Multiple signals can trigger the transcription of the COX-2 gene by binding transcription factors at specific binding sites. These binding sites include cyclic AMP response element (CRE), and binding sites for nuclear factor B (NF B), nuclear factor interleukin-6 (NF-IL6), Myb and Ets factors. Serum, growth factors like PDGF, oncogene products and Tumor necrosis factor (TNF) also plays a critical role in the upregulation of COX-2. PGE_2 acts as a mitogen that stimulates the growth of epithelial cells promoting cell growth and proliferation in the presence of EGF.COX-2 has a large active site, so substrates other than AA can bind. COX-2 overexpression may lead to a detectable change in cAMP and ca^{2+}levels which then helps in the tumor cells to invade the neighboring host's normal cell. This contact with the neighboring cells is very important in the survival and progression of the tumor cell. This then results in metastasis.COX-2 reduces the apoptosis of tumor cell by a mechanism of activation of the PI3 kinase/Akt pathway by PGE2, which results in the upregulation of antiapoptotic molecules that ultimately reduces the cellular level of arachidonic acid, which is an inducer of apoptosis [13]. PG_s also reduce the growth of immune cells leads to immunosuppression.

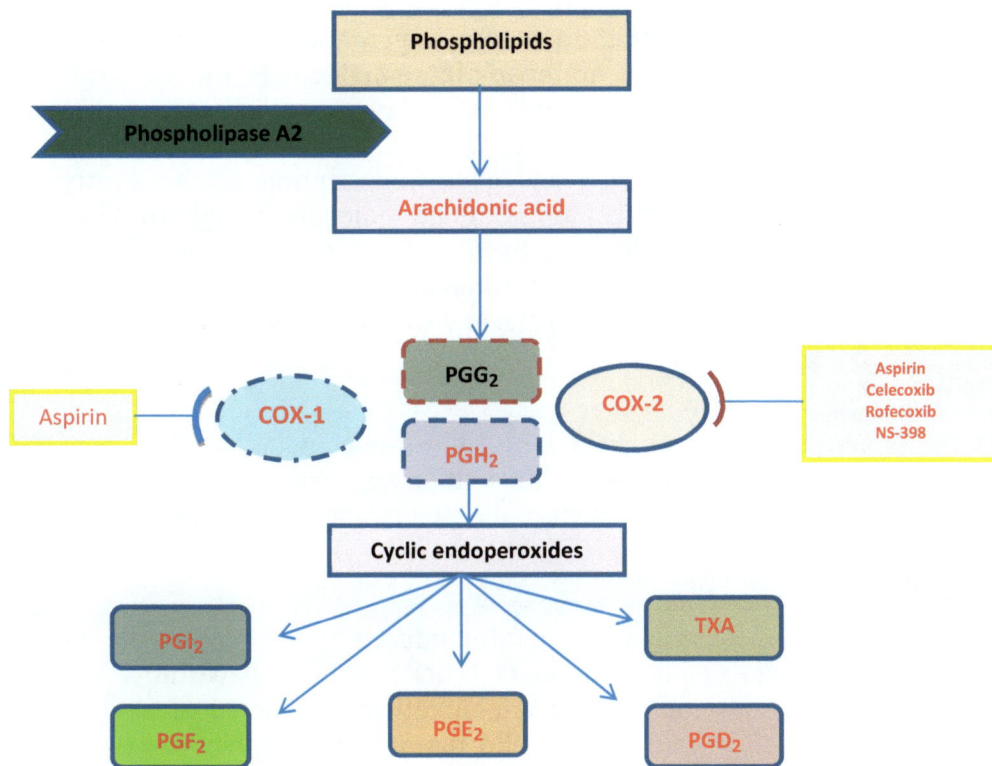

Fig. (3). Mechanism of Cyclooxygenase (COX).

COX-2 Inhibitors

COX-2 became a useful target in the prevention and treatment of many epithelial cancers including breast cancer. COX-2 inhibitors are generally anti-inflammatory drugs with anti-inflammatory, antipyretic and pain-relieving actions. these drugs are termed as nonsteroidal anti-inflammatory drugs (NSAIDs). This mainly includes nonselective COX inhibitors like aspirin, diclofenac, indomethacin, ketoprofen, naproxen, ibuprofen, phenylbutazone, and meclofenamate, exerts a similar effect on both COX-1 and COX-2. A subgroup of preferential COX-2 inhibitors does exist among the nonselective COX inhibitors group. These include celecoxib, rofecoxib, lumiracoxib, etoricoxib, valdecoxib, and nimesulide [14].

LYSYL OXIDASE

Mechanism of Action

Inception property of the copper-dependent amine oxidase enzyme exhibits oxidation of primary amine substrates to responsive (proactive) aldehydes. Lysyl oxidase (LOX) belongs to this group of enzymes. Maintenance of the tensile (mechanical) strength of tissues in different organs is the chief role of LOX. This family of enzymes crosslinks the tissue collagen and/or elastin present in the extracellular matrix producing aminoadipic semialdehyde by the oxidation of peptidyl lysine. Aminoadipic semialdehyde plays a critical role in the tensile strength of the tissues [15]. Other intracellular functions include regulation of cell differentiation, motility/migration and gene transcription. LOX, a polypeptide contains 417 amino acids with an approximate molecular weight of 32 kDa. The signal peptide is present in the first twenty amino acid residues. The catalytic active site is present in the carboxyl-terminal of the polypeptide which constitutes the active copper (II) ion, lysine, tyrosine and cysteine residues (Fig. **4**).

Genetics

In homo sapiens chromosome 5q23.3-31.2 location contains the LOX gene. LOX, a polypeptide contains 417 amino acids with an approximate molecular weight of 32 kDa. The signal peptide is present in the first twenty amino acid residues.

Tumorigenesis

The expression of LOX protein is generally induced by the hypoxia condition as a result of the hypoxia response element (HRE) embedded within its promoter sequence. Hypoxia is a condition in which cells are depleted of oxygen, it is a salient feature of solid tumors. The solid tumors characterized by dysregulated growth leading to oxygen depletion [16]. The propeptide is cleaved by the action of procollagen – C – proteinase into two distinct parts: the mature form and the LOX propeptide, this mature LOX protein modifies the extracellular matrix, facilitating the formation of cancer niches where cancer (tumor) develops and eventually undergoes metastasis. The LOX propeptide on the other hand has tumor inhibiting action [17, 18]. The correlation between tumor progression and LOXL2 expression depends upon the type of tissue. The elevated level of LOX mRNA expression is observed in highly invasive/metastatic breast cancer [19 - 21]. FAK/Src signaling pathway is involved *via* the hydrogen peroxide–intermediated mechanism for LOX regulated adhesion formation. LOXL-2 activity is very high in basal-like carcinoma cells, which affects the tight junction and cell polarity complexes by a mechanism of downregulating the involved genes.

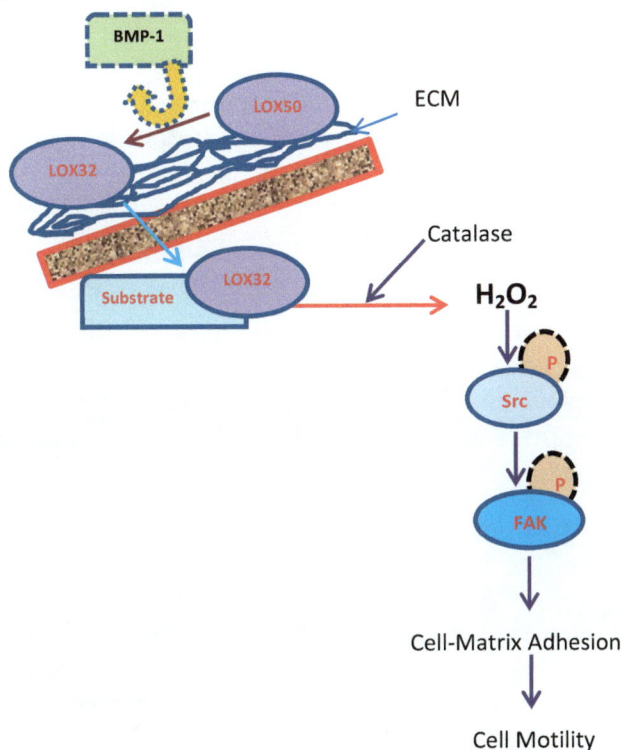

Fig. (4). Mechanism of LOX in promoting tumor adhesion and mobility.

Mechanism of Action

Copper ions play a crucial role in the biochemical pathway of living organisms which is a major active component of LOX. The redox activity of copper offers effective functionalities as well as toxic activity to proteins. LOX is the one responsible for adhesion and mobility of tumor cells. An inactive proenzyme is cleaved into active LOX by Bone Morphogenetic Protein-1 (BMP1). An inactive proenzyme is cleaved into active LOX by Bone Morphogenetic Protein-1 (BMP1). Active LOX enters by translocation into the cell thereupon intercommunicate by enzymatic pathway whereby producing hydrogen peroxide as a by-product. The produced by-product facilitates the FAK/Src signaling pathway. Src and FAK are the key elements in cell adhesion and mobility. Upon tumor progression (proliferative immortality, angiogenesis and metastasis) clinical trials have also revealed that tumor tissue and blood serum contain an increased level of copper content [22 - 24] in breast cancer patients.

HYDROXYSTEROID DEHYDROGENASES

Ovaries are the main and primary source of estrogen production for the pre-menopausal women but it is a different physiological manifestation for menopausal women where peripheral tissues play a major role [25]. Relatively a major section of the breast cancer (tumor) patients (60-80%) express high levels of estrogen where the cytosol-protein concentration of equal to or more than 10 fmol/mg carried out by ligand binding assay (Gold standard) is considered estrogen positive, hence a proliferative effect is observed [26].

Mechanism of Action

Hydroxysteroid dehydrogenases (HSDs) are a group of alcohol oxidoreductases that catalyze the interconversion of alcohol and carbonyl functional groups which play a major role in steroid biosynthesis and metabolism. There are about four different classes of HSDs namely, 3β-, 11β-, 17β-, and 20α-HSDs of which the 3β and 17β- HSDs are reported to play a crucial role in the regulation of hormone-dependent breast cancers. This section gives a brief about these two enzymes.

17β- Hydroxysteroid Dehydrogenase

HSD17B (EC 1.1.1.51) is a prominent oxidoreductase enzyme that catalyses both the reduction of 17-ketosteroids and the dehydrogenation of 17β- hydroxysteroids during steroidogenesis and steroid metabolism. Of all the reactions catalysed by this enzyme, the interconversion of estrone and estradiol appears to be the most prominent one in cases of breast carcinomas [27]. The conversion of adrenal androgen to estrogen in women post-menopause by this enzyme serves as a major stimulator of breast carcinoma cells and 14 isozymes of HSD17B have been identified to promote this activity. Despite the relatively low sequence homogeneity (only about 20-30%) observed among the isozymes of HSD17, Intersection of activity at a significant level has been observed in the enzymatic biocatalytic activity among HSD17B1, 3, 5, 7,12 and 2, 4, 14. Reductive as well as oxidative biocatalytic activity of these enzymes differs upon the corresponding substrate on which they act upon and their pattern of expression [28]. The main role of HSD17B1 is to mediate the reduction of E1(Estrone) to E2(Estradiol), Dehydroepiandrosterone (DHEA) to androstenediol, which leads to reduced androgenic and increased estrogenic activity and metabolizes DHT into 3β-diol and 3α-diol both of which have a much higher affinity for estrogen receptor (ER). On the other hand, HSD17B2 catalyzes the oxidation of E2 to E1, testosterone to androstenedione and androstenediol to DHEA [29]. In a brief, both these isozymes perform contrasting functions thereby regulating the proliferation, metastasis and inhibition of tumor cells. The control of the relative expression levels of HSD17B isozymes in human breast carcinomas is thought to play a pivotal role in the supply of estradiol to estrogen receptor (ER) containing breast cancer cells [30]. E2 binds to estrogen receptors (ERα and ERβ) or the G protein-coupled membrane receptor (GPR30), after which it recruits promoters of several genes related to proliferation, thus stimulating cell growth [31] (Fig. **5**).

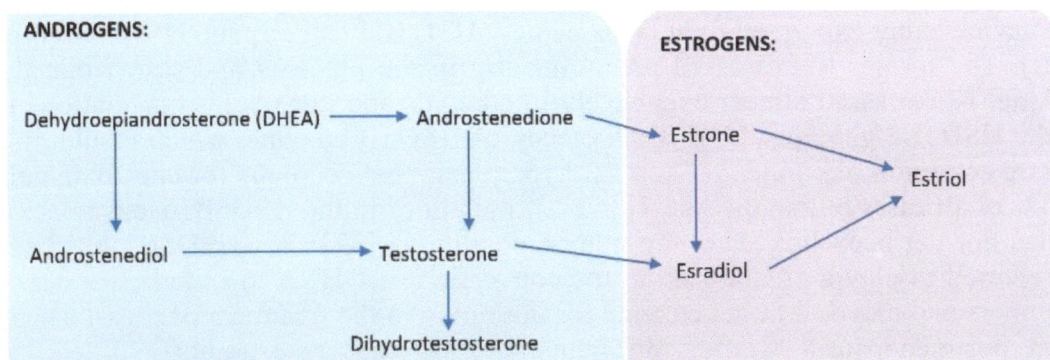

Fig. (5). HSD17B Catalyzed reactions.

From the breast cancer patient biochemical laboratory analysis perspective, a high ratio of [E2]/[E1] in the pathological cellular milieu contributes very much in a very significant way for breast cancer cell proliferation. For breast cancer patients in the menopausal group intra-tumoral [E2]/[E1] is very significant. Breast cancer therapy leading to disease-free survival should be aimed at decreasing the production of E2 thereby reducing the ratio of [E2]/[E1] should be worked upon by the biotechnology-based start-up companies [32].

Recent studies in MCF-7 breast cancer cell lines demonstrated varied gene expression modulation at the micro (mRNA) and also at macro (protein) levels carried out by HSD17B which also involved in cell proteome of breast cancer. It also facilitates the migration of breast cancer cells even when it exhibits positive regulation of the anti-metastatic gene NM23 which is further correlated in the stimulation of breast cancer cell growth thereupon confirming the targeted therapy for breast cancer [33]. Irrespective of the ER status be it +ve or –ve, HSD17B1 mRNA can be considered as a prognostic marker for BC progression, where the breast cancer patient tumors expressing HSD17B1 mRNA manifested shorter disease-free survival proving higher fatality [34]. These factors add up to the prominence of HSD17B in breast cancer studies.

3β- Hydroxysteroid Dehydrogenase

HSD3B or Δ^{5-4} isomerase (EC 1.1.1.145) is a member of the oxidoreductase family and is a bifunctional enzyme that catalyses the oxidation of Delta (5)-en--3-beta-hydroxy steroids and the oxidation of ketosteroids. It is involved in the conversion of progesterone from pregnenolone, 17α-hydroxyprogesterone from 17α-hydroxy pregnenolone, and androstenedione from dehydroepiandrosterone (DHEA) thus playing a crucial role in the biosynthesis of all classes of hormonal steroids. Though not being a member of cytochrome P450 class, this enzyme participates in the adrenal pathway of corticosteroid synthesis making it unique among the other members of the HSD family [35]. There are two isozymes of this enzyme being expressed in humans namely HSD3B1 (type I) and HSD3B2 (type II). The former is expressed predominantly in the placenta and skin while the latter is expressed almost exclusively in adrenals and gonads [36]. Mutations in the HSD3B2 gene results in a deficiency of HSD3B enzyme, which results in a rare condition of congenital adrenal hyperplasia that accounts for approximately 1% of all cases of that disease. However, mutations in the HSD3B1 isozyme gene has not yet been linked to any human conditions [37]. Yet HSD3B1 has been reported to play a crucial role in the conversion of DHEA to estradiol in breast tumors and maybe a target enzyme for inhibition in the treatment of breast cancer in postmenopausal women. In human breast cancer, estradiol-17β (E2) is produced in the tumor of breast cancer patients by the circulating dehydroepiandrosterone-sulfate (DHEA-S) present in the adrenal gland which is converted by 3β-HSD1. This conversion happens in the presence of other enzymes also namely, steroid sulfatase, aromatase and 17β-HSD.

This E2 acts as a precursor of estrogen ultimately leading to the increased proliferation of breast carcinoma [38, 39]. Significant estrogenic biocatalytic activity is exhibited by 5α-androstane-3β,17β-diol (3β-diol) which is produced in the inactive state by converting the prospective androgen DHT. Hence upon, the HSD3B enzyme is considered for the modulatory performance in the production or inactivation of active sex steroid which plays a dominant function for the subtype, hormone-dependent breast cancer [40, 41]. This gives the enzyme great significance both as a clinical and prognostic marker in the study of BCs in addition to currently used biomarkers such as HER-2, PgR and lymph node status, invasive tumor size and histological grade [42]. Recent studies also reveal that HSD3B1 expression in breast cancer is associated with an increased risk of cancer recurrence and that genetic or pharmacologic inhibition of HSD3B1 suppresses malignant cancer cell phenotypes both *in vitro* and *in vivo* [43] (Fig. **6**).

Pregnenolone ⟶ 17α-OH -Pregnenolone ⟶ DHEA

3β- HYDROXYSTEROID DEHYDROGENASE

Progesterone   Cortisol   Androstenedione ⟶ Testosterone

17β- estradiol

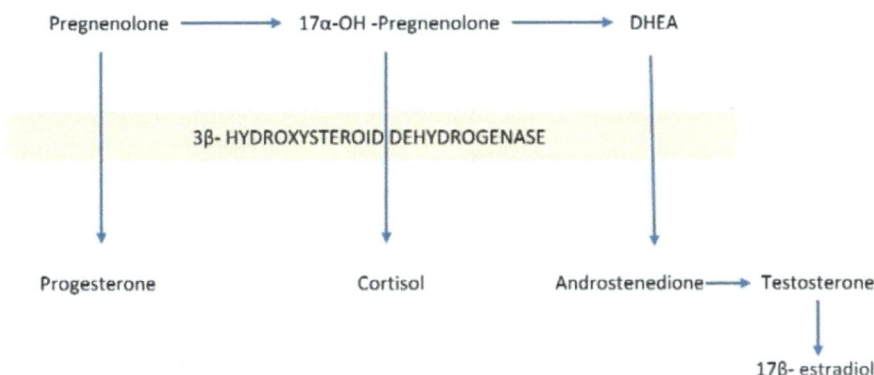

Fig. (6). Reactions catalyzed by HSD3B.

Tousled-like Kinases

Breast cancer subtypes have five categories namely, luminal A, luminal B, triple-negative or basal-like, HER2 Over Expressed and normal-like. Further to subtyping, histological grading and proliferation indexing provide significant prognostic information for breast cancer therapy and cure. Endocrine therapy is the most common medication regimen as an estrogen receptor (ERβ) overexpression inducing breast cancer being the majority occurrence. Physiological and psychological as a clinical outcome differ significantly among different breast cancer patients. With the corroboration of cancer genome atlas, the associated gene encoding for TLK1 and TLK2 [44] was identified and the encoded protein kinases were found to be nuclear serine/threonine kinases. From the genetic perspective, these kinases are involved (S-phase and G-phase) [45] in cell morphogenesis by promoting chromatin assembly during S-phase of cell morphogenesis and also during mitosis for chromosome segregation. In terms of the molecular reaction mechanism, TLKs namely, TLK1 and TLK2 interact coercively further involving ASF1 histone binding and DNA damage response and as of date exact mechanism at the molecular level is not yet to be elucidated. TLKs specifically TLK2 inhibition by the prescription of TLK2 inhibitors yield decreased cell viability and enhanced breast cancer cellular apoptosis [46]. Luminal A and luminal B subtyping are the subdivisions of ERβ breast cancer category of which luminal B is very belligerent in the pathological conditions of worse tumor grade, tumor size being larger and also with a higher proliferation index, hence this type of breast cancer also being therapy-resistant till date remains a great clinical confrontation and challenge for treatment and disease-free survival. Convergence technology through integrative genomic data analysis identified and distinguished a kinase enzyme Tousled-like kinase 2 (TLK2) as a target site for inhibitor action negating the overexpression of TLK2 leading to the apoptosis of the patient breast cancer cells as the ensuing overexpression of TLK2 associates and correlates with worst-case clinical endpoints [47, 48]. TLK2 overexpression is of copy number amplifications rather than an expression of qualitative nature. This type of copy number amplification is observed in patients with intellectual disability (ID). Hence technology derived medications are of a very high necessity to develop a very efficacious therapeutic drug regimen to negate the underlying genetic aberrations associated with TLK2 overexpression aggressive tumors.

The latest clinical trials in advanced breast cancer patients result have shown the in-vivo success of the CDK4/6-specific inhibitors invoking cell cycle kinases (TLKs) as a potential drug targeting site which will be of a great clinical and pharmaceutical commercial breakthrough. Resulting reports based on the preclinical xenograft tumor model has shown TLK2 inhibition drastically improved progression-free survival exhibiting a feasible therapeutic significance in TLK2-amplified luminal sub-type breast tumors.

Drugs Used to Inhibit Enzymes Involved in Breast Cancer

Novel nano-drug delivery systems for controlled and targeted delivery [49] would be the cutting edge and forefront technology for feminine oncological indications manifested at breast, cervix and ovaries. Path-breaking polymeric nanocarriers developed by biotechnology-based start-up companies [50, 51] would be the progressive track and as a matter of fact in a fast-track facility for FDA approval. The process can be innovative or the drug of choice for which the drug has been identified (Metformin) [52] to be tried for a different oncological especially feminine indication.

Anastrazole

Anastrazole is a type 2, a non-steroidal inhibitor of aromatase. Docking studies were carried out for the interplay of different forces between anastrozole and aromatase by three-dimensional modeling. Basic essential coordination was among anastrozole's triazole group and heme (iron) moiety. Thereupon interaction of anastrozole with aromatase's active site is through hydrogen bonding. Aromatase's T310 residue and triazole group (N-one atom) and Anastrozole's D309 residue and cyano group get involved in the interaction. Inhibitory studies were carried out demonstrating a mutant version situated in B'-C loop and β4 sheet influence binding interaction which supported the logical interference of the docking studies involving the cyanoisopropyl group with β4 sheet and B'-C loop. The market reach of anastrozole is provided as follows.

According to IMS data, 2009 sales of the branded product in the United States were approximately $916.8 million, with approximately 105 million tablets sold annually.

Side-effects

Food and Drug Administration (FDA) approved anastrozole for the treatment of breast cancer in 1996 for Astrazeneca as an oral pill. https://www.accessdata. fda.gov/drugsatfda_docs/label/2011/020541s026lbl.pdf. A single dose of anastrozole that results in life-threatening symptoms has not been established. Dosage of a single dose of 60 mg to a healthy volunteer did not show any clinical manifestation as given in the FDA access data portal. Clinical trials and patient data showed anastrozole lowers the estrogen body levels; hence less estrogen reaches the bone cells leading to osteoporosis as carried out by AstraZeneca. (NCT008 https://clinicaltrials.gov/ct2/show/NCT00849030). General side effects include hot flashes, general body weakness and pain in the joints. Serious side effects will be: Elevation of blood cholesterol levels, skin lesions leading to irritation and itching, mucosal ulcer and inflammation of the liver.

Exemestane

Aromatase inhibition can be carried out by producing a reactive electrophile by hydroxylation of exemestane at C19 irreversibly to the corresponding active site of aromatase. Recently a novel mechanism has been reported called the clamping mechanism whereby involving heme, I helix, B'-C loop, and the β-4 sheet [53]. The following residues are involved between E302 and T310), containing I133 and F134 and containing S478 and H480 for heme, I helix, B'-C loop, and the β-4 sheet correspondingly.

Side-effects

Approval of the drug Aromasin (generic name, exemestane) to treat advanced-stage breast cancer in postmenopausal patients was given by FDA in 1999 (https://www.pfizermedicalinformation.com/en-us/aromasin) as an oral pill based on the clinical trial called Intergroup Exemestane Study (https://clinicaltrials. gov/ct2/show/NCT00003418). General side effects include visual disturbances, nausea, arthralgia and hot flushes. https://www.accessdata.fda.gov/drugsatfda_ docs/label/2018/020753s020lbl.pdf

Celecoxib

Celecoxib trade name Celebrex™ belongs to the category of non-steroidal anti-inflammatory drugs (NSAIDs) and a COX-2 inhibitor particularly suppressing and inhibiting COX-2 without any action on COX-1 but shows side effects such as stomach ulceration in the stomach and bleeding [54]. After the administration of celecoxib, its action of inhibiting COX-2 is 10-20 more than COX-1 inhibition. Its mechanism involves the binding of its polar sulfonamide side chain to a hydrophilic side pocket region close to the active COX-2 binding site. this selectivity allows celecoxib and other COX-2 inhibitors to reduce inflammation (and pain) while minimizing adverse drug reactions that are common with nonselective NSAIDs [55]. It also reduces cancer progression by binding with cadherin-11(which is overexpressed in breast cancer). Celecoxib is available in the market mainly for arthritis, acute pain and musculoskeletal pain.

Side-effects

Celecoxib to be used as an analgesic, anti-inflammatory, and antipyretic drug was approved by FDA in 1999 which was marketed by Pfizer, Inc. Several clinical trials have been conducted and various clinical trials are being pursued using celecoxib for breast cancer therapy.

Drug Inhibitors of HSD17B

Technology development for prospective inhibitors for HSD17B is very much in the preclinical and clinical trials since HSD17B plays a greater role in the regulation of E2 levels in tumors of breast cancer patients. Clinical commercialization of bedside translation of HSD17B inhibitors is only at the preclinical and not even at the early clinical trial stages [56] even being with the huge success of other steroidogenic enzymes namely aromatase and steroid sulphatase at the clinical trials. One of the earliest inhibitors used in this context is flavonoids such as apigenin, chrysin, genistein, and naringenin which reduces the conversion of estrone to estradiol thereby exhibiting anti-HSD17B activity. This inhibition is further stabilized as the flavonoids act as estrogen-mimicking molecules and as antagonists of estrogen receptors [57]. When T-47D breast cancer cell lines were used, products of E1 as inhibitors with 2-ethyl- and 2-methoxy (E1) were reported to inhibit HSD17B1 activity [58]. Furthermore, many patented steroidal and non-steroidal inhibitors of HSD17Bs have also been reported in the literature. These include E1 pyrazole N-ethyl derivatives and C15-derivatives which are best among the steroidal inhibitors and pyrimidinone derivatives, fluorinated derivatives and methylated compounds as potent non-steroidal inhibitors of HSD17B1. Inhibition of HSD17B2 is exerted by phenylthiofurane derivatives which are patented compounds of non-steroidal origin [59]. Other promising treatments include the use of hydroxybenzothiazole and PBRM [3-(2-bromoethyl)-16β-(m-carbamoylbenzyl)-17β-hydroxy-1,3,5(10)-estratriene] as effective non-steroidal and steroidal inhibitors of estrogen-dependent BC respectively. However, there are currently no inhibitors developed against isozymes HSD17B7 and HSD17B14 which are also reported in cases of BC with only one recent study demonstrating the use of INH7 as a selective inhibitor for HSD17B7. INH7 significantly suppresses the regulation of 78 kDa glucose-regulated protein (GRP78) and anti-apoptosis factor Bcl-2 *via* 17βHSD7 inhibition. This led to decreased proliferation and increased apoptosis in breast cancer cell lines [60].

Drug Inhibitors of HSD3B

Current approaches to treat breast cancer through inhibition of HSD3B mainly proceed through the selective knockdown of HSD3B1 isozyme as it is reported to have a major effect in hormone-dependent breast cancers. In this context, the most widely used drugs are trilostane and epostane both of which can competitively inhibit purified human 3β-HSD1 with much higher affinity compared to human 3β-HSD2 [61]. Trilostane competitively inhibits HSD3B, which is responsible for converting pregnenolone to progesterone in the adrenal cortex. This leads to a decrease in the production of cortisol as well as, aldosterone and other sex hormones (estrogen) thereby inhibiting the proliferation and prognosis of BCs. HSD3B by acting on the estrogen and growth factor-dependent pathways simultaneously thereby cell proliferation is stimulated by estradiol. It is also an allosteric modulator. Several clinical trials are currently ongoing where initial reports suggest trilostane as an effectual therapy for patients with relapsed breast cancer. Currently, efforts are being made in examining the potential for the use of trilostane in pre-menopausal breast cancer [62]. Trilostane is sold under the trade names Desopan, Modrastane, and Modern. Epostane is a derivative of trilostane and exhibits similar inhibitory properties for HSD3B as the latter. It is predominantly used as an anti-progestogenic drug while trilostane functions as an anti-estrogenic drug [63]. Epostane is sold under the trade name GyMiso.

Potential Drugs as TLK2 Inhibitors

Various laboratory preclinical level testing has shown that Staurosporine and Nocardiopsis have exhibited total abrogation of TLK2 activity towards breast cancer cells. Traditional Chinese medicine cocktail formulation by the name of Danggui Longhui Wan (Chinese herb) in which Indirubin is the active pharmaceutical ingredient (API) has shown very promising results for treating chronic myelocytic leukemia patients [64, 65] rendering this molecule and its derived molecules as a potential TLK2 inhibitor, which showed a substantial reduction in TLK2 activity. For sure more in-depth further experimentations using breast cancer patient-derived xenograft tumor models are needed to substantiate the TLK2 inhibitory activities of these potential molecules. Finally clinical trials are mandatory to prove the therapeutic effect of these molecules as TLKs inhibitors for FDA approval and further commercialization bring these molecules to the breast cancer patient bedside as a therapeutic agent for treatment and cure.

CONCLUSION

Feminine cancer leads to several 100,000s deaths annually around the world. Nearly 600,000 women have died of breast cancer alone in 2017 globally. Other feminine cancers being cervical and ovarian add up to several 100,000s. Convergence technology will and should be able to eradicate breast cancer from the world by the next two decades using innovative polymers as drug delivery carriers. Hence this chapter on breast cancer-related enzyme inhibitors can give a major perspective on the early diagnosis, treatment and cure of feminine cancer.

AUTHOR'S CONTRIBUTION

Hariharan J and Praveen Kumar PK are equally contributed as first authors for this manuscript.

CONSENT FOR PUBLICATION

Not applicable.

CONFLICT OF INTEREST

The author(s) confirms that there is no conflict of interest.

ACKNOWLEDGMENT

I acknowledge Mr. Srihari S, a native English language expert from the Department of Humanities and Social Sciences, Sri Venkateswara College of Engineering, Sriperumbudur for thoroughly extensively editing the book chapter for grammar, punctuation, spelling and overall style.

REFERENCES

[1]   Bray F, Ferlay J, Soerjomataram I, Siegel RL, Torre LA, Jemal A. Global cancer statistics 2018: GLOBOCAN estimates of incidence and mortality worldwide for 36 cancers in 185 countries. CA Cancer J Clin 2018; 68(6): 394-424.
[http://dx.doi.org/10.3322/caac.21492] [PMID: 30207593]

[2]   McPherson K, Steel CM, Dixon JM. ABC of breast diseases. Breast cancer-epidemiology, risk factors, and genetics. BMJ 2000; 321(7261): 624-8.
[http://dx.doi.org/10.1136/bmj.321.7261.624] [PMID: 10977847]

[3]   Top 10 Pharmaceutical Companies 2018 - Oncology - IgeaHub nd 2018. https://www.igeahub.com/2018/06/27/top-10-pharmaceutical-companies-2018-oncology/ (Accessed December 27, 2018).

[4]   Kim A, Balis FM, Widemann BC. Sorafenib and sunitinib. Oncologist 2009; 14(8): 800-5.

[http://dx.doi.org/10.1634/theoncologist.2009-0088] [PMID: 19648603]

[5]   Rouzier R, Perou CM, Symmans WF, *et al.* Breast cancer molecular subtypes respond differently to preoperative chemotherapy. Clin Cancer Res 2005; 11(16): 5678-85.
[http://dx.doi.org/10.1158/1078-0432.CCR-04-2421] [PMID: 16115903]

[6]   Chan HJ, Petrossian K, Chen S. Structural and functional characterization of aromatase, estrogen receptor, and their genes in endocrine-responsive and -resistant breast cancer cells. J Steroid Biochem Mol Biol 2016; 161: 73-83.
[http://dx.doi.org/10.1016/j.jsbmb.2015.07.018] [PMID: 26277097]

[7]   Chumsri S, Howes T, Bao T, Sabnis G, Brodie A. Aromatase, aromatase inhibitors, and breast cancer. J Steroid Biochem Mol Biol 2011; 125(1-2): 13-22.
[http://dx.doi.org/10.1016/j.jsbmb.2011.02.001] [PMID: 21335088]

[8]   Singh-Ranger G, Mokbel K. The role of cyclooxygenase-2 (COX-2) in breast cancer, and implications of COX-2 inhibition. Eur J Surg Oncol 2002; 28(7): 729-37.
[http://dx.doi.org/10.1053/ejso.2002.1329] [PMID: 12431470]

[9]   Singh B, Lucci A. Role of cyclooxygenase –2 in breast cancer journal of surgical research 2002; 108: 173-9.

[10]   Denkert C, Winzer KJ, Hauptmann S. Prognostic impact of cyclooxygenase-2 in breast cancer. Clin Breast Cancer 2004; 4(6): 428-33.
[http://dx.doi.org/10.3816/CBC.2004.n.006] [PMID: 15023244]

[11]   Brueggemeier RW, Richards JA, Petrel TA. Aromatase and cyclooxygenases: enzymes in breast cancer. J Steroid Biochem Mol Biol 2003; 86(3-5): 501-7.
[http://dx.doi.org/10.1016/S0960-0760(03)00380-7] [PMID: 14623550]

[12]   Tanabe T, Tohnai N. Cyclooxygenase isozymes and their gene structures and expression. Prostaglandins Other Lipid Mediat 2002; 68-69: 95-114.
[http://dx.doi.org/10.1016/S0090-6980(02)00024-2] [PMID: 12432912]

[13]   Lee JS, Choi YD, Lee JH, *et al.* Expression of cyclooxygenase-2 in adenocarcinomas of the uterine cervix and its relation to angiogenesis and tumor growth. Gynecol Oncol 2004; 95(3): 523-9.
[http://dx.doi.org/10.1016/j.ygyno.2004.08.036] [PMID: 15581957]

[14]   Regulski M, Regulska K, Prukała W, Piotrowska H, Stanisz B, Murias M. COX-2 inhibitors: a novel strategy in the management of breast cancer. Drug Discov Today 2016; 21(4): 598-615.
[http://dx.doi.org/10.1016/j.drudis.2015.12.003] [PMID: 26723915]

[15]   Cox TR, Gartland A, Erler JT. Lysyl Oxidase, a Targetable Secreted Molecule Involved in Cancer Metastasis. Cancer Res 2016; 76(2): 188-92.
[http://dx.doi.org/10.1158/0008-5472.CAN-15-2306] [PMID: 26732355]

[16]   Levental KR, Yu H, Kass L, *et al.* Matrix crosslinking forces tumor progression by enhancing integrin signaling. Cell 2009; 139(5): 891-906.
[http://dx.doi.org/10.1016/j.cell.2009.10.027] [PMID: 19931152]

[17]   Johnston KA, Lopez KM. Lysyl oxidase in cancer inhibition and metastasis. Cancer Lett 2018; 417: 174-81.
[http://dx.doi.org/10.1016/j.canlet.2018.01.006] [PMID: 29309816]

[18]   Moon HJ, Finney J, Ronnebaum T, Mure M. Human lysyl oxidase-like 2. Bioorg Chem 2014; 57: 231-41.
[http://dx.doi.org/10.1016/j.bioorg.2014.07.003] [PMID: 25146937]

[19]   Gilkes DM, Semenza GL. Role of hypoxia-inducible factors in breast cancer metastasis. Future Oncol 2013; 9(11): 1623-36.
[http://dx.doi.org/10.2217/fon.13.92] [PMID: 24156323]

[20]   Peinado H, Del Carmen Iglesias-de la Cruz M, Olmeda D, *et al.* A molecular role for lysyl oxidase-

like 2 enzyme in snail regulation and tumor progression. EMBO J 2005; 24(19): 3446-58.
[http://dx.doi.org/10.1038/sj.emboj.7600781] [PMID: 16096638]

[21]   Payne SL, Fogelgren B, Hess AR, *et al.* Lysyl oxidase regulates breast cancer cell migration and adhesion through a hydrogen peroxide-mediated mechanism. Cancer Res 2005; 65(24): 11429-36.
[http://dx.doi.org/10.1158/0008-5472.CAN-05-1274] [PMID: 16357151]

[22]   Blockhuys S, Wittung-Stafshede P. Roles of copper-binding proteins in breast cancer. Int J Mol Sci 2017; 20:18(4): pii: E871.

[23]   Barker HE, Cox TR, Erler JT. The rationale for targeting the LOX family in cancer. Nat Rev Cancer 2012; 12(8): 540-52.
[http://dx.doi.org/10.1038/nrc3319] [PMID: 22810810]

[24]   Krawetz SA. The origin of lysyl oxidase, comparative biochemistry, and physiology. Europe PMC Plus 1994; 108(1): 117-9.

[25]   Agneta Jansson A. 17 Beta-hydroxysteroid dehydrogenase enzymes and breast cancer. Drug Discov Today 2016; 21(4): 598-615.
[PMID: 26723915]

[26]   Bouchard-Fortier A, Provencher L, Blanchette C, Diorio C. Prognostic and predictive value of low estrogen receptor expression in breast cancer. Curr Oncol 2017; 24(2): e106-14.
[http://dx.doi.org/10.3747/co.24.3238] [PMID: 28490933]

[27]   Labrie F, Luu-The V, Lin SX, *et al.* The key role of 17 beta-hydroxysteroid dehydrogenases in sex steroid biology. Steroids 1997; 62(1): 148-58.
[http://dx.doi.org/10.1016/S0039-128X(96)00174-2] [PMID: 9029730]

[28]   Nagasaki S, Miki Y, Akahira J, Suzuki T, Sasano H. 17beta-hydroxysteroid dehydrogenases in human breast cancer. Ann N Y Acad Sci 2009; 1155: 25-32.
[http://dx.doi.org/10.1111/j.1749-6632.2008.03682.x] [PMID: 19250189]

[29]   Hilborn E, Stål O, Jansson A. Estrogen and androgen-converting enzymes 17β-hydroxysteroid dehydrogenase and their involvement in cancer: with a special focus on 17β-hydroxysteroid dehydrogenase type 1, 2, and breast cancer. Oncotarget 2017; 8(18): 30552-62.
[http://dx.doi.org/10.18632/oncotarget.15547] [PMID: 28430630]

[30]   Aka JA, Zerradi M, Houle F, Huot J, Lin SX. 17beta-hydroxysteroid dehydrogenase type 1 modulates breast cancer protein profile and impacts cell migration. Breast Cancer Res 2012; 14(3): R92.
[http://dx.doi.org/10.1186/bcr3207] [PMID: 22691413]

[31]   Aka JA, Mazumdar M, Chen CQ, Poirier D, Lin SX. 17β-hydroxysteroid dehydrogenase type 1 stimulates breast cancer by dihydrotestosterone inactivation in addition to estradiol production. Mol Endocrinol 2010; 24(4): 832-45.
[http://dx.doi.org/10.1210/me.2009-0468] [PMID: 20172961]

[32]   Zhang CY, Chen J, Yin DC, Lin SX. The contribution of 17beta-hydroxysteroid dehydrogenase type 1 to the estradiol-estrone ratio in estrogen-sensitive breast cancer cells. PLoS One 2012; 7(1)e29835
[http://dx.doi.org/10.1371/journal.pone.0029835] [PMID: 22253796]

[33]   Aka JA, Zerradi M, Houle F, Huot J, Lin SX. 17beta-hydroxysteroid dehydrogenase type 1 modulates breast cancer protein profile and impacts cell migration. Breast Cancer Res 2012; 14(3): R92.
[http://dx.doi.org/10.1186/bcr3207] [PMID: 22691413]

[34]   Oduwole OO, Li Y, Isomaa VV, *et al.* 17β-hydroxysteroid dehydrogenase type 1 is an independent prognostic marker in breast cancer. Cancer Res 2004; 64(20): 7604-9.
[http://dx.doi.org/10.1158/0008-5472.CAN-04-0446] [PMID: 15492288]

[35]   Cravioto MD, Ulloa-Aguirre A, Bermudez JA, *et al.* A new inherited variant of the 3 beta-hydroxysteroid dehydrogenase-isomerase deficiency syndrome: evidence for the existence of two isoenzymes. J Clin Endocrinol Metab 1986; 63(2): 360-7.
[http://dx.doi.org/10.1210/jcem-63-2-360] [PMID: 3088022]

[36]   Simard J, Ricketts ML, Gingras S, Soucy P, Feltus FA, Melner MH. Molecular biology of the 3beta-hydroxysteroid dehydrogenase/delta5-delta4 isomerase gene family. Endocr Rev 2005; 26(4): 525-82.
       [http://dx.doi.org/10.1210/er.2002-0050] [PMID: 15632317]

[37]   Rhéaume E, Simard J, Morel Y, *et al.* Congenital adrenal hyperplasia due to point mutations in the type II 3 beta-hydroxysteroid dehydrogenase gene. Nat Genet 1992; 1(4): 239-45.
       [http://dx.doi.org/10.1038/ng0792-239] [PMID: 1363812]

[38]   Penning TM, Burczynski ME, Jez JM, *et al.* Human 3alpha-hydroxysteroid dehydrogenase isoforms (AKR1C1-AKR1C4) of the aldo-keto reductase superfamily: functional plasticity and tissue distribution reveals roles in the inactivation and formation of male and female sex hormones. Biochem J 2000; 351(Pt 1): 67-77.
       [PMID: 10998348]

[39]   Pasqualini JR. The selective estrogen enzyme modulators in breast cancer: a review. Biochim Biophys Acta 2004; 1654(2): 123-43.
       [PMID: 15172700]

[40]   Simard J, Ricketts ML, Gingras S, Soucy P, Feltus FA, Melner MH. Molecular biology of the 3beta-hydroxysteroid dehydrogenase/delta5-delta4 isomerase gene family. Endocr Rev 2005; 26(4): 525-82.
       [http://dx.doi.org/10.1210/er.2002-0050] [PMID: 15632317]

[41]   Hanamura T, Ito T, Kanai T, *et al.* Human 3β-hydroxysteroid dehydrogenase type 1 in human breast cancer: clinical significance and prognostic associations. Cancer Med 2016; 5(7): 1405-15.
       [http://dx.doi.org/10.1002/cam4.708] [PMID: 27139182]

[42]   Morabito A, Magnani E, Gion M, *et al.* Prognostic and predictive indicators in operable breast cancer. Clin Breast Cancer 2003; 3(6): 381-90.
       [http://dx.doi.org/10.3816/CBC.2003.n.002] [PMID: 12636883]

[43]   Chang YC, Chen CK, Chen MJ, *et al.* Expression of 3β-hydroxysteroid dehydrogenase type 1 in breast cancer is associated with poor prognosis independent of estrogen receptor status. Ann Surg Oncol 2017; 24(13): 4033-41.
       [http://dx.doi.org/10.1245/s10434-017-6000-6] [PMID: 28744792]

[44]   Segura-Bayona S, Knobel PA, González-Burón H, *et al.* Differential requirements for Tousled-like kinases 1 and 2 in mammalian development. Cell Death Differ 2017; 24(11): 1872-85.
       [http://dx.doi.org/10.1038/cdd.2017.108] [PMID: 28708136]

[45]   Bruinsma W, van den Berg J, Aprelia M, Medema RH. Tousled-like kinase 2 regulates recovery from a DNA damage-induced G2 arrest. EMBO Rep 2016; 17(5): 659-70.
       [http://dx.doi.org/10.15252/embr.201540767] [PMID: 26931568]

[46]   Kim JA, Tan Y, Wang X, *et al.* Comprehensive functional analysis of the tousled-like kinase 2 frequently amplified in aggressive luminal breast cancers. Nat Commun 2016; 7: 12991.
       [http://dx.doi.org/10.1038/ncomms12991] [PMID: 27694828]

[47]   Mortuza GB, Hermida D, Pedersen AK, *et al.* Molecular basis of Tousled-Like Kinase 2 activation. Nat Commun 2018; 9(1): 2535.
       [http://dx.doi.org/10.1038/s41467-018-04941-y] [PMID: 29955062]

[48]   Garrote AM, Redondo P, Montoya G, Muñoz IG. Purification, crystallization and preliminary X-ray diffraction analysis of the kinase domain of human tousled-like kinase 2. Acta Crystallogr F Struct Biol Commun 2014; 70(Pt 3): 354-7.
       [http://dx.doi.org/10.1107/S2053230X14002581] [PMID: 24598926]

[49]   Sun B, Ranganathan B, Feng SS. Multifunctional poly(D,L-lactide-co-glycolide)/montmorillonite (PLGA/MMT) nanoparticles decorated by Trastuzumab for targeted chemotherapy of breast cancer. Biomaterials 2008; 29(4): 475-86.
       [http://dx.doi.org/10.1016/j.biomaterials.2007.09.038] [PMID: 17953985]

[50]   Ramakrishnan R, Gimbun J, Samsuri F, Narayanamurthy V, *et al.* Needleless electrospinning

technology–an entrepreneurial perspective. Indian J Sci Technol 2016; 9(15): 1-11.
[http://dx.doi.org/10.17485/ijst/2016/v9i15/91538]

[51]   Khargonekar P, Sinskey A, Miller C, Ranganathan B. Convergence revolution – piloting the third scientific revolution through start-ups for breast cancer cure. Cancer Science & Research 2017; 4(1): 1-6.
[http://dx.doi.org/10.15226/csroa.2017.00130]

[52]   Prabhakaran S, Thirumal D, Gimbun J, Ranganathan B. Metformin-A panacea pharmaceutical agent through convergence revolution initiative. J Nat Rem 2018; 17(3): 69-79.

[53]   Hackett JC, Brueggemeier RW, Hadad CM. The final catalytic step of cytochrome p450 aromatase: a density functional theory study. J Am Chem Soc 2005; 127(14): 5224-37.
[http://dx.doi.org/10.1021/ja044716w] [PMID: 15810858]

[54]   Kismet K, Akay MT, Abbasoglu O, Ercan A. Celecoxib: a potent cyclooxygenase-2 inhibitor in cancer prevention. Cancer Detect Prev 2004; 28(2): 127-42.
[http://dx.doi.org/10.1016/j.cdp.2003.12.005] [PMID: 15068837]

[55]   Fanun M. Solubilization of celecoxib in microemulsions based on mixed nonionic surfactants and peppermint oil. J Dispers Sci Technol 2010; 31(8): 1140-9.
[http://dx.doi.org/10.1080/01932690903224565]

[56]   Joanna M. Day, Helena J Tutill, Alan (Atul) Purohit, Michael J Reed. Design and validation of specific inhibitors of 17 -hydroxysteroid dehydrogenases for therapeutic application in breast and prostate cancer, and in endometriosis. Endocr Relat Cancer 2008; 15(3): 665-92.
[http://dx.doi.org/10.1677/ERC-08-0042] [PMID: 18541621]

[57]   Le Bail JC, Laroche T, Marre-Fournier F, Habrioux G. Aromatase and 17beta-hydroxysteroid dehydrogenase inhibition by flavonoids. Cancer Lett 1998; 133(1): 101-6.
[http://dx.doi.org/10.1016/S0304-3835(98)00211-0] [PMID: 9929167]

[58]   Purohit A, Tutill HJ, Day JM, et al. The regulation and inhibition of 17beta-hydroxysteroid dehydrogenase in breast cancer. Mol Cell Endocrinol 2006; 248(1-2): 199-203.
[http://dx.doi.org/10.1016/j.mce.2005.12.003] [PMID: 16414180]

[59]   Audet-Walsh É, Bellemare J, Lacombe L, et al. The impact of germline genetic variations in hydroxysteroid (17-beta) dehydrogenases on prostate cancer outcomes after prostatectomy. Eur Urol 2012; 62(1): 88-96.
[http://dx.doi.org/10.1016/j.eururo.2011.12.021] [PMID: 22209174]

[60]   Wang X-Q, Aka JA, Li T, Xu D, Doillon CJ, Lin SX. Inhibition of 17beta-hydroxysteroid dehydrogenase type 7 modulates breast cancer protein profile and enhances apoptosis by down-regulating GRP78. J Steroid Biochem Mol Biol 2017; 172: 188-97.
[http://dx.doi.org/10.1016/j.jsbmb.2017.06.009] [PMID: 28645527]

[61]   James L. Thomas, Kevin M. Bucholtz, and Balint Kacsoha. Selective inhibition of human 3β-hydroxysteroid dehydrogenase type 1 as a potential treatment for breast cancer. Steroid Biochem Mol Biol 2011; 125(1-2): 57-65.
[http://dx.doi.org/10.1016/j.jsbmb.2010.08.003]

[62]   Puddefoot JR, Barker S, Vinson GP. Trilostane in advanced breast cancer. Expert Opin Pharmacother 2006; 7(17): 2413-9.
[http://dx.doi.org/10.1517/14656566.7.17.2413] [PMID: 17109615]

[63]   Birgerson L, Odlind V, Johansson ED. Effects of Epostane on progesterone synthesis in early human pregnancy. Contraception 1986; 33(4): 401-10.
[http://dx.doi.org/10.1016/0010-7824(86)90103-4] [PMID: 3731777]

[64]   Hoessel R, Leclerc S, Endicott JA, et al. Indirubin, the active constituent of a Chinese antileukaemia medicine, inhibits cyclin-dependent kinases. Nat Cell Biol 1999; 1(1): 60-7.
[http://dx.doi.org/10.1038/9035] [PMID: 10559866]

[65] Blažević T, Heiss EH, Atanasov AG, Breuss JM, Dirsch VM, Uhrin P. Indirubin and indirubin derivatives for counteracting proliferative diseases. Evid Based Complement Alternat Med 2015; 2015654098
[http://dx.doi.org/10.1155/2015/654098] [PMID: 26457112]

228 *Frontiers in Enzyme Inhibition*, 2020, *Vol. 1*, 228-262

CHAPTER 12

Enzyme Inhibition Applications in Treatment of Human Viral Diseases

Subasree Sekar, P.K. Praveen Kumar* and Arthi Udhayachandran

Department of Biotechnology, Sri Venkateswara College of Engineering (Autonomous), Sriperumbudur Tk – 602117, Tamilnadu, India

Abstract: Enzyme inhibitor molecules are used for the development of antiviral drugs. Understanding the mechanisms of enzyme inhibitors are needed for the treatment of HIV, Chikungunya, Dengue, Ebola, Influenza, and Nipah viral diseases. Inhibition of viral entry and its replication in the host cell was the most prominent mode of action against these viruses. In this chapter, the detailed list of plant compounds to be used as drugs for the treatment of above viral diseases through targeting of enzymes, reverse transcriptase, and RNA-dependent RNA polymerase is explained. Recent advancements such as emerging technologies, Next Generation Sequencing, and CRISPR used as an effective approach for the diagnosis, treatment, and alleviation of viral disease progression, are explained.

Keywords: Antiviral drugs, CRISPR, Enzyme inhibitors, Next Generation Sequencing, Reverse transcriptase, RNA-dependent RNA polymerase.

INTRODUCTION

Enzymes are biocatalyst that accelerates the chemical reactions. Enzyme inhibitors are chemical compounds with a low molecular weight that may scale back or completely inhibit the enzyme catalytic activity reversibly or irreversibly (permanently). An enzyme-inhibitor complex is formed once the enzyme is bound to the inhibitor. However, the complex is not formed if the enzyme is not bound. The presence of naturally occurring enzyme inhibitors, like antitrypsin, anti-thrombin, and anti-pepsin, controls the activity of an enzyme in the human body and under physiological circumstances ensures their extracellular and intracellular action [1].

Most commonly, competitive enzyme inhibitors are utilized as pharmaceutical drugs or agents. Competitive inhibition is an analog to biochemical substrates that

* **Corresponding author Praveen Kumar P.K:** Department of Biotechnology, Sri Venkateswara College of Engineering (Autonomous), Sriperumbudur Tk – 602117, Tamilnadu, India; Tel: +919444495008; Fax: +914427162462; Email: praveenpk@svce.ac.in

G. Baskar, K. Sathish Kumar & K. Tamilarasan (Eds.)

compete with the natural substrate for an enzyme's active site and prevent unwanted metabolic products [2]. Furthermore, additional inhibitors target enzymes that use bi - substrate only after a transition to the active site has occurred due to the binding either of the two reaction substrates. Such uncompetitive inhibitors bound to the substrate and hinder the enzyme catalysis [3]. For example, mycophenolic acid, which inhibits the inosine 5′-monophosphate dehydrogenase (IMDH) enzyme, is used in treating cancer and viral diseases [4].

Antiviral medicines are used in particular to treat viral infections. Most antivirals required activation by viral and cellular enzymes before antiviral use [5]. Moreover, some viruses have protease enzymes, which involve in cleavage of viral protein chains. Substantial research was carried out to identify HIV protease inhibitors as drugs for the treatment of HIV attacks in humans [6]. Protease inhibitors were found efficient in the 1990s but later it developed side effects [7]. Protease inhibitors development from natural sources is focused in the present era. For example, Shiitake mushroom (*Lentinus edodes*) possess protease inhibitors that have shown antiviral activity in *in vitro* [8, 9].

MECHANISM OF ANTIVIRAL DRUGS

In the following section, the drugs used for viral diseases are discussed and are shown in Fig. (**1**).

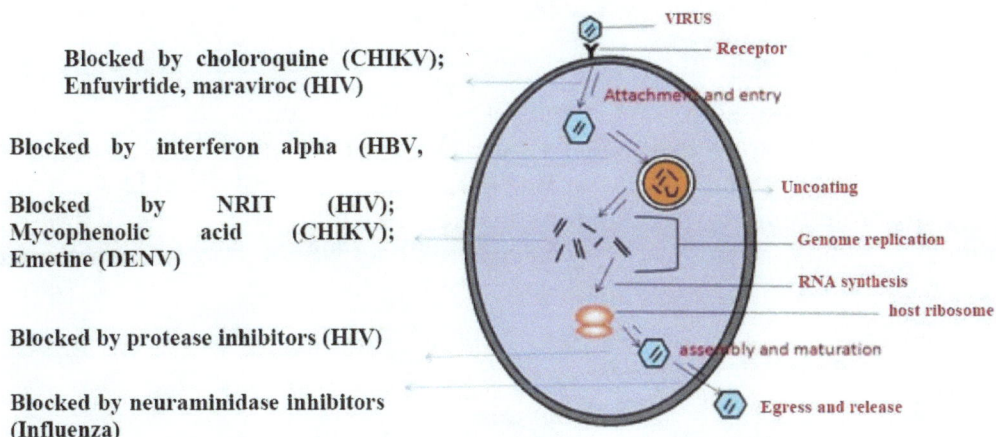

Blocked by choloroquine (CHIKV); Enfuvirtide, maraviroc (HIV)

Blocked by interferon alpha (HBV,

Blocked by NRIT (HIV); Mycophenolic acid (CHIKV); Emetine (DENV)

Blocked by protease inhibitors (HIV)

Blocked by neuraminidase inhibitors (Influenza)

VIRUS
Receptor
Attachment and entry
Uncoating
Genome replication
RNA synthesis
host ribosome
assembly and maturation
Egress and release

Fig. (1). Mechanism of action of antiviral drugs.

Attachment

Virus infections are initiated when virus capsid or envelope-related viral proteins bind to the specific host cell membrane receptors on host cells. The HIV envelope glycoproteins allow binding of the virus to bind to CD4+ T lymphocytes expressing the chemokine receptor 5 (CCR5) and/or the C-X-C motif chemokine receptor 4 (CXCR4) [10].

Entry

Viruses enter across host cell membranes into the cytoplasm. For example, the host cell (CD_4+ T lymphocyte) membranes are promoted by the HIV envelope protein (gp41) [11]. A virus-mediated fusion of the HIV envelope with a plasma membrane of the host CD4+ T lymphocytes is currently being offered as a virus - entry blocker.

Uncoating

Uncoating of viral entry involves the removal/degradation of nucleocapsids. Structural modification of nucleocapsids causes the viral genome to be released into the host cell cytoplasm and transported to the host cell nuclear nuclei. Currently, there are viral uncoating inhibitors that block influenza A virus M2 proton channel and prevent virus matrix protein dissociation dependent on pH [12].

Transcription and Translation

After the uncoating step, the gene expression of viral nucleic acid was determined, *i.e.*, the transcription of viral RNA or DNA into mRNA and mRNA translation into proteins, and viral polyproteins proteolytic cleavage into individual protein units, become available. Existing viral gene expression inhibitors presently disrupt HCV – related functional NS3/4A protease expression [13].

Replication

The ribonucleoside triphosphates or deoxyribonucleoside triphosphates generation is required for viral genome replication. In the host cell cytoplasm, most RNA viruses replicate their genomes. In the host cell nucleus, however, DNA viruses are replicating their genomes. Nucleoside analogue inhibitor phosphorylates viral cellular kinase is incorporated into the growing viral genome. Inhibitors of non-nucleoside polymerase type inhibit RNA or DNA polymerases directly [14].

Assembly & Maturation

This formation of immature viruses is the next step in the cycle of viral life. The maturation process occurs after the assembly step. New virions turn into infectious in this step of the viral life cycle. This process majorly focuses on proteolytic cleavage by virus proteases. Currently, viral maturation inhibitors used are HIV protease inhibitors. Moreover, the virus egresses from the infected host cells through cell lysis or budding through the cell membrane.

Release

However, certain virions need an additional step of discharge in the host cytoplasm. Influenza A and B viruses require neuraminidase in host cell membranes to affect the extracellular surface release. Existing viral release inhibitors or inhibitors of neuraminidase prevent the release of host cell detaches of new influenza A and B virions [15]. The mechanism of action of anti-viral drugs is shown in the schematic diagram, Fig. (**1**).

In the following section, we will discuss the enzyme inhibition activity of some antiviral drugs against the respective diseases.

HUMAN VIRAL DISEASES

Chikungunya Virus (CHIKV)

Chikungunya virus (CHIKV) is a re-emerged mosquito-borne viral disease that has become a prominent life-threatening disease in many areas of the world. CHIKV belongs to the Alphavirus genera of the Togaviridae family. This enveloped, single-stranded, positive-sense RNA virus has two open reading frames (ORFs) and icosahedral symmetry, possessing an 11.8 kb genome. The genomic RNA is translated to 50 ORF which forms the viral replicate with four non-structural proteins (nsP1, nsP2, nsP3, and nsP4), while the subgenomic RNA translated to 30 ORF, forms capsid protein (C) [16], with two surface envelope glycoproteins (E1 and E2) [17] and two small peptides named (E3 and K6) [18]. *Aedes aegypti* and *A. albopictus* mosquitos generally transmit CHIKV. The mechanism of the action of CHIKV drugs is shown in Fig. (**2**).

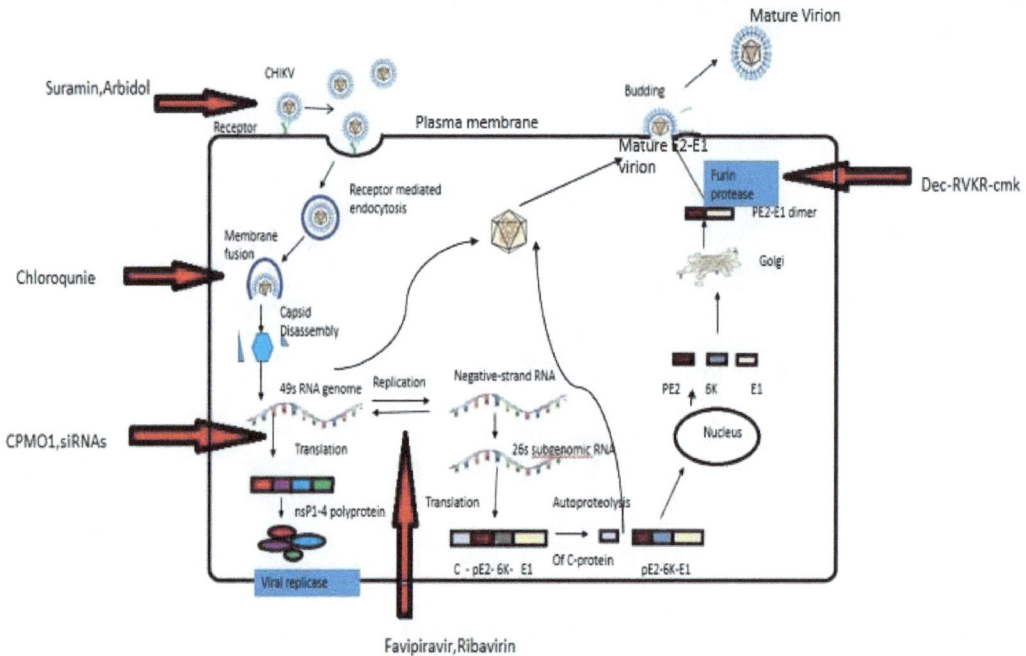

Fig. (2). Mechanism of action of drugs against CHIKV.

CHIKV has shown to modulate the pro-survival of PI3K/AKT/mTOR pathway, pattern recognition receptor (PRR) mediated innate immune pathways and endoplasmic reticulum (ER) stress response pathways [19].

The A226V mutation in the envelope glycoprotein of CHIKV suggests an increase in the activity of CHIKV in *A.albopictus* as well as improving transmissibility of the virus over *A.albopictus* [20]. Presently no licensed antiviral vaccines, antiviral drugs and non-steroidal anti-inflammatory drugs (*e.g.,* corticosteroids, *etc.*) have been developed for the treatment of CHIKV infection in humans. It shows no symptomatic relief of arthralgia and myalgia [21].

A summary of plant compounds with anti-CHIKV properties and its mode of action are listed in Table **1**.

Table 1. Plant Compounds with inhibitory activities against CHIKV.

Plant Name	Active Compounds	Toxicity	Enzyme Inhibitor	Mode of Action	References
Rhizoma coptidis	Isoquinoline alkaloid	+++ (Acute toxicity: 2.95 g/kg in mice)	Mitogen-activated protein kinase (MAPK), mitogen-activated protein kinase(MAPKK)	Inhibits viral entry	[23]
Cephalotaxus harringtonia	Harringtonine and Homoharringtonine alkaloids	+++ (Acute toxicity: 2 g/kg in mice)	RNA-dependent RNA polymerase	Inhibits transcription and translation	[25]
Silybum marianum	Flavonolignan	+++ (Acute toxicity: 1 g/kg in mice)	RNA-dependent RNA polymerase proteins, puromycin acetyltransferase.	Inhibits viral entry, RNA synthesis.	[26]
Curcuma longa	Diarylheptanoid	+ (Acute toxicity:>5 g/kg in mice)	RNA-dependent RNA polymerase	Inhibits protein synthesis	[27]
Trigonostemon cherrieri	Trigocherrins and Trigocherriolides	+ (Acute toxicity:>5 g/kg in mice)	RNA-dependent RNA polymerase	Inhibits viral entry	[28]

In this study, few plant compounds Berberine, Harringtonine, Silymarin, Curcumin, Daphnane diterpenoids and its mechanism of action was studied for anti-chikungunya activity.

Berberine

Berberine, an isoquinoline alkaloid present in *Rhizoma coptidis* has significant anti-viral activity including CHIKV [22, 23]. Berberine activates Adenosine Monophosphate-Activated Protein Kinase (AMPK) enzyme while inhibiting Protein-Tyrosine Phosphatase 1B (PTP1B). Berberine is effective in various CHIKV strains. The structure of Berberine is shown in Fig. (**3**).

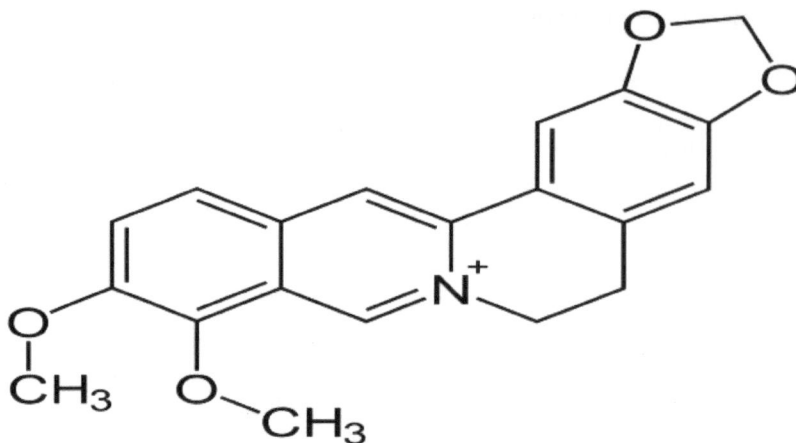

Fig. (3). Structure of Berberine (PubChem Id: 2353).

Mechanism of Action

CHIKV infection is majorly elicited by Extracellular Signal-Regulated Kinase (ERK), mitogen-activated protein kinase (MAPK), p38 and Janus Kinase (JNK) signaling pathways. The typical MAPK signaling pathway involves a three-tiered cascade of activating kinases, MAPK kinase (MAPKKK), MAPK kinase (MAPKK) and MAP kinase (MAPK) to phosphorylate and activate an embarrassment of cytoplasmic and nuclear substrates. These transcription factors play a crucial role in cell growth, differentiation, proliferation, migration, and apoptosis. Berberine, an organic compound verified to be efficient at reducing each viral non-structural and structural protein levels at this high MOI (Multiplicity of infection) conditions. However, the impact of berberine on virus entry was ruled out by luciferase signals.

Harringtonine

Harringtonine, a cephalotaxine organic compound derived from *Cephalotaxus harringtonia* is known to inhibit the primary cycle of the elongation phases of eukaryotic translation [24]. Extra methyl in the side chain of the Homoharringtonine increases the stability of the compound. Viral genomic ribonucleic acid transfection showed harringtonine inhibiting the pre-CHIKV replication cycle. Fig. (**4**) shows the structure of Harringtonine.

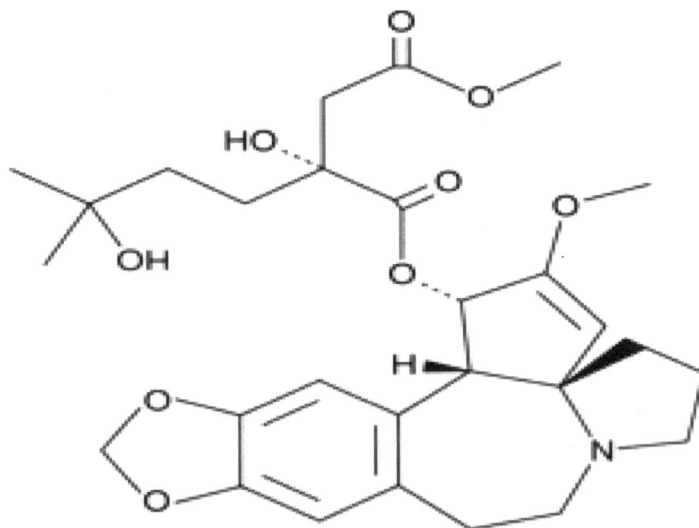

Fig. (4). Structure of Harringtonine (PubChem Id: 276389).

Mechanism of Action

Harringtonine acts as an inhibitor for protein synthesis; it suppresses the structural and non-structural proteins of CHIKV by inhibiting its eukaryotic ribosomal unit. Replicase levels decrease because of inhibition of non-structural protein3 (nsP3). Henceforth, levels of negative-sense RNA strands tend to decrease including the synthesis of the positive-sense RNA strands [25]. Thus, the decrease in levels of RNA leads to host cell translation inhibition by harringtonine, inhibiting viral structural protein production.

Silymarin

Silymarin is a flavonolignan that is extracted from the seeds and fruit of *Silybum marianum*. It is a combination of 3 structural isomer components: silydianine, silychristine and silibinin. Fig. (**5**) shows the structure of Silymarin.

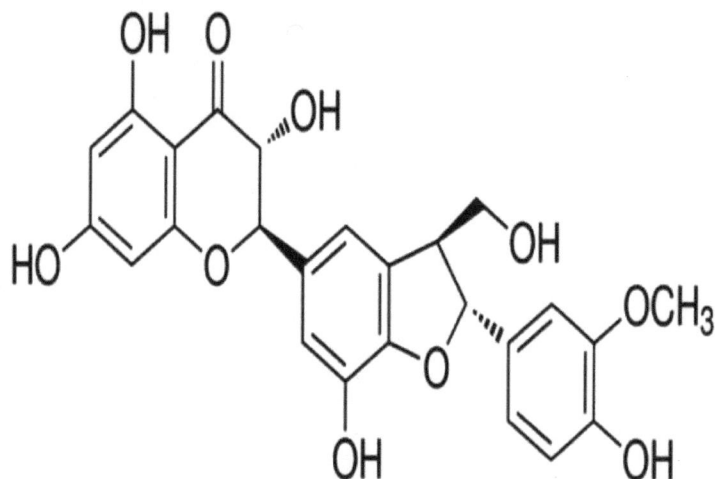

Fig. (5). Structure of Silymarin (PubChem Id: 7073228).

Mechanism of Action

Silymarin is an inhibitor for CHIKV entry and blocks cell-to-cell viral infection both *in vitro* and *in vivo*. The cell line of CHIKV replicon contains CHIKV non-cytotoxic replicon and virus replicase proteins. Puromycin acetyltransferase, EGFP, and Rluc markers are expressed in this protein. Quantitatively, 100μg/ml of silymarin suppresses the CHIKV infection [26].

Curcumin

Curcumin, the most important active compound from turmeric (*Curcuma longa*), possesses antiviral properties. Recent researches eluded the effectiveness of curcumin against the CHIKV. Chemically, curcumin is a diarylheptanoid of the curcuminoids, which is responsible for its yellow color. It is a tautomeric compound in organic solvents over the enolic form and in water as a keto form. The bioavailability efficacy and aqueous solubility of curcumin have been reported against several infectious diseases [27]. Fig. (**6**) shows the structure of Curcumin.

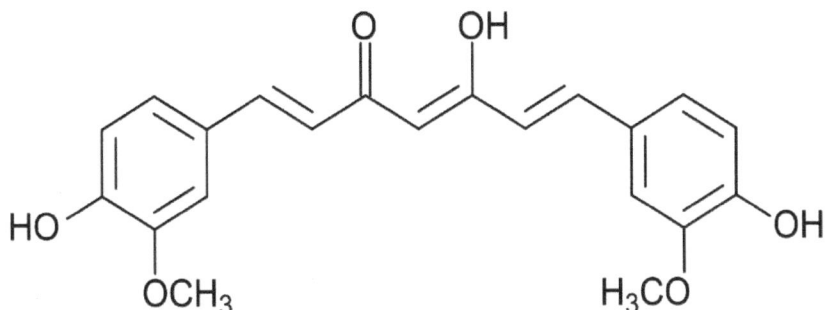

Fig. (6). Structure of Curcumin (PubChem Id: 969516).

Mechanism of Action

Curcumin acts as an inhibitor for protein synthesis and its spin-off demethoxycurcumin shows higher inhibitory efficiency because of lipophilic properties against viruses.

Daphnane Diterpenoids

Purification was performed on *Trigonostemon cherrieri* bark extracts to target chikungunya and dengue virus using RNA-dependent RNA polymerase NS5 protein inhibitors.

Mechanism of Action

Trigocherrins A, B, F, and trigocherriolides A, B and C are potent CHIKV replicase inhibitors. These compounds suppress the cytopathic effect (CPE) caused by CHIKV. However, mechanisms of action of these diterpenoids against CHIKV are yet to be unleashed [28].

Dengue Virus

Dengue viruses are spread by the mosquitos of the Flaviviridae family and are considered global human pathogens. Non-structural protein 3 (NS3) acts as a helicase, serine protease, and nucleoside triphosphatase (NTPase). Non-structural protein 5 (NS5) is a protein complex comprising the activities of methyltransferase (MTase) and RNA-dependent RNA polymerase (RdRp) [29]. The mechanism of action of dengue virus drugs is shown in Fig. (**7**).

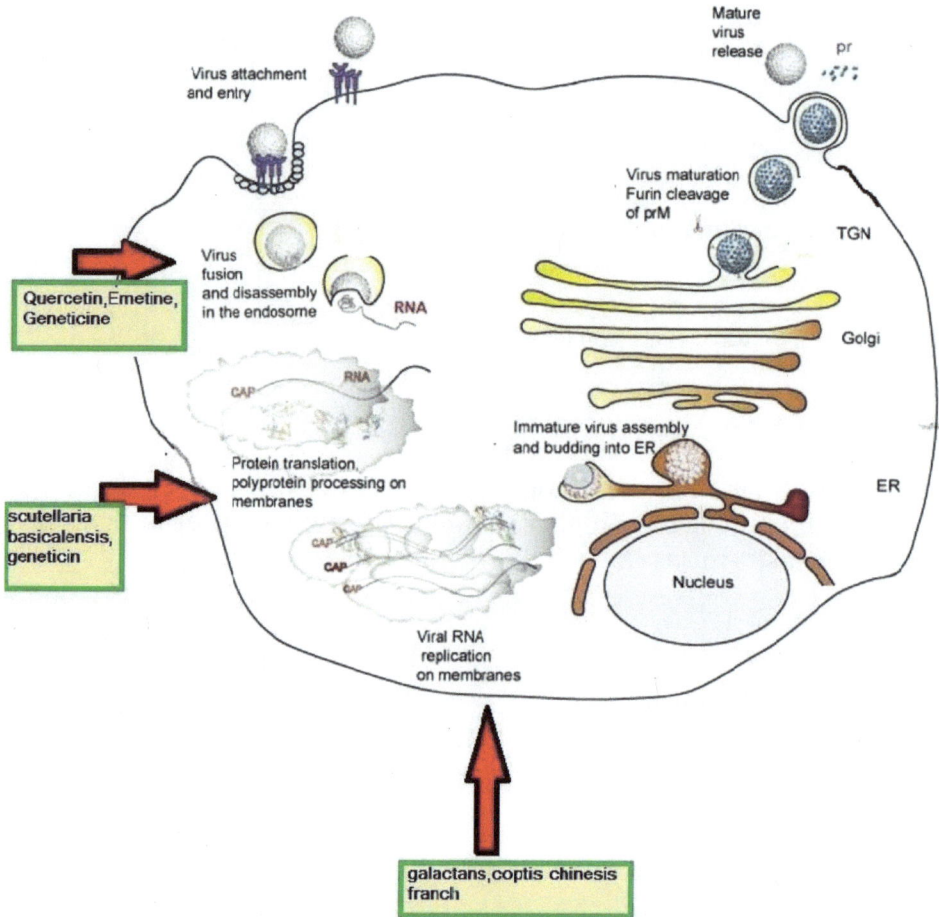

Fig. (7). Mechanism of action of drugs against Dengue Virus.

No effective antiviral vaccines and therapies are presently available to treat or prevent DENV infection. Antivirals decrease DENV infection mortality by more than a 10-fold viral load. Inhibition of host cellular functions required to replicate or activate DENV in the host cell should impair viral replication. Indeed, one of the major advantages of targeting host functions compared to targeting viral functions is the lower likelihood that drug-resistant mutants will emerge [30]. A summary of anti-dengue activity compounds exhibited is listed in Table **2**.

Table 2. Compounds with inhibitory activities against dengue virus.

Plant Name	Active Compound	Plant Part Extract	Mode of Action	References
Houttuynia cordata	Flavonoids	Leaf and root extract	Inhibits the viral RNA synthesis	[34]
Boesenbergia rotunda	Flavonoids	Root extract	Inhibits of protease enzyme activity	[35]
Fucoidan	Sulfated polysaccharide	Brown seaweed extract	Inhibits virus infection	[34]
Cryptomeria crenulata	Flavonoids	Leaf and grape extract	Inhibits the RNA polymerase	[37]
Geneticin	Alkaloids	Root extract	Inhibits RNA and protein synthesis	[39]
Coptis chinensis Franch	alkaloid	Root extract	Inhibits the viral replication	[36]

Houttuynia Cordata

Houttuynia Cordata is a medicinal vegetable used to promote health and regulate the inflammatory reactions that have been consumed by the people in East and Southeast Asia. The bioactive compound, Hyperoside obtained from aqueous extract of *H.cordata* is non-toxic and it acts effectively against dengue infection [31]. The structure of Hyperoside is shown in Fig. (**8**).

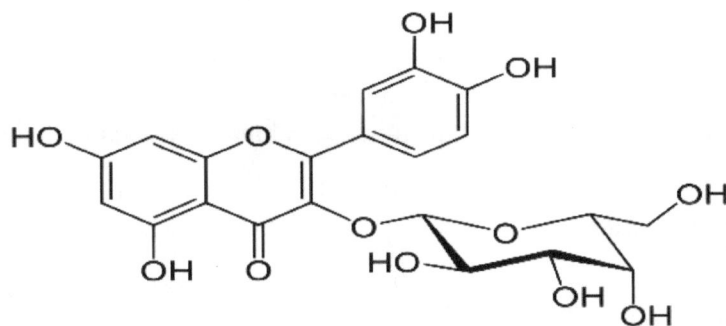

Fig. (8). Structure of Hyperoside (PubChem Id: 5281643).

Mechanism of Action

The aqueous extract of *H. Cordata* in HepG2 cells showed an inhibitory effect on the production of DENV-2 RNA. The higher concentration of *H. Cordata* was effective in protecting HepG2 cells from infection with DENV-2 (protective mode), decreasing intracellular RNA synthesis (treatment mode) and inactivating the virus (direct blocking) [32].

The aqueous extract exhibited a protective effect on the release of virions for LLC-MK2 cells. The major component of the flavonoid hyperoside was identified by HPLC (High-Performance Liquid Chromatography). The detected antiviral activity of dengue is expected to be associated with the flavonoid hyperoside. It protects viral entry and after adsorption has anti-virus activity by inhibiting viral replication and suppressing intracellular RNA synthesis.

Fucoidan

Cladosiphon okamuranus is a marine algae brown seaweed. Fucoidan is one of the polysaccharides isolated from *C.okamuranus*. The fucose containing sulfated polysaccharide (FCSP) provides various effective bioactive functions including inhibiting dengue virus type 2 (DENV-2) infections for humans. Fig. (**9**) shows the structure of fucoidan and its derivatives.

Mechanism of Action

Fucoidan is bound to the envelope glycoprotein (EGP) of DEN2. The fucoidan derivatives have been generated by desulfation, which led to derivative 1, 2 and 4 that proved significant suppression of inhibitory activity. Sulfation is required for the antiviral activity of glycosaminoglycan. Compound 3 has also decreased the capability to prevent serotype 2 viruses [32]. The glucuronic acid and sulfate residue of fucoidan is critical to the action of the DEN V-2 inhibitors, but the exact molecular mechanisms of their inhibitors have not been clarified.

Boesenbergia Rotunda

Boesenbergia rotunda is a Chinese and Southeast Asian medicinal and cooking herb commonly known as Chinese keys, finger root or Chinese ginger. Cyclohexenyl chalcone derivatives of 4-hydroxypanduratin A and Panduratin A of the *Boesenbergia rotunda* (L.) had competitive inhibitory activities towards dengue 2 NS3 protease, whereas pinostrobin and cardamonin have been non-competitive [33].

Fucoidan (**1**) and derivatives **2–4**.

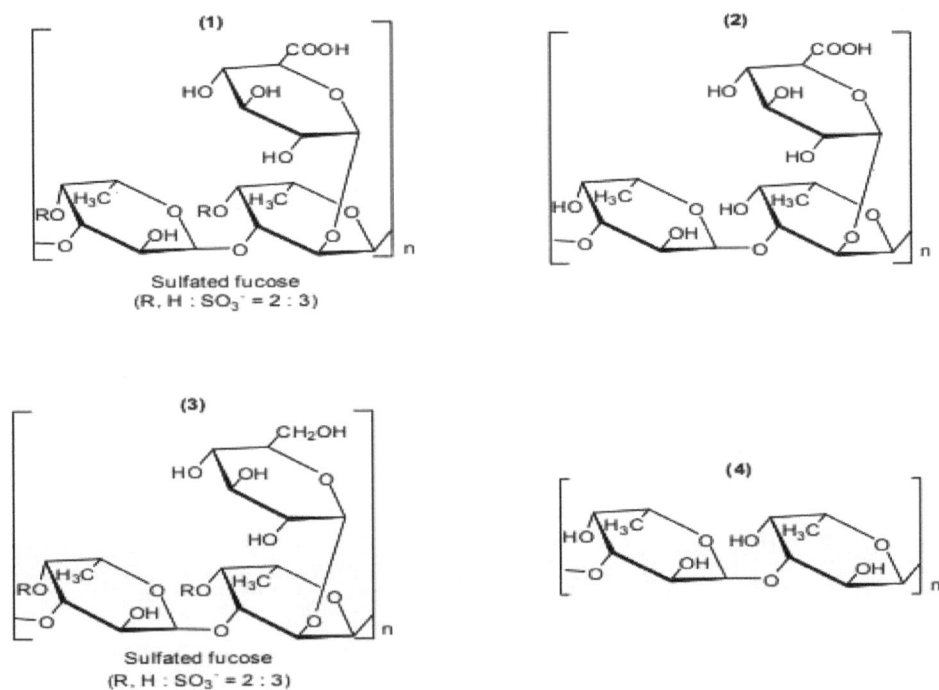

Fig. (**9**). Structure of Fucoidan and derivatives (PubChem Id: 92023653).

Mechanism of Action

4-Hydroxyphanuratin A compound has greater potential inhibitors for dengue-2 NS2B/NS3 protease.

Quercetin

Quercetin is a flavonoid isolated from red seaweed *cryptonemia crenulata*. It is a powerful selective inhibitor of diverse strains of DENV-2 in Vero cells. This is designed to block viral binding and host cell entry. Fig. (**10**) shows the structure of Quercetin.

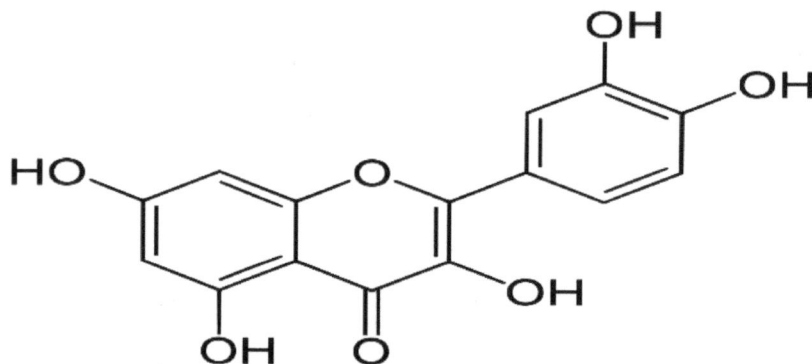

Fig. (10). Structure of Quercetin (PubChem Id: 5280343).

Mechanism of Action

The mechanism of action of quercetin includes improving the attachment of virus, entry of the virus, interferon's antiviral activity, binding to viral proteins, and interfering with the synthesis of viral nucleic acid by binding to viral polymerases [34, 35].

Palmatine

Coptidis rhizome (CR), a medicinal plant is a dried rhizome of *Coptis Chinensis* used to treat the virus, bacterial, fungal and other diseases. The coptidis alkaloid compound, palmatine was screened in *in-vitro* studies for DENV-2 anti-virus activity. Palmatine inhibits non-structural 2B and non-structural 3 protease of the West Nile Virus.

Mechanism of Action

The palmatine is a non - competitive inhibitor of the viral reproduction by the protease enzyme. The mechanism used to inhibit the virus proteases by palmatine has not yet been understood [36].

Geneticin

Geneticin is an analog of aminoglycoside neomycin. Geneticin inhibits the function of 80S ribosomes and protein synthesis in eukaryotic cells. DENV-2 replication and translation were inhibited and have shown greater activity than other pathogens like YFV for this virus. Moreover, Geneticin's has no effect on DENV infection with structural analogs like kanamycin, gentamicin, and guanidylation geneticin. The structure of Geneticin is shown in Fig. (**11**).

Structure of the aminoglycoside geneticin

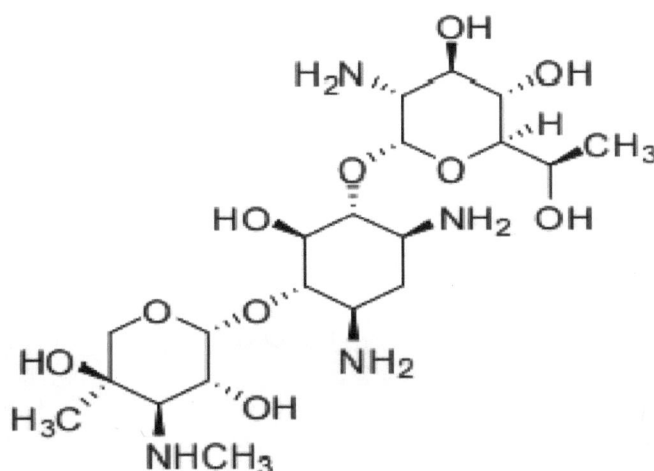

Fig. (11). Structure of Geneticin (PubChem Id: 123865).

Mechanism of Action

Geneticin could inhibit NS4B viral translation and RNA replication by protease enzyme [37]. Viral E protein was used as a marker of DENV-2 translation to provoke the translation of viral proteins. Geneticin treatment inhibited E protein by 80%. Geneticin is still unclear about the mechanism of antiviral activity but is likely to inhibit viral translation, leading to decreased viral RNA synthesis. Additionally, Geneticin inhibits viral folding of RNA, prevents viral RNA accumulation and therefore reduces the production of viral proteins.

Mycophenolic Acid (MPA)

A non - nucleoside analog, MPA is a powerful, non - competitive inhibitor of guanine nucleotide biosynthesis of inosine monophosphate dehydrogenase

(IMPDH). IMPDH is a key enzyme required for the biosynthesis of guanine nucleotides. Consequently, IMPDH inhibition is expected to inhibit not only eukaryotic cell proliferation but also DNA and RNA virus replication. It is approved as an inhalation medicine for treating the respiratory syncytial virus and for treating hepatitis C virus (HCV) infections orally together with alpha-interferon [38]. The mycophenolic acid structure is shown in Fig. (**12**).

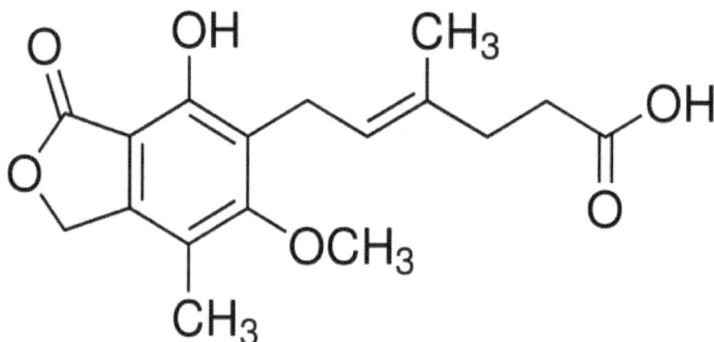

Fig. (12). Structure of Mycophenolic acid (PubChem Id: 446541).

Mechanism of Action

The MPA acts as an inhibitor for the synthesis of viral RNA and the release of infectious virions due to a drop in the guanosine nucleotide intracellular pool. It shows the antiviral activity of MPA by inhibition of IMPDH and causes the error-prone RNA - dependent RNA polymerase (RdRP) to produce a defective genome to be incorrectly incorporated into nucleotides. Therefore, DENV-2 polyprotein processing, replicates complex assembly and NS5 levels would eventually be reduced because of the reduction of viral RNA templates.

Iminosugars

The α-glucosidases I and II competitive inhibitors are glucose imitators represented by deoxynojirimycin (DNJ), an Iminosugar. Iminosugar has antiviral activities inhibiting endoplasm reticulum - resident (N - linked) oligosaccharides on glycoprotein precursors in the sequence of hydrolysis of glucose residues of asparagine [39]. Several iminosugars, including dengue virus (DENV), were therefore investigated as therapeutic agents for treating infections. Fig. (**13**) shows the structure of deoxynojirimycin.

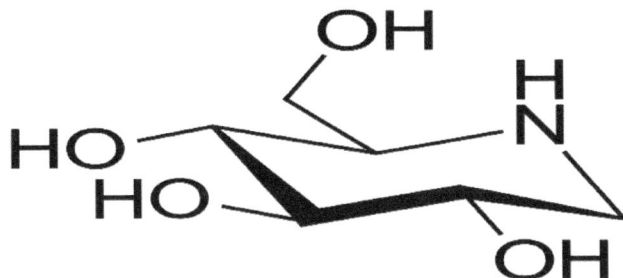

Fig. (13). Structure of Deoxynojrimycin.

Mechanism of Action

The virion assembly and secretion of many enveloped viruses, including DENV, is selectively inhibited by iminosugar derivatives. However, only recently in a genome-wide siRNA knockdown effort to identify host factors required for or to limit the infection, the essential role of cellular α-glucoside 2 in DENV infection has been established [40]. Host glycosylation inhibitors can result in the aberration and interference of viral glycoproteins. Iminosugars are monosaccharides that can imitate glucosidase functions and thus compete with glycoprotein.

Attachment of a preformed oligosaccharide structure with 3 terminal glucose residues (Glc3) is the first step of N - linked protein glycosylation. The three-terminal glucose residues are removed sequentially with α-glucosidases I and II during the maturation of glycoproteins. The retention of glucose residues on non-connecting glycoprotein oligosaccharides, leads to disruption and subsequent degradation of glycoprotein, by the inhibition of α-glucosidases [41]. Due to the misfolded degradation of glycoprotein, free oligosaccharides (FOS) are produced, which contain mono-, bi- and tri - glucose residues. FOS mono - glucosylated to analyze the impact of inhibition of glucosidase on the mice treated with iminosugar. A corresponding nonglycosylated free FOS was used as a control, resulting from the normal metabolism of proteins and being abundant and readily detectable. The absence of tri-glucosylated FOS after inhibitor treatment suggests that there may be insufficient cellular concentrations at the doses used to inhibit α-glucosidase I.

Dasatinib

Dasatinib is powerful adenosine triphosphate and competitive inhibitor of tyrosine kinases. They are efficient Src and Abl competitive inhibitors that associate hydrogen - bonds with the ATP binding site and inhibition of kinase activity. Dasatinib is an oncogenic tyrosine kinase BCR - ABL inhibitor and has more potency than imatinib. The structure of Dasatinib is shown in Fig. (**14**).

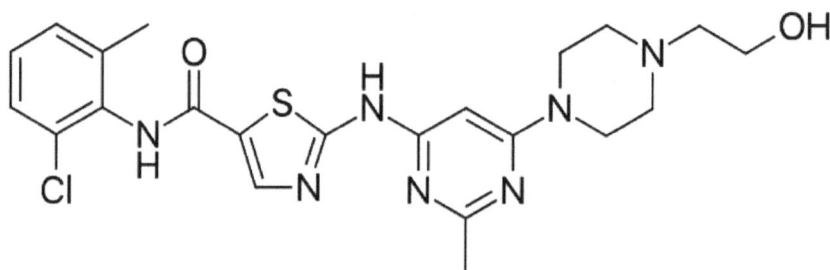

Fig. (14). Structure of Dasatinib (PubChem Id: 3062316).

Mechanism of Action

Dasatinib involves depletion of Fyn, Lyn or Src kinases by strong inhibition of DENV2 RNA. Dasatinib treatment results in a reduction of luciferase signal and a decreased steady-state DENV2 RNA [42]. RNA stability and transition decline are unable to synthesize viral RNA due to GDD catalytic triad mutation in the viral RNA polymerase.

Nipah Virus (NiV)

A zoonotic Nipah virus belongs to the family of Paramyxoviridae and the genus of the henipavirus. NiV first spread from a village on the Malaysian peninsula, Kumpung Sungai Nipah, where pig growers became ill with or due to encephalitis. No approved vaccines or therapeutics can be used in humans despite the pathogenicity of henipaviruses. HeV subunit was approved for use as a vaccine for horses is effective in several animal models and appears safe for human use in Australia as a veterinary vaccine for horses. In animal models for post-exposure prophylaxis, monoclonal antibodies aimed at viral envelope proteins and also have shown to be effective and were used safely by people under compassionate use, even though their efficacy for treating human illness is unknown. In an open-label study with a 36 percent reduction in mortality, a broad-spectrum ribavirin antiviral was initially used in the Malaysian outbreak. However, several animal study experiments have repeatedly shown that monotherapy with ribavirin has not been effective in reducing the mortality of henipavirus infections, together with combined therapy with chloroquine. Fig. (**15**) illustrates the drug mechanism for NiV action.

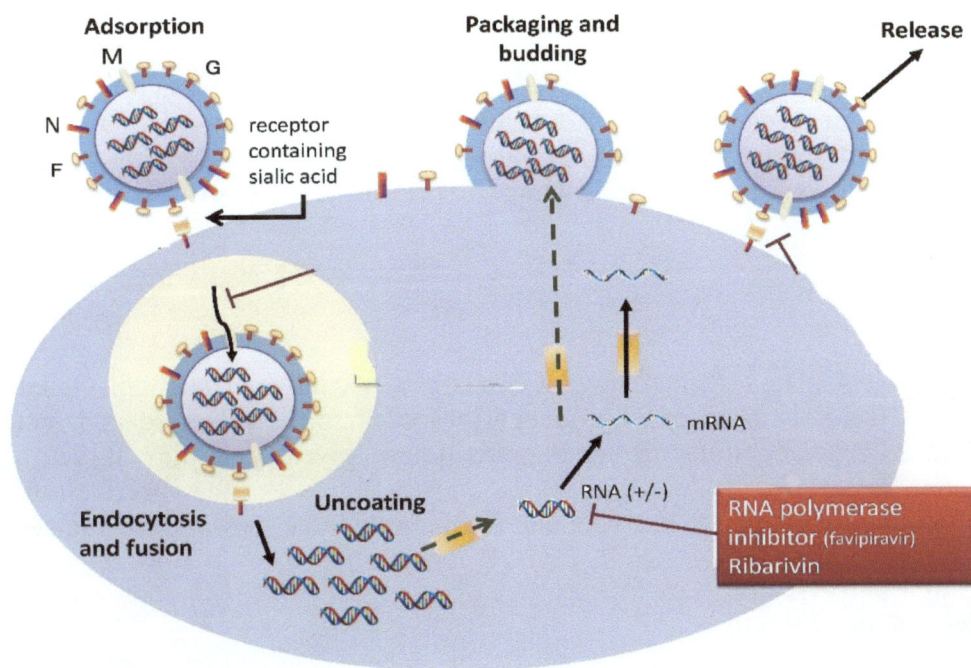

Fig. (15). Mechanism of action of drugs against NiV.

Favipiravir

The Toyama Chemical Company developed FaviPiravir (T-705) as a flu antiviral. It has been licensed for phase 3 clinical trials. Likewise, Favipiravir has proven effective against many other RNA viruses. In the hamster model, Favipiravir inhibited NiV. Similarly, the replication and transcription studies to micromolar concentrations of Nipah virus were inhibited by Favipiravir in *in-vitro* [43].

<u>*Mechanism of Action*</u>

Favipiravir is an alternative substrate for the viral polymerase and potent inhibitor of RNA synthesis (RNAs) or viral mutagenesis. Inclusion of purine or pyrimidine nucleosides inhibits the Nipah virus.

Remdesivir

The drug Remdesivir is intended to deliver nucleoside monophosphate into the cell. Nucleoside analogues target viral DNA or RNA polymerase inhibiting viral infections.

Mechanism of Action

Remedesivir targets viral RNA - dependent polymerase and therefore inhibits viral transcription and replication. Besides, the GS-5734 was able to reduce the NiV at much lower concentrations using the prodrug's in its active form after the release [44].

Galectin-1

Galectin-1 is isolated from C*inachyrella sp,* a marine ball sponge. Galectins proteins selectively bind to oligosaccharides containing galactose and lactose attached as repeated units in asparagines-linked glycan complex serves as a common target for NiV [45]. Gal-1 also increases the production of proinflammatory cytokines like NiV and Ebola virus infections in dendritic cells.

Mechanism of Action

Galectin-1 is an endogenous carbohydrate-binding protein that binds selectively to glycan on NiV - F to inhibit endothelial cell fusion, a consequence that can reduce pathophysiology sequelae of the NiV infection [45].

HIV (Human Immunodeficiency Virus)

Human Immunodeficiency Virus (HIV) is an enveloped virus of the retroviridae family. Reverse Transcriptase of all retroviruses plays a major role in the process of reverse transcription [46, 47].

Human Immunodeficiency type-1 Virus (HIV-1) life cycle involves multiple enzyme interventions for effective molecular targets with acquired immunodeficiency syndrome (AIDS). Of course, HIV-1 integrase (IN) has been identified as a promising therapeutic target as it helps in catalyzing the viral and host DNA strands integration, and this process is necessary for virus replication.

The HIV genome and proteins (human immunodeficiency virus) were the key topics of extensive research. The virus was believed to be a form of the human T - cell leukemia virus (HTLV) that affects the human immune system and causes certain leukemia. Each virion consists of a viral envelope and a capsid matrix containing two copies of the single-stranded RNA genome and several enzymes [48, 49]. Fig. (**16**) shows the mechanism of HIV action.

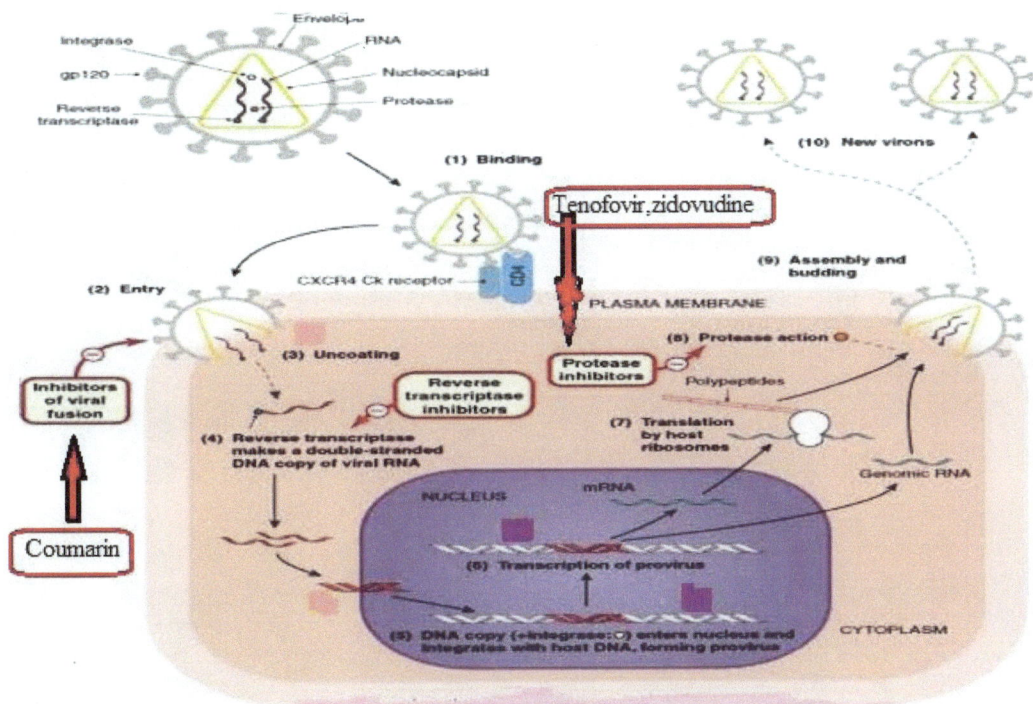

Fig. (16). Mechanism of action of drugs against HIV.

For many people infected with HIV, the use of ART (antiretroviral therapy) reduced the HIV count in the blood by helping in increasing their CD4 cell counts. There are three commonly known primary HIV target enzymes: they are reverse integrase (IN), Protease (PR) and transcriptase (RT). Table **3** shows the search for more and better anti - HIV agents for natural products, despite chemical medicines, such as toxicity, lack of curative and multifaceted effects.

Table 3. Compounds with inhibitory activities against HIV.

Plant Name	Active Compounds	Plant Part/Extract Type	Mode of Action	References
Dipteryx odorata	Coumarin	Tonka bean	Inhibits post-HIV-1 entry	[50]
Agastache rugosa	Rosmarinic acid	Aqueous methanolic extract of roots	Inhibits virus integrase enzyme	[52]
Calophyllum brasiliense	Apetalic acid, Calanolides B and C	The hexane extract of the leaves	Inhibitory effect on reverse transcriptase	[53]
Rhizophora mucronata	Polysaccharide	Alkaline extract of barks	Inhibits the viral binding to the cell	[54]
Ricinus communis	Lectin	Methanolic extract of leaves	Inhibits reverse transcriptase and the N-glycohydrolases	[55]

Coumarin

Coumarin is an organic, aromatic chemical compound found in the chemical class of benzopyrone. It is a naturally occurring substance found in many plants and is colorless crystalline. The structure of Coumarin is shown in Fig. (**17**).

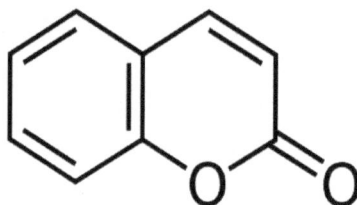

Fig. (17). Structure of Coumarin (PubChem Id: 323).

In the last two decades, coumarins have been discovered and developed as anti - HIV agents. Inhibitory activity on reverse transcriptase, anti-integrase, and antiprotease activity has also been described [50]. Recent requirements for potential anti - HIV agents are increasing, and this also applies for both natural and synthetic Coumarins, to the adequate definition of the mechanism for action and the definition of toxic effects.

Mechanism of Action

Dicamphanoyl-khellactone (DCK), a coumarin derivative has potent anti-HIV activity. DCK is different from medications currently used in AIDS therapy in terms of chemical structure and mechanism of action. DCK is ineffective against both the nucleoside RT and NNRTI resistant strain HIV-1 but it is a potent inhibitor for many HIV-1 isolates. The synthesis and effective control of the multiple - RT inhibitor - resistant strain was demonstrated by several compounds with structural modifications on the DCK khellactone ring to enhance the drug-resistant profile. DCK primarily detects RNA polymerase dependent DNA activities of HIV-1 RT [50].

Rosmarinic Acid

A caffeic ester of 4-dihydroxyphenyllactic acid is Rosmarinic acid. It is found naturally in *Boraginaceae* species and the *Nepetoideae* subfamily of the *Lamiaceae*. The polyphenolic substance in edibles such as mint (*Mentha arvense L.*), perilla (*Perilla frutescens L.*), basil (*Ocimum basilicum L.*) and sage (*Salvia officinalis L.*) are polyphenolic substances. The structure of Rosmarinic acid is shown in Fig. (**18**).

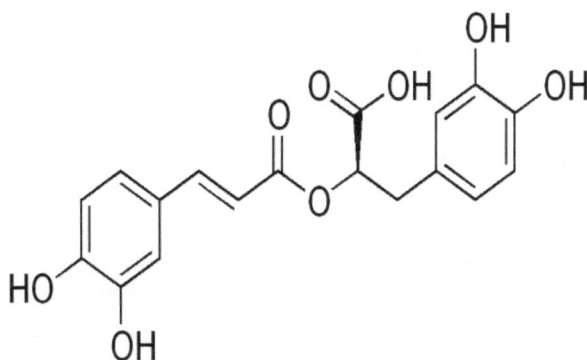

Fig. (18). Structure of Rosmarinic acid (PubChem Id: 5281792).

Mechanism of Action

Without increasing cellular toxicity, the nitration of rosmarinic acid significantly improves the anti-integration and the antiviral activity. Rosmarinic acid was active in submicromolar compounds as inhibitors of HIV-1. The viral replication is inhibited in MT-4 cells with similar selectivity indexes [51].

Calanolides

Calophyllum brasiliense (guanandi) is native to subtropical and tropical regions of the American. Calanolide A and Calanolide B compounds are extracted from the latex and leaves of Calophyllum as AIDS inhibitors. It has a range of medications for the treatment of ulcer and gastritis to avoid prostate harm for skin scarification. Calanolide B, Calanolide C and Apetalic Acid structures shown in Fig. (**19**).

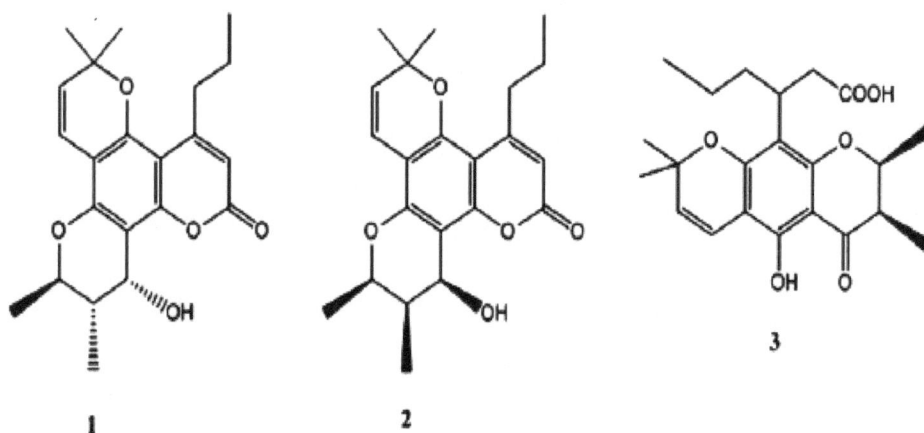

Fig. (19). Structure of (1) Calanolide B, (2) Calanolide C and (3) Apetalic acid.

Mechanism of Action

Type 1 reverse transcriptase (HIV-1 RT) Calophyllum species are the sources of calanolides that inhibit human immunodeficiency virus. *Calophyllum brasiliense* leaves extracts of hexane, acetone and methanol are the result of high RT HIV-1 inhibition, low cytotoxicity to cells of MT2 and high HIV-1 IIIb / LAV inhibition. Three anti - HIV1 dipyranocoumarins are isolated: calanolides A and B, and attrolides of the soul. In contrast, HIV-1 RT inhibitor activity was not present in other isolated compounds such as apetalic acid, isolated acid, isopetalic acid, a structural isomer of insulin, friedelin, canophyllol, and amentoflavone. Calanolide C was also obtained as a natural product and has moderate inhibitory properties [52].

Tenofovir

Tenofovir, an anti - HIV drug belongs to nucleoside analogues. The structure of Tenofovir is shown in Fig. (**20**).

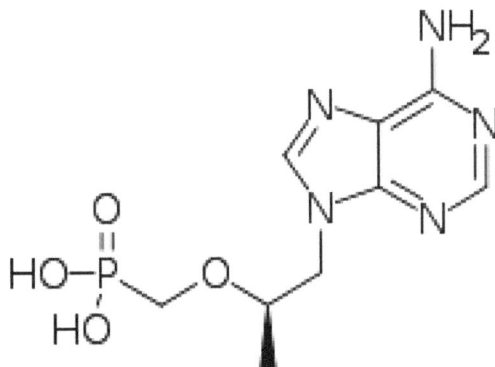

Fig. (20). Structure of Tenofovir (PubChem Id: 464205).

Tenofovir inhibits reverse transcriptase, an enzyme used to produce new viruses by HIV - infected cells. Tenofovir inhibits or reduces the enzyme's activity, causes the slower or higher production of new viruses by HIV infected cells [54].

This is used with another anti - HIV medicines and sometimes other classes of medications, including protease inhibitors. These combinations are known as ART or antiretroviral therapy. Tenofovir disoproxil fumarate (DF) is the initial inhibitor for the treatment of HIV-1 infection approved for use in combination with other HIV-1 antiretroviral agents [55].

Mechanism of Action

Nucleotide Reverse Transcriptase Inhibitors (NRTIs) block a reverse transcriptase enzyme called the HIV enzyme. NRTIs can prevent the multiplication of HIV by blocking reverse transcriptase, which can reduce the amount of HIV in the body.

Abacavir

Abacavir is a reverse transcriptase inhibitor analog of guanosine. This reduces the risks of developing AIDS. The Abacavir structure is shown in Fig. (**21**).

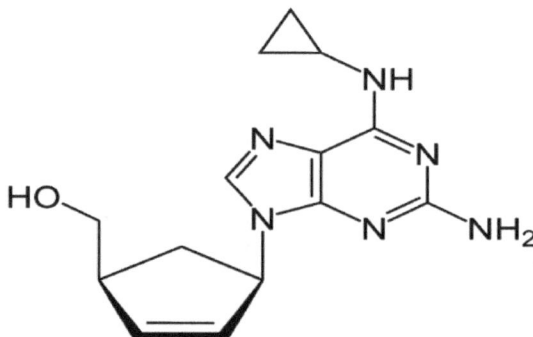

Fig. (21). Structure of Abacavir (PubChem Id: 441300).

Abacavir is a carbocyclic nucleoside analog converted to its active metabolite Carbovir triphosphate that also inhibits the HIV transcriptase effects. It is metabolized by glucuronyl transferase, alcohol dehydrogenase. Abacavir is commonly used as a 2-NRTI backbone in combination with lamivudine and is available with lamivudine or with zidovudine and lamivudine as a fixed-dose combination.

Mechanism of action

Abacavir inhibits the reverse transcriptase enzyme, which is used to produce new viruses in HIV - infected cells. This drug inhibits HIV - infected cells, leading to fewer viruses [56].

INFLUENZA

Influenza often called flu, is an infectious disease with mild to severe symptoms. The four anti-influenza drugs in use are the two proton channel inhibitors of virus uncoating are amantadine and rimantadine, whilst oseltamivir and zanamivir are inhibitors of neuraminidase (NA) that inhibit the release of viruses. The viral fusion protein hemagglutinin (HA) has been one of the most advanced anti-influenza drug targets available. Table **4** shows compounds with inhibitory influenza action.

Table 4. Compounds with inhibitory activities against Influenza.

Plant Type	Active Compounds	Extract Type	Mode of Action	References
Mangifera indica	flavonoid	Methanolic Extract	Inhibits the replication of the virus	[60]
Aloe barbadensis	Anthraquinones, Aloe-emodin	Hot glycerine extract	Partial destruction of the viral envelope	[61]
Camellia sinensis	Catechin	Aqueous extract	Inhibits virus replication and hemagglutinin	[62]
Commelina communis	Alkaloids	Methanolic Extract	Inhibits the virus growth and reduce viral titers in lungs	[63]
Scutellaria baicalensis	Isoscutellarei-8 -methyl ether	Aqueous Extract	Inhibit viral replication	[63]

Isoquercetin

The flavonoid, a chemical compound type, is isoquercetin. The quercetin is the 3-O - glucoside can be isolated from the noble rhubarb and the mango. The structure of Isoquercetin is shown in Fig. (**22**).

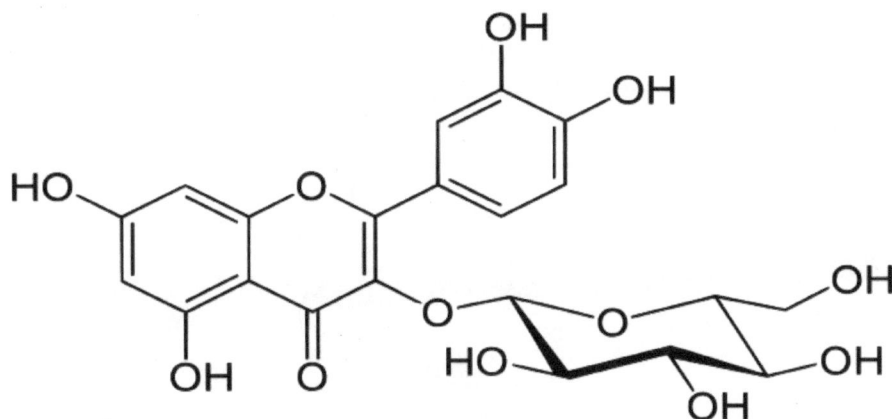

Fig. (22). Structure of Isoquercetin (PubChem Id: 5280804).

Some polyphenols, such as resveratrol and epigallocatechin gallate, have recently been identified to show significantly *in vitro* and/or *in vivo* anti-influenza activity.

Isoquercetin among polyphenols at low effective concentration suppresses the replication of both influenza A and B viruses. Synergistic effects of treatment were observed with patients treated with isoquercetine and amantadine for the treatment of viral infection [57].

The serial virus transit in the presence of isoquercetin only did not lead to the development of resistant viruses; however, adding isoquercetine to the treatment of amantadine or oseltamivir suppressed the development of a virus that is resistant to amantadine or oseltamivir. In influenza - based mouse model intraperitoneally administered isoquercetine to injected mice with human influenza A virus reduces virus titers and pathological changes in the lungs significantly.

Mechanism of Action

In the initial stage of infection with the virus, the glycoprotein hemaggelutinin plays an important role and is an active target in the development of anti-influenza therapy. These mechanisms have identified the interaction of quercetin with the subunit HA2 to inhibit H5N1 entry *via* a drug - screening system based on pseudoviruses [58].

Oseltamivir

Oseltamivir is a medication used to treat and prevent influenza A and influenza B (flu), which is sold under the brand name Tamiflu. It is used to prevent and treat influenza caused by influenza A and B viruses as neuraminidase inhibitors. Fig. (**23**) shows the structure of Oseltamivir.

Fig. (23). Structure of Oseltamivir (PubChem Id: 65028).

Two inhibitors of influenza virus neuraminidase (NA) - zanamivir and oseltamivir—provide both antiviral effects and clinical benefits to humans with influenza. The acquisition of resistance to NA inhibitors in *vitro* studies occurs through the reduction of the virus's dependence on NA activity due to changes in the hemagglutinin (HA), and acquisition of NA resistance due to changes in the enzyme itself [59].

Mechanism of action

Oseltamivir inhibits the neuraminidase enzyme by cleaving the sialic acid found on glycoproteins on the surface of human cells that helps new virions to exit the cell [60].

CONCLUSION

Even though various herbal medicines are available for treating various human viral diseases like Chikungunya, Dengue, NiV and HIV, there are some limitations to overcome in the treatment. The main limitations of herbal medicine are the lack of reproducibility and standardization of plant-derived products. Even, vaccines have not been developed yet for treating NiV and HIV and so are not yet available in the market.

Apart from the existing technologies, researchers have now been developing new and effective technologies for the development of antivirals for human viral diseases targeting viral enzymes. Some of the current technologies for antiviral drug development include Clustered Regularly Interspaced Short Palindromic Repeats (CRISPR) gene-editing followed by next-generation sequencing. CRISPR-Cas system can be used to prevent viral infection or replication by virus-host interactions and by the nucleic acid cleavage rules or gene activation [64].

HIV and poliovirus encode their genetic information in RNA rather than DNA that they repurpose to create new viruses. The latest study reveals the inhibition of HIV-1 replication by binding to the T cells with Cas9 and antiviral gRNAs by the non-homologous end-joining pathway [65]. Massive parallel sequencing of the next-generation sequencers (NGS) platforms has changed the discovery field of treatment of HIV and NiV.

CONSENT FOR PUBLICATION

Not applicable.

CONFLICT OF INTEREST

The author(s) confirms that there is no conflict of interest.

ACKNOWLEDGEMENT

I acknowledge Mr. Shrihari S, a native English language expert from the Department of Humanities and Social Sciences, Sri Venkateswara College of Engineering, Sriperumbudur for thoroughly extensively editing the book chapter for grammar, punctuation, spelling and overall style.

REFERENCES

[1]   Bjelaković G, Stojanović I, Bjelaković GB, Pavlović D, *et al.* Competitive inhibitors of enzymes and their therapeutic application. Med Biol (Milano) 2002; 9: 201-6.

[2]   Mohan C, Long KD, Mutneja M. An introduction to inhibitors and their biological applications. EMD Millipore Corp. 2013; pp. 3-13.

[3]   Copeland RA, Harpel MR, Tummino PJ. Targeting enzyme inhibitors in drug discovery. Expert Opin Ther Targets 2007; 11(7): 967-78.
        [http://dx.doi.org/10.1517/14728222.11.7.967] [PMID: 17614764]

[4]   Digits JA, Hedstrom L. Species-specific inhibition of inosine 5′-monophosphate dehydrogenase by mycophenolic acid. Biochemistry 1999; 38(46): 15388-97.
        [http://dx.doi.org/10.1021/bi991558q] [PMID: 10563825]

[5]   Kaur R, Taheam N, Anil K. Sharma. Important Advances on Antiviral Profile of Chromone Derivatives. RJPBCS 2013; 4(2): 73-93.

[6]   Anderson J, Schiffer C, Lee SK, Swanstrom R. Viral protease inhibitors InAntiviral strategies. Berlin, Heidelberg: Springer 2009; pp. 85-110.

[7]   Flint OP, Noor MA, Hruz PW, *et al.* The role of protease inhibitors in the pathogenesis of HIV-associated lipodystrophy: cellular mechanisms and clinical implications. Toxicol Pathol 2009; 37(1): 65-77.
        [http://dx.doi.org/10.1177/0192623308327119] [PMID: 19171928]

[8]   Odani S, Tominaga K, Kondou S, *et al.* The inhibitory properties and primary structure of a novel serine proteinase inhibitor from the fruiting body of the basidiomycete, *Lentinus edodes.* Eur J Biochem 1999; 262(3): 915-23.
        [http://dx.doi.org/10.1046/j.1432-1327.1999.00463.x] [PMID: 10411656]

[9]   Suzuki H, Okubo A, Yamazaki S, Suzuki K, Mitsuya H, Toda S. Inhibition of the infectivity and cytopathic effect of human immunodeficiency virus by water-soluble lignin in an extract of the culture medium of *Lentinus edodes mycelia* (LEM). Biochem Biophys Res Commun 1989; 160(1): 367-73.
        [http://dx.doi.org/10.1016/0006-291X(89)91665-3] [PMID: 2469420]

[10]   Levroney EL, Aguilar HC, Fulcher JA, *et al.* Novel innate immune functions for galectin-1: galectin-1 inhibits cell fusion by Nipah virus envelope glycoproteins and augments dendritic cell secretion of proinflammatory cytokines. J Immunol 2005; 175(1): 413-20.
        [http://dx.doi.org/10.4049/jimmunol.175.1.413] [PMID: 15972675]

[11]   Shivendra K, Sunil K, Ishita R, Pradeep D, *et al.* Molecular herbal inhibitors of dengue virus: an update. Int J Med Arom Plants 2012; 2(1): 1-21.

[12]   Li JZ, Coen DM. Pharmacology of viral infections.Principles of Pharmacology: The pathophysiologic basis of drug therapy. 4th ed.. Philadelphia, PA: Wolters Kluwer. 2017; pp. 694-722.

[13]   Akhtar J, Shukla D. Viral entry mechanisms: cellular and viral mediators of herpes simplex virus entry. FEBS J 2009; 276(24): 7228-36.
        [http://dx.doi.org/10.1111/j.1742-4658.2009.07402.x] [PMID: 19878306]

[14]   Sugar AM. Fungi The Merck manual of diagnosis and therapy Robert S. 19th ed. Porter, Merck Sharp

& Dohme Corp 2011; pp. 1319-35.

[15]   Anthony JT, Bertram GK, Marieke MKH, Susan BM. Pharmacology Examination & Board Review. 11th ed. Mc Graw Hill Education publishers 2015; pp. 1-593.

[16]   King AMQ, Adams MJ, Leowitz EJ. Classification and Nomenclature of Viruses. Ninth Report of the International Committee on Taxonomy of Viruses 2011.

[17]   Simizu B, Yamamoto K, Hashimoto K, Ogata T. Structural proteins of Chikungunya virus. J Virol 1984; 51(1): 254-8.
       [http://dx.doi.org/10.1128/JVI.51.1.254-258.1984] [PMID: 6726893]

[18]   Solignat M, Gay B, Higgs S, Briant L, Devaux C. Replication cycle of chikungunya: a re-emerging arbovirus. Virology 2009; 393(2): 183-97.
       [http://dx.doi.org/10.1016/j.virol.2009.07.024] [PMID: 19732931]

[19]   Thaa B, Biasiotto R, Eng K, *et al.* Differential phosphatidylinositol-3-kinase-Akt-mTOR activation by Semliki Forest and chikungunya viruses is dependent on nsP3 and connected to replication complex internalization. J Virol 2015; 89(22): 11420-37.
       [http://dx.doi.org/10.1128/JVI.01579-15] [PMID: 26339054]

[20]   Rashad AA, Mahalingam S, Keller PA. Chikungunya virus: emerging targets and new opportunities for medicinal chemistry. J Med Chem 2014; 57(4): 1147-66.
       [http://dx.doi.org/10.1021/jm400460d] [PMID: 24079775]

[21]   Queyriaux B, Simon F, Grandadam M, Michel R, Tolou H, Boutin JP. Clinical burden of chikungunya virus infection. Lancet Infect Dis 2008; 8(1): 2-3.
       [http://dx.doi.org/10.1016/S1473-3099(07)70294-3] [PMID: 18156079]

[22]   Lou T, Zhang Z, Xi Z, *et al.* Berberine inhibits inflammatory response and ameliorates insulin resistance in hepatocytes. Inflammation 2011; 34(6): 659-67.
       [http://dx.doi.org/10.1007/s10753-010-9276-2] [PMID: 21110076]

[23]   Varghese FS, Thaa B, Amrun SN, *et al.* The antiviral alkaloid berberine reduces chikungunya virus-induced mitogen-activated protein kinase signaling. J Virol 2016; 90(21): 9743-57.
       [http://dx.doi.org/10.1128/JVI.01382-16] [PMID: 27535052]

[24]   Fresno M, Jiménez A, Vázquez D. Inhibition of translation in eukaryotic systems by harringtonine. Eur J Biochem 1977; 72(2): 323-30.
       [http://dx.doi.org/10.1111/j.1432-1033.1977.tb11256.x] [PMID: 319998]

[25]   Kaur P, Thiruchelvan M, Lee RC, *et al.* Inhibition of chikungunya virus replication by harringtonine, a novel antiviral that suppresses viral protein expression. Antimicrob Agents Chemother 2013; 57(1): 155-67.
       [http://dx.doi.org/10.1128/AAC.01467-12] [PMID: 23275491]

[26]   Lani R, Hassandarvish P, Chiam CW, *et al.* Antiviral activity of silymarin against chikungunya virus. Sci Rep 2015; 5: 11421.
       [http://dx.doi.org/10.1038/srep11421] [PMID: 26078201]

[27]   Subudhi BB, Chattopadhyay S, Mishra P, Kumar A. Current strategies for inhibition of chikungunya infection. Viruses 2018; 10(5): 235.
       [http://dx.doi.org/10.3390/v10050235] [PMID: 29751486]

[28]   Kaur P, Chu JJ. Chikungunya virus: an update on antiviral development and challenges. Drug Discov Today 2013; 18(19-20): 969-83.
       [http://dx.doi.org/10.1016/j.drudis.2013.05.002] [PMID: 23684571]

[29]   Zou J, Xie X, Wang QY, *et al.* Characterization of dengue virus NS4A and NS4B protein interaction. J Virol 2015; 89(7): 3455-70.
       [http://dx.doi.org/10.1128/JVI.03453-14] [PMID: 25568208]

[30]   Chang J, Schul W, Butters TD, *et al.* Combination of α-glucosidase inhibitor and ribavirin for the

treatment of dengue virus infection *in vitro* and *in vivo*. Antiviral Res 2011; 89(1): 26-34.
[http://dx.doi.org/10.1016/j.antiviral.2010.11.002] [PMID: 21073903]

[31] Leardkamolkarn V, Sirigulpanit W, Phurimsak C, Kumkate S, *et al*. The inhibitory actions of *Houttuynia cordata* aqueous extract on dengue virus and dengue□infected cells. J Food Biochem 2012; 36(1): 86-92.
[http://dx.doi.org/10.1111/j.1745-4514.2010.00514.x]

[32] Teixeira RR, Pereira WL, Oliveira AF, *et al*. Natural products as source of potential dengue antivirals. Molecules 2014; 19(6): 8151-76.
[http://dx.doi.org/10.3390/molecules19068151] [PMID: 24941340]

[33] Kiat TS, Pippen R, Yusof R, Ibrahim H, Khalid N, Rahman NA. Inhibitory activity of cyclohexenyl chalcone derivatives and flavonoids of fingerroot, *Boesenbergia rotunda* (L.), towards dengue-2 virus NS3 protease. Bioorg Med Chem Lett 2006; 16(12): 3337-40.
[http://dx.doi.org/10.1016/j.bmcl.2005.12.075] [PMID: 16621533]

[34] Schwartz A, Sutton SL, Middleton E Jr. Quercetin inhibition of the induction and function of cytotoxic T lymphocytes. Immunopharmacology 1982; 4(2): 125-38.
[http://dx.doi.org/10.1016/0162-3109(82)90015-7] [PMID: 6211417]

[35] Zandi K, Teoh BT, Sam SS, Wong PF, Mustafa MR, Abubakar S. Antiviral activity of four types of bioflavonoid against dengue virus type-2. Virol J 2011; 8(560): 560.
[http://dx.doi.org/10.1186/1743-422X-8-560] [PMID: 22201648]

[36] Meng FC, Wu ZF, Yin ZQ, Lin LG, Wang R, Zhang QW. *Coptidis rhizoma* and its main bioactive components: recent advances in chemical investigation, quality evaluation and pharmacological activity. Chin Med 2018; 13(1): 13.
[http://dx.doi.org/10.1186/s13020-018-0171-3] [PMID: 29541156]

[37] Zhang XG, Mason PW, Dubovi EJ, *et al*. Antiviral activity of geneticin against dengue virus. Antiviral Res 2009; 83(1): 21-7.
[http://dx.doi.org/10.1016/j.antiviral.2009.02.204] [PMID: 19501253]

[38] Takhampunya R, Ubol S, Houng HS, Cameron CE, Padmanabhan R. Inhibition of dengue virus replication by mycophenolic acid and ribavirin. J Gen Virol 2006; 87(Pt 7): 1947-52.
[http://dx.doi.org/10.1099/vir.0.81655-0] [PMID: 16760396]

[39] Chang J, Wang L, Ma D, *et al*. Novel imino sugar derivatives demonstrate potent antiviral activity against flaviviruses. Antimicrob Agents Chemother 2009; 53(4): 1501-8.
[http://dx.doi.org/10.1128/AAC.01457-08] [PMID: 19223639]

[40] Sessions OM, Barrows NJ, Souza-Neto JA, *et al*. Discovery of insect and human dengue virus host factors. Nature 2009; 458(7241): 1047-50.
[http://dx.doi.org/10.1038/nature07967] [PMID: 19396146]

[41] Helenius A, Aebi M. Roles of N-linked glycans in the endoplasmic reticulum. Annu Rev Biochem 2004; 73(1): 1019-49.
[http://dx.doi.org/10.1146/annurev.biochem.73.011303.073752] [PMID: 15189166]

[42] McEvoy GK. American Hospital formulary service drug information 2000. Bethesda, MD: American Society of Hospital Pharmacists Inc. 2000.

[43] Dawes BE, Kalveram B, Ikegami T, *et al*. Favipiravir (T-705) protects against Nipah virus infection in the hamster model. Sci Rep 2018; 8(1): 7604.
[http://dx.doi.org/10.1038/s41598-018-25780-3] [PMID: 29765101]

[44] Lo MK, Jordan R, Arvey A, *et al*. GS-5734 and its parent nucleoside analog inhibit Filo-, Pneumo-, and Paramyxoviruses. Sci Rep 2017; 7: 43395.
[http://dx.doi.org/10.1038/srep43395] [PMID: 28262699]

[45] Garner OB, Aguilar HC, Fulcher JA, *et al*. Endothelial galectin-1 binds to specific glycans on nipah virus fusion protein and inhibits maturation, mobility, and function to block syncytia formation. PLoS

Pathog 2010; 6(7)e1000993
[http://dx.doi.org/10.1371/journal.ppat.1000993] [PMID: 20657665]

[46]   Miller MD, Anton KE, Mulato AS, Lamy PD, Cherrington JM. Human immunodeficiency virus type 1 expressing the lamivudine-associated M184V mutation in reverse transcriptase shows increased susceptibility to adefovir and decreased replication capability *in vitro.* J Infect Dis 1999; 179(1): 92-100.
[http://dx.doi.org/10.1086/314560] [PMID: 9841827]

[47]   Gu Z, Gao Q, Fang H, *et al.* Identification of a mutation at codon 65 in the IKKK motif of reverse transcriptase that encodes human immunodeficiency virus resistance to 2′,3′-dideoxycytidine and 2′,3′-dideoxy-3′-thiacytidine. Antimicrob Agents Chemother 1994; 38(2): 275-81.
[http://dx.doi.org/10.1128/AAC.38.2.275] [PMID: 7514855]

[48]   Wainberg MA, Miller MD, Quan Y, *et al. In vitro* selection and characterization of HIV-1 with reduced susceptibility to PMPA. Antivir Ther (Lond) 1999; 4(2): 87-94.
[PMID: 10682153]

[49]   Srinivas RV, Fridland A. Antiviral activities of 9-R-2-phosphonomethoxypropyl adenine (PMPA) and bis(isopropyloxymethylcarbonyl)PMPA against various drug-resistant human immunodeficiency virus strains. Antimicrob Agents Chemother 1998; 42(6): 1484-7.
[http://dx.doi.org/10.1128/AAC.42.6.1484] [PMID: 9624498]

[50]   Huang L, Yuan X, Yu D, Lee KH, Chen CH. Mechanism of action and resistant profile of anti-HIV-1 coumarin derivatives. Virology 2005; 332(2): 623-8.
[http://dx.doi.org/10.1016/j.virol.2004.11.033] [PMID: 15680427]

[51]   Dubois M, Bailly F, Mbemba G, *et al.* Reaction of rosmarinic acid with nitrite ions in acidic conditions: discovery of nitro- and dinitrorosmarinic acids as new anti-HIV-1 agents. J Med Chem 2008; 51(8): 2575-9.
[http://dx.doi.org/10.1021/jm7011134] [PMID: 18351727]

[52]   Huerta-Reyes M, Basualdo MdelC, Abe F, Jimenez-Estrada M, Soler C, Reyes-Chilpa R. HIV-1 inhibitory compounds from Calophyllum brasiliense leaves. Biol Pharm Bull 2004; 27(9): 1471-5.
[http://dx.doi.org/10.1248/bpb.27.1471] [PMID: 15340243]

[53]   Miller MD, Margot NA, Hertogs K, Larder B, Miller V. Antiviral activity of tenofovir (PMPA) against nucleoside-resistant clinical HIV samples. Nucleosides Nucleotides Nucleic Acids 2001; 20(4-7): 1025-8.
[http://dx.doi.org/10.1081/NCN-100002483] [PMID: 11562951]

[54]   Margot NA, Isaacson E, McGowan I, Cheng AK, Schooley RT, Miller MD. Genotypic and phenotypic analyses of HIV-1 in antiretroviral-experienced patients treated with tenofovir DF. AIDS 2002; 16(9): 1227-35.
[http://dx.doi.org/10.1097/00002030-200206140-00004] [PMID: 12045487]

[55]   Fung HB, Stone EA, Piacenti FJ. Tenofovir disoproxil fumarate: a nucleotide reverse transcriptase inhibitor for the treatment of HIV infection. Clin Ther 2002; 24(10): 1515-48.
[http://dx.doi.org/10.1016/S0149-2918(02)80058-3] [PMID: 12462284]

[56]   Baum PD, Sullam PM, Stoddart CA, McCune JM. Abacavir increases platelet reactivity *via* competitive inhibition of soluble guanylyl cyclase. AIDS 2011; 25(18): 2243-8.
[http://dx.doi.org/10.1097/QAD.0b013e32834d3cc3] [PMID: 21941165]

[57]   Kim Y, Narayanan S, Chang KO. Inhibition of influenza virus replication by plant-derived isoquercetin. Antiviral Res 2010; 88(2): 227-35.
[http://dx.doi.org/10.1016/j.antiviral.2010.08.016] [PMID: 20826184]

[58]   Wu W, Li R, Li X, *et al.* Quercetin as an antiviral agent inhibits influenza A virus (IAV) entry. Viruses 2015; 8(1): 6.
[http://dx.doi.org/10.3390/v8010006] [PMID: 26712783]

[59]   Stiver G. The treatment of influenza with antiviral drugs. CMAJ 2003; 168(1): 49-56.
[PMID: 12515786]

[60]   Gubareva LV, Kaiser L, Hayden FG. Influenza virus neuraminidase inhibitors. Lancet 2000;
355(9206): 827-35.
[http://dx.doi.org/10.1016/S0140-6736(99)11433-8] [PMID: 10711940]

[61]   Henegar KE, Cebula M. Process development for (S, S)-reboxetine succinate *via* a Sharpless
asymmetric epoxidation. Org Process Res Dev 2007; 11(3): 354-8.
[http://dx.doi.org/10.1021/op700007g]

[62]   Wang C, Takeuchi K, Pinto LH, Lamb RA. Ion channel activity of influenza A virus M2 protein:
characterization of the amantadine block. J Virol 1993; 67(9): 5585-94.
[http://dx.doi.org/10.1128/JVI.67.9.5585-5594.1993] [PMID: 7688826]

[63]   Hamed SA, Abdellah MM. The relationship between valproate induced tremors and circulating
neurotransmitters: a preliminary study. Int J Neurosci 2017; 127(3): 236-42.
[http://dx.doi.org/10.1080/00207454.2016.1181631] [PMID: 27161592]

[64]   Chen S, Yu X, Guo D. CRISPR-cas targeting of host genes as an antiviral strategy. Viruses 2018;
10(1): 40.
[http://dx.doi.org/10.3390/v10010040] [PMID: 29337866]

[65]   Wang G, Zhao N, Berkhout B, Das AT. CRISPR-Cas9 can inhibit HIV-1 replication but NHEJ repair
facilitates virus escape. Mol Ther 2016; 24(3): 522-6.
[http://dx.doi.org/10.1038/mt.2016.24] [PMID: 26796669]

CHAPTER 13

Enzyme Inhibition Applications in Treatment of Neurological Disorders

D. Lokapriya and **P.K. Praveen Kumar**[*]

Department of Biotechnology, Sri Venkateswara College of Engineering (Autonomous), Sriperumbudur Tk, 602117, Tamilnadu, India

Abstract: Enzyme inhibitors are widely prescribed drugs for many diseases including neurological disorders and today most of the drugs are enzyme inhibitors and are in the clinical/pre-clinical phase of the drug discovery. For many of the neurological disorders, especially neurodegenerative disorders, only symptomatic therapy is available rather than the therapy based on an understanding of the mechanism of these diseases. In this case, enzyme inhibitors become the solution as they inhibit the action of enzymes whose abnormal activity may be one of the causes of the disease. They also halt the progression of the disease and alleviate the symptoms. This chapter focuses on some of the enzyme inhibitors that have been prescribed as drugs for neurological disorders, their mechanism of action and discuss some inhibitors that are still in their research level of development.

Keywords: Drug discovery, Enzyme Inhibitors, Neurological disorders, Symptomatic therapy.

INTRODUCTION

Enzymes are protein molecules that act as catalysts in various biochemical reactions. Substances that inhibit the catalytic activity of enzymes are called enzyme inhibitors. Being the low molecular weight chemical compounds, enzyme inhibitors reduce the enzyme activity either reversibly or irreversibly [1]. Nowadays, chemo drugs are based on reducing the activity of overactive enzymes which in turn declines the progression of the disease and alleviates symptoms [2].

Generally, competitive enzyme inhibition is the mechanism used for employing enzyme inhibitors as pharmaceutical agents in which inhibitors structurally similar to normal biochemical substrates are used to compete with the natural substrate for the active site of the enzyme and it results in blocking the formation

[*] **Corresponding author Praveen Kumar P.K:** Department of Biotechnology, Sri Venkateswara College of Engineering (Autonomous), Sriperumbudur Tk, 602117, Tamilnadu, India; Tel: +919444495008; Fax: +914427162462; Email: praveenpk@svce.ac.in

G. Baskar, K. Sathish Kumar & K. Tamilarasan (Eds.)

of undesirable metabolic products [2]. Hence, enzyme inhibitors constitute a significant portion of the clinical usage of oral therapeutic drugs. Currently, enzymes are attractive targets for drug discovery and efforts are being made on identifying and optimizing drug candidates that specifically inhibit enzyme targets in the field of drug discovery and drug development [3].

Diseases that affect the brain, spine and nerves are called neurological disorders. Nearly, 600 common diseases of the nervous system are occurring in humans mainly: Stroke, Alzheimer's and, Parkinson, *etc.* The current population is aging. Hence, these disorders are very prevalent nowadays as aging is one of the significant risk factors of neurodegenerative disorders along with inflammation, apoptosis, and excitotoxicity [4]. Most of the drugs that are enzyme inhibitors for neurological disorders have been approved by the FDA and some enzyme inhibitors are still in the clinical and pre-clinical phase of the drug development. This chapter gives an overview of some enzyme inhibitors that have been used as drugs for neurological disorders along with their mechanisms.

NEURODEGENERATIVE DISEASES

Alzheimer's Disease (AD)

AD is one of the generally occurring neurodegenerative diseases. Symptomatic therapy is chosen for the treatment of AD. Dementia along with impaired learning and cognition is the primarily noticed clinical symptom but irritability, confusion, and behavioral changes occur later as the disease progresses. Peptides like β-amyloid (Aβ) plaques and neurofibrillary tangles are the pathological hallmarks of AD. They accumulate in the brains of patients of AD. Due to the mutations in amyloid precursor protein (APP), β-secretase (APP cleaving enzyme) does increase the proteolysis and it leads to the increased formation of Aβ plaques by γ-secretases. The accumulated Aβ plaques have destructive effects on neurons especially neuronal loss as it generates free radicals in the brain. Hence, neurotoxicity in the brain due to AD is directly proportional to aggregates formed by Aβ plaques [4].

There are five major types of drugs used for treating AD: cholinergic treatment, anti-glutamatergic treatment, nonsteroidal anti-inflammatory drugs (NSAIDs), vitamins and antioxidants, and pharmacological management of neuropsychiatric symptoms. Among these, acetylcholinesterase inhibitors (AChEIs) are widely used for treating AD [5].

Inhibition of Cholinesterases

Acetylcholine (Ach) is a neurotransmitter responsible for the conduction of electrical impulses among neurons. When the acetylcholinesterase (AChE) enzyme (EC 3.1.1.7) hydrolyses Ach, it's level decreases [6]. ACh is generated in neurons by the action of choline acetyltransferase concentrated in vesicles, and released from the presynaptic cell, which is shown in Fig. (**1**).

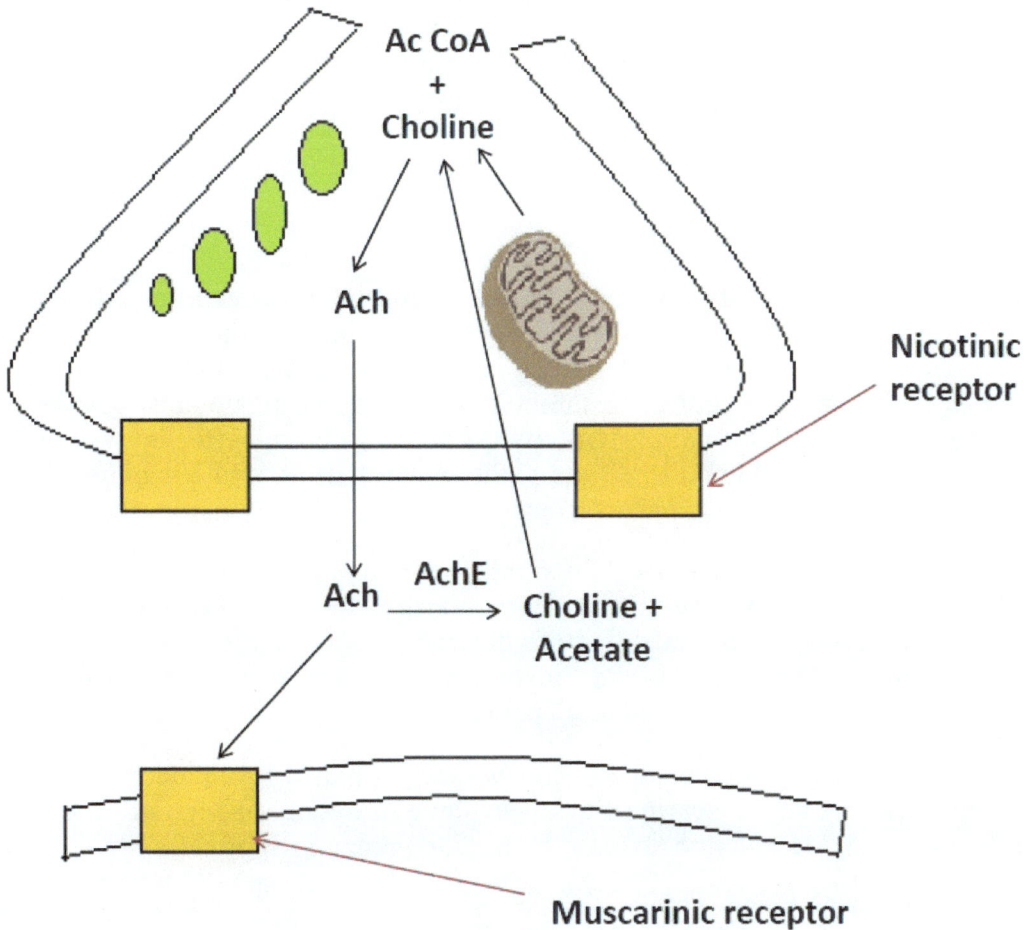

Fig. (1). Acetylcholine biosynthesis, and synaptic transmission.

ACh is released *via* nicotinic and muscarinic cholinergic receptors on post- and presynaptic cells. Once released, it interacts with the muscarinic receptor on postsynaptic cells. The interaction of Ach with muscarinic receptors leads to the hydrolysis of Ach by AChE. When AChE is inhibited, neurotransmission can be prolonged for the time as ACh molecules remain unhydrolyzed and still interact with muscarinic receptors. AChE inhibitors (AChEIs) reduce the degradation of ACh and therefore maintain neurotransmission. Clinical AChEIs should not interfere with the synthesis of ACh [5].

Drugs

Despite patients showing hepatotoxicity in the clinical trials, Tacrine was originally prescribed for AD and it was the first AChEI drug approved by the FDA. Currently, donepezil, galantamine and rivastigmine are prescribed as they prove to be effective and safer in clinical trials [5]. Possible side effects of these drugs are nausea, diarrhea, vomiting, muscle cramps, tiredness, headache, weight loss and bruising.

Parkinson's Disease (PD)

PD is the second most common neurodegenerative disorder after AD. Patients experience tremor and display difficulty in bodily movements, rigidity, hypokinesia, and impaired balance as brains affected with PD lose striatal and dopaminergic neurons in the substantia nigra which coordinate motor movements. Lewy bodies are abnormal protein clumps found in brains affected with PD and hence they are considered to be the pathological hallmark of progression of PD [4].

Similar to AD, PD is also based on symptomatic therapy. There are no drugs that can prevent neuronal loss. At an earlier stage, reduced olfactory sensitivity, autonomic dysfunction, and affective disorder are markable symptoms, but the most apparent symptoms of Parkinson's disease are due to the depletion of dopamine (DA) in the nigrostriatal pathway [7].

Pharmacological treatment of PD can be accomplished by levodopa, dopamine agonists (DAs), antimuscarinics or amantadine, Catechol-O-Methyl- Transferase inhibitors and, Monoamine Oxidase B (MAO-B) enzyme inhibitors [8].

Inhibition of Monoamine Oxidase (MAO)

MAO (EC 1.4.3.4) is a mitochondrial enzyme highly expressed in neuronal tissues. Oxidative deamination of amines catalyzed by MAO is shown in Fig. (**2**). It also plays a major role in metabolization of neurotransmitters *i.e.* noradrenaline, serotonin (5-HT) and DA and detoxification of amines. Drugs that inhibit MAO are currently in clinical use for the treatment of PD [7, 9].

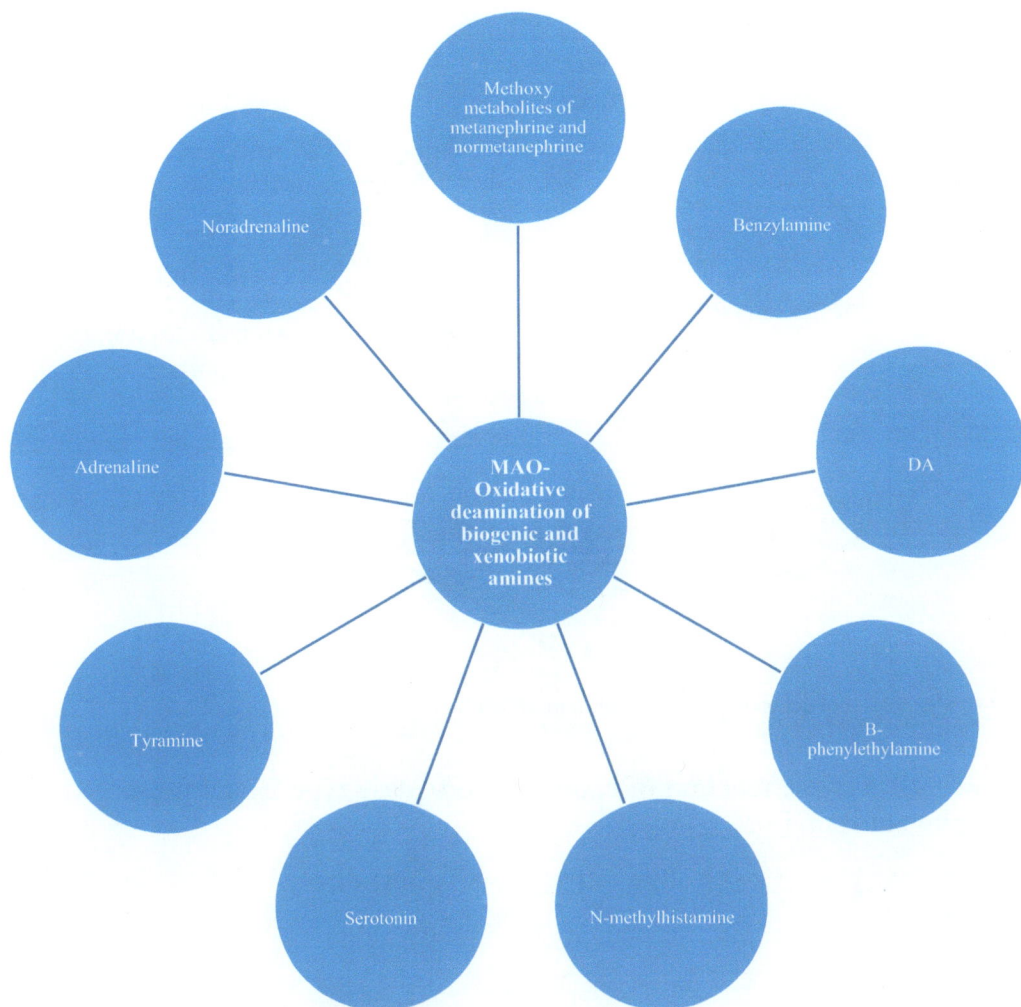

Fig. (2). Oxidative deamination of amines catalyzed by MAO.

The mechanism of enzyme inhibition of MAO is shown in Fig. (3). Originally, MAO inhibitors were used for treating depression diseases as it inhibits degradation of 5-HT and noradrenaline which consequently reduce amine levels in their receptors. Brains of patients affected with PD have reduced levels of DA and hence inhibition of DA oxidative metabolism is effective in returning to normal neurotransmitter levels [7].

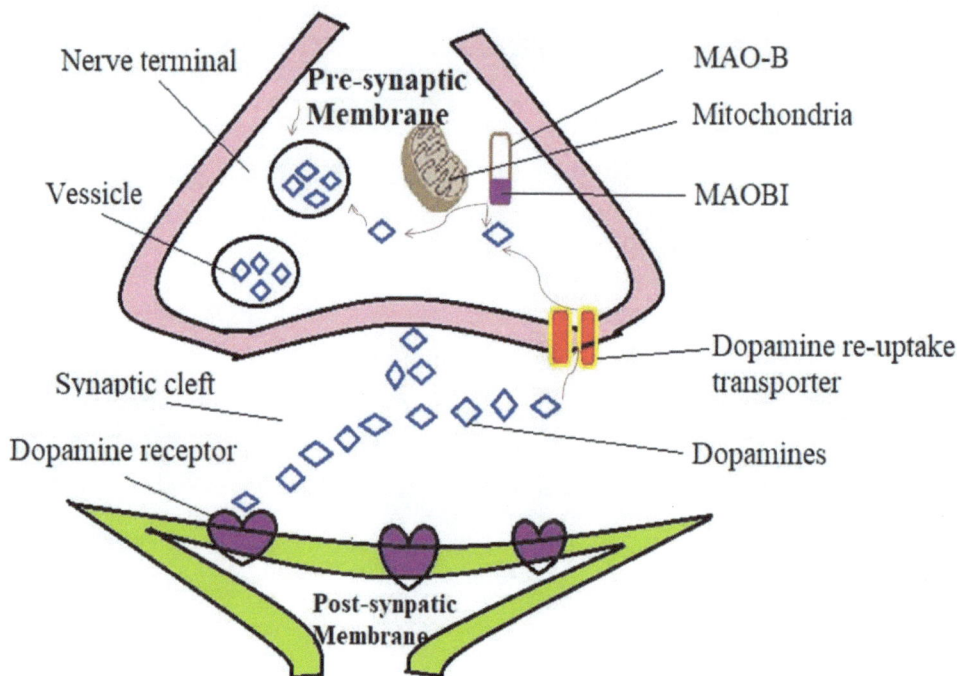

Fig. (3). Mechanism of action of enzyme inhibition of MAO-B.

The overall enzyme reaction of Monoaminooxidase type B is represented in the following equation [9]:

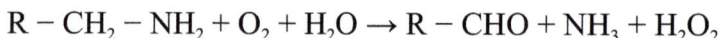

$$R - CH_2 - NH_2 + O_2 + H_2O \rightarrow R - CHO + NH_3 + H_2O_2$$

Drugs

Rasagiline is a selective inhibitor of MAO type B and to inactivate the enzyme, it binds covalently with the N5 nitrogen of the flavin residue of MAO [7]. Selegiline is another MAO-B inhibitor that prevents enzyme action by the mechanism of irreversible noncompetitive inhibition. Hence, it reduces the dopamine level and results in a free radical generation. In Parkinson's disease pathogenesis, the production of free radicals damages lipids and proteins [10]. Common side effects of this Rasagiline drug are dizziness, joint pain, headache, depression, nausea, fever, muscle pain, vomiting, impotence, *etc.*

Stroke

Stroke ranks next to AD in neurological disorders. Stroke is the largest killer in general diseases next to heart disease and cancer. Types of stroke are shown in Fig. (**4**). Hypertension, diabetes, atherosclerosis, and heart disease are well-known risk factors of stroke.

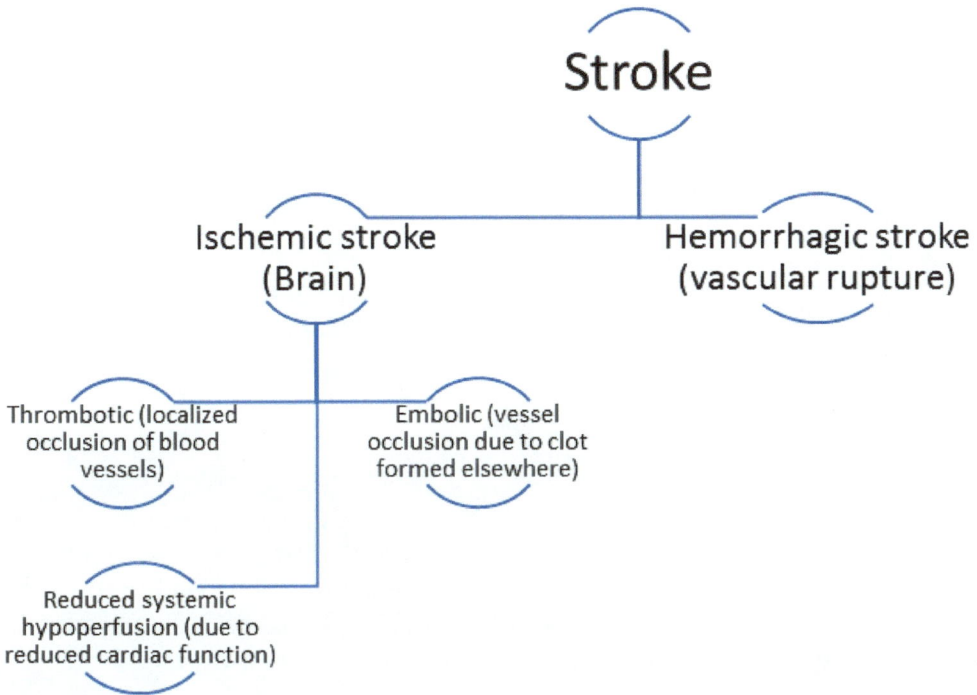

```
                              Stroke

        Ischemic stroke              Hemorrhagic stroke
          (Brain)                     (vascular rupture)

  Thrombotic (localized      Embolic (vessel
  occlusion of blood         occlusion due to clot
     vessels)                 formed elsewhere)

  Reduced systemic
  hypoperfusion (due to
  reduced cardiac function)
```

Fig. (4). Types of Stroke.

The mechanism of stroke is shown in Fig. (**5**). The main pathology of the stroke is a vascular occlusion in the brain. Blood vessels are compromised not to supply blood to the brain and hence oxygen and glucose supply is reduced. Tissues in the brain rely on aerobic metabolism. The brain parenchyma suffers death immediately compared to the surrounding areas because of no oxygen supply and due to the lack of necessary respiratory reserve [4].

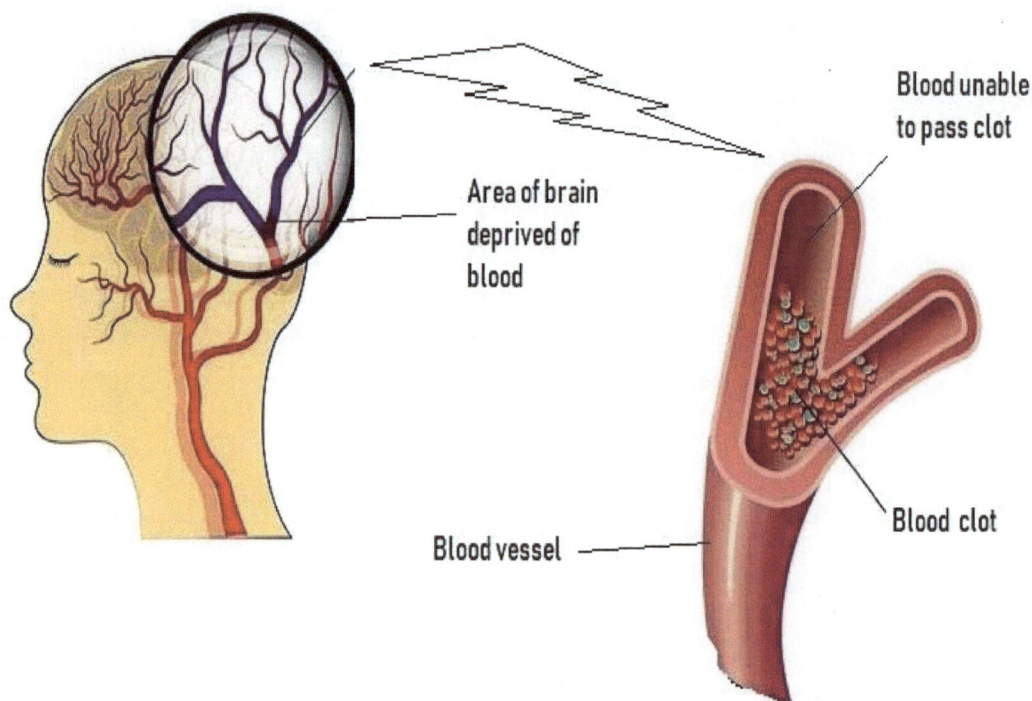

Fig. (5). Mechanism of Stroke.

Antiplatelet agents are commonly preferred pharmacologic treatment to prevent stroke. Antithrombotic agents like clopidogrel, dipyridamole, ticlopidine, extended-release warfarin, and aspirin are also used in the prevention of ischemic stroke. Aspirin is the first prescribed drug for stroke since it is affordable and there is no need to monitor it extensively [11].

Inhibition of Cyclooxygenase (COX)

Drug: Aspirin

Aspirin inhibits the effects of platelets hence preventing vascular complications. Within 30 minutes of ingestion the drug exhibits its antiplatelet effects and persists for the platelet lifespan. Aspirin inhibits the prostanoid TXA2 production as TXA2 is responsible for platelet aggregation and vasoconstriction, by inactivating COX (EC 1.14.99.1) enzymes. Aspirin also affects haemostasis and thrombogenesis. Aspirin is an efficient anti-thrombogenic agent because it inhibits platelets, promotes fibrinolysis and suppresses plasma coagulation factors. Among all, its ability to inactivate COX-1 is essential for preventing arterial thrombosis [11]. Common side effects of Aspirin are rash, gastrointestinal ulcerations, abdominal pain, heartburn, drowsiness, headache, and cramping.

Brain Tumor

Generally, tumor/cancer is a disease/state where cells divide abnormally and destroy healthy cells. A brain tumor is a collection, or mass, of abnormal cells inside the skull which when grows causes damage to the brain. There are two different types of brain tumors: they are primary and secondary. Primary brain tumors are benign as they originate in the brain itself whereas secondary brain tumors are metastatic as they are spread to the brain from different organs [12]. The types of brain tumors are represented in Fig. (**6**).

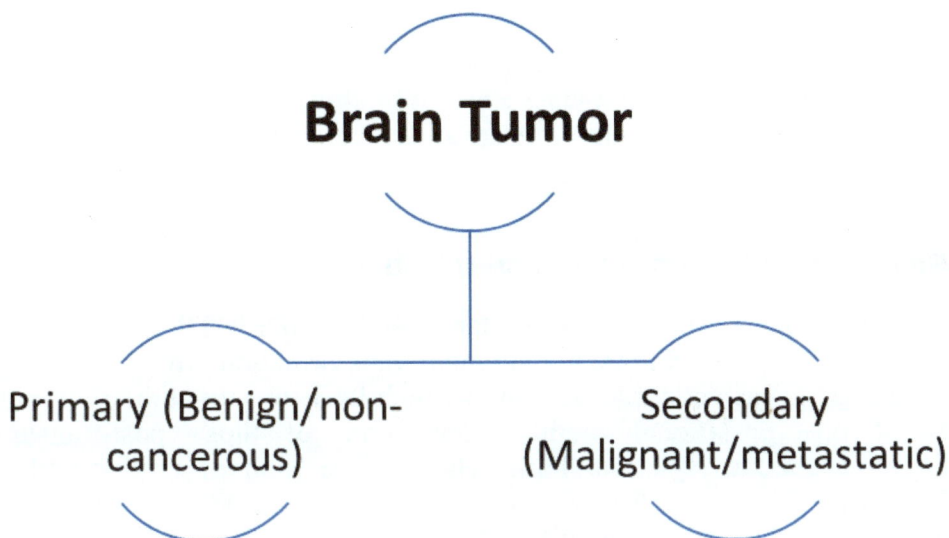

Brain Tumor

Primary (Benign/non-cancerous)

Secondary (Malignant/metastatic)

Fig. (6). Types of Brain tumor.

Inhibition of Matrix Metalloproteinases (MMPs)

MMPs (EC 3.4.24.24) are involved in the degradation of connective tissues like monocytes, fibroblasts, macrophages, endothelial cells and metastatic tumor cells. They include matrilysin, gelatinases, collagenases, and stromelysins [13]. These enzymes are dependent on zinc and calcium and they share similar amino acid sequences up to 40-50%. MMPs degrade the protein components including collagen, proteoglycan, fibronectin and laminin. Hence, in pathological conditions like tumor metastasis, the deregulation of MMP leads to the uncontrolled breakdown of the extracellular matrix. Substances that inhibit MMP inhibitors can be used as therapeutic drugs for tumors as they reduce the rate of degradation of connective tissues. Drugs used under this inhibition of MMPs are marimastat, metastat, and prinomastat.

Inhibition of Phosphoinositide 3-Kinase (PI3k)

The PI3Ks [EC 2.7.1.137] are lipid kinases that phosphorylate phosphatidy-linositol 4, 5-bisphosphate to phosphatidylinositol 3, 4, 5-trisphosphate. This phosphorylation in turn activates Akt which plays a major role in carcinogenesis. Hence, PI3K is the most often activated pathway in cancer metabolism [14] and is an important anticancer drug target. Currently interest in the development of PI3K inhibitors is very high.

Drug: Idelalisib

Idelalisib is the approved first-class PI3K inhibitor in the USA and Europe for cancer therapy. Other PI3K inhibitors are still in clinical trials including BKM120 and ZSTK474.

Inhibition of Ras: Farnesyl Transferase (FTase)

Ras (rat sarcoma), a low molecular weight GDP/GTP-binding guanine triphosphatase plays a vital role in malignant transformation, invasion, and spread of gliomas [15]. Ras undergoes mutation which leads to activation of gliomas by series of posttranslational modifications. One of those post-translational modifications is farnesylation. Farnesylation is a kind of lipid modification catalyzed by FTase (EC 2.5.1.58). This process depends on the recognition of a specific amino acid sequence with a carboxyl-terminal called CAAX by FTase.

CAAX is the abbreviated for

C – Cystine

AA - Aliphatic amino acids

X - Amino acid, preferably methionine or serine

This farnesylation process enables Ras to anchor to the cell membrane which is the essential step for Ras to activate cancer.

Drug: Lonafarnib/Sarasar

It is an oral non-inhibitor of FTase and is active against tumors with Ras proteins. Rho, Rheb, and CENP-E, F proteins are some of the G proteins which can undergo farnesylation. These proteins are targeted by farnesyltransferase inhibitors as they are potential substrates for farnesylation. These inhibitors are even able to inhibit P-glycoprotein, a product of a multi-drug resistant gene that plays a major role in resisting chemotherapy; the development of resistance to chemotherapy by cancer cells.

Drug: Tipifarnib

It is a non-peptidomimetic methyl-quinolone originally developed for antifungal infections. It is also a selective non-peptidomimetic inhibitor of FTase. It induces postmitotic necrotic cell death of radioresistant tumor cell lines as a radiosensitizer.

Inhibition of Proteasome

Proteasomes are potential targets for cancer drugs because they are the reason for the degradation of proteins involved in the cell cycle, DNA transcription and repair, apoptosis, angiogenesis, and cell growth [15].

Drug: Velcade

Velcade is working against cells that have been transformed into cancer by inducing apoptosis in the G2-M phase.

Inhibition of COX-2

The cyclooxygenase pathways are actively present in both normal and tumoral astrocytes [15]. The COX-2 enzyme is highly expressed in cancer cells, and its expression is important for carcinogenesis. High-grade human gliomas are found with an upregulated COX-2 enzyme. By inhibiting the COX-2 enzyme, proliferation and migration of human glioblastoma cell lines can be reduced as they suppress growth and induce apoptosis.

Drug: Celecoxib

Celecoxib is an NSAID that inhibits prostaglandin synthesis by specifically inhibiting the COX-2 enzyme.

Glioblastoma

Glioblastomas are like malignant brain tumors. It is different from normal brain tumors in a way it origins from star-shaped glial cells such as astrocytes and oligodendrocytes which support the health of the nerve cells. They also act as immune cells within the brain [16].

Like a common brain tumor, the blood supply nourishes the growth of abnormal gliomas. Though the tumor majorly consists of abnormal astrocytic cells, it also contains different cell types including blood vessels and areas of dead cells. Glioblastomas can easily pass through one hemisphere of the brain to another (metastasis inside the brain) *via* the corpus callosum, the connection fibers act as a bridge between two fibers. Glioblastomas spread rarely outside the brain [17].

Inhibition of Protein Kinase C (PKC)

PKC (EC 2.7.11.13) is cytoplasmic threonine/serine kinases regulate cell survival, stimulation, and apoptosis of tumor cell [18]. Important proteins in several cellular signaling pathways are phosphorylated by PKC. It also plays an important role in regulating angiogenesis through VEGF. Thus, PKC is a target receptor for the discovery and development of neurological disorder treatment therapies.

Drugs

Tamoxifen is a lipid-soluble, nonsteroidal agent and blood-brain barrier permeability drug used for breast cancer. It increases the apoptosis of cancer cells by inhibition. If supplied at a high dose, it can counteract chemoresistance. Enzastaurin is another small lipid-soluble molecule like Tamoxifen having good blood-brain barrier permeability that inhibits the PKCb receptor.

Inhibition of Histone Acetylation

Histone acetylation is an intrinsic role in epigenetic modification and gene regulation [18]. Lysine residues present on histone core proteins are modified by histone acetylases and histone deacetylases enzymes. Thus during acetylation, positively charged proteins are converted to negative ones. Histone acetylating leads to chromatin relaxation and subsequently transcription. Moreover, histone acetylases and histone deacetylases involved in cell survival, differentiation, proliferation, and apoptosis.

Drug: Vorinostat

Vorinostat, a linear suberoylanilide hydroxamic acid acts on histone deacetylases silencing gene expression as this drug tightly coils the chromatin.

Multiple Sclerosis (MS)

MS is the most common inflammatory condition of the brain, spinal cord and nerves. It is a neurodegenerative disorder, where the immune system destroys the white matter covering nerves. Hence, it is also an autoimmune disorder [19].

Inhibition of Histone Deacetylases (HDAC)

During gene expression, gene sequences which code for important proteins such as enzymes and neurotransmitters are made accessible to transcription factors by enzymes. One of those enzymes is HDAC (EC 3.5.1.98) [20]. Inside the nucleus, the DNA and histone proteins assembled to form nucleosomes. The lysine residues present in the histone proteins are acetylated which is necessary for transcriptional activation. The enzyme histone acetyltransferase mediates the addition of acetyl groups whereas their removal by HDACs. Hence, the balance between these two enzymes is indispensable to express specific sets of genes. Their imbalance leads to an imbalance in acetylation and deacetylation and hence inflammations as MS arise. Especially decrease in histone acetylation contributes to several disease states. It can be counteracted by inhibiting HDAC as inhibitors of HDACs show therapeutic benefits in treating epilepsy, mood disorders, immune diseases like MS, *etc.*

Other than core histone proteins, the targets of HDAC include nuclear hormone receptors, transcription factors, and cytoskeletal elements.

Drugs

The most commonly prescribed inhibitors of HDAC include suberoylanilide hydroxamic acid, hydroxamate compounds trichostatin A, aliphatic acids, sodium butyrate, and valproic acid. Table **1** shows the types of neurological diseases, enzymes and the drugs used for the treatment.

Table 1. Disease, Inhibited enzyme and prevalent drugs.

Disease	Enzyme	Drugs	Reference
AD	AChE	Tacrine	[5]
		Donepezil	
		Rivastigmine	
		Galantamine	
PD	MAO	Rasagiline	[7]
		Selegiline	[10]
Stroke	COX-1	Aspirin	[11]
Brain tumor	MMP	Marimastat	[13]
		Metastat	
		Prinomastat	
	PI3k	Idelalisib	[14]
	Ftase	Lonafarnib/Sarasar	[15]
		Tipifarnib	
	Proteasome	Velcade	
	COX-2	Celecoxib	
Glioblastoma	PKC	Tamoxifen	[18]
		Enzastaurin	
	Histone Acetylase	Vorinostat	
MS	HDAC	Trichostatin A	[20]
		Suberoylanilide hydroxamic acid	
		Valproic acid	
		Sodium butyrate	

ENZYME INHIBITORS FROM PLANTS

Plant derived chemical substances have been used to treat neurological diseases since the origin of medicines. Most of the chemo drugs developed in the past two decades are from natural products. Advancements in technological innovations have renewed the interest in natural products as drugs in medicine. There are only traditional methods to prove the efficacy of various natural inhibitors. Their potential has to be validated *via* scientific research [21].

Many of the plant compounds that are used as enzyme inhibitors in clinical and pre-clinical trials of drug discovery are shown in Table **2**.

Besides this, numerous plant compounds possess natural cholinesterase [30, 31] and Monoamine Oxidase B [29, 32 - 35] inhibitory properties. More extracts of many plants have been found to possess enzyme inhibitory properties.

PLA2 inhibitors are also present in extracts of curcumin, *Centella Asiatica and Ginkgo biloba* and they have been studied for treating neurological disorders in the research level [36].

Rosemary (Satapatrika) contains natural COX-2 inhibitors like Apigenin, thymol, carvacrol, oleanolic acid, eugenol, and ursolic acid [30].

Table 2. Disease, Inhibited enzyme and prevalent drugs.

Compound	Enzyme Inhibition	Disease	Origin	Reference
Huperzine A	AChe	AD	*Huperzia serrata*	[22]
Zt-1	AChe	AD	*Huperzia serrata*	[22]
Physostigmine	AChe	AD and Myasthenia Gravis	*Physostigma venenosum* L	[23]
Curcumin	MAO-B	PD and other Neurological Disorders	*Curcuma longa*	[24]
Ellagic Acid	MAO-B	PD and other Neurological Disorders	Wide range of plant species	[24]
Genistein	Tyrosine Kinases	Amyotrophic Lateral Sclerosis [ALS]	Wide range of plant species	[25]
Apocynin	NADPH-Oxidase [NOX]	Ischemic Stroke	Canadian hemp (*Apocynum cannabinum*), *Picrorhiza kurroa*	[25]
Honokiol	NADPH-Oxidase [NOX]	Ischemic Stroke	*Magnolia Officinalis*	[26]
Plumbagin	NOX	Ischemic Stroke	*Plumbago zeylanica*	[26]
Hypericin	MAO	MS	*Hypericum perforatum* L	[27]
Arecaidine, Arecoline, Guvacine	MAO-A	Depression	*Areca catechu*	[27]
Galanthamine	AChe	AD	*Galanthus nivalis*	[28]
Gastrodin, Vanillin	GABA Transaminase	PD	*Gastrodia elata*	[28]

OTHER ENZYMES

Beta-site Amyloid Precursor Protein Cleaving Enzyme-1 [BACE-1]

BACE-1(EC 3.4.23.46) is a protease enzyme that processes amyloid precursor protein (APP) to form the protein-soluble APPβ and N-terminus of Aβ peptides [37]. BACE-1 initiates the generation and aggregation of Aβ plaques, which in turn gives rise to neurodegenerative disorders especially AD. Inhibition of BACE-1 is a new concept for treating AD as it prevents the generation and accumulation of Aβ rather than treating the symptoms.

γ-secretase

Based on the amyloid hypothesis, γ-secretase (EC 3.4.23) is responsible for forming Aβ [38]. Hence, compounds that inhibit γ-secretase are potential therapeutics for AD. Inhibitors of γ-secretase are being developed for the treatment of AD.

Sirtuin 2 (SIRT2) Deacetylase

Inhibition of SIRT2 deacetylase (EC 3.5.1) mediates protective effects on cells and invertebrate models of PD and Huntington's disease (HD). SIRT2 is a highly abundant protein in the adult brain, and it is expressed in oligodendrocytes and neurons, although the protein function remains elusive [39].

Thiazole-containing inhibitors of the SIRT2 deacetylase with neuroprotective activity have been developed [40].

Phospholipases A2 (PLA2)

PLA2 (EC 3.1.1.4) hydrolyzes membrane phospholipids into lysophospholipids and arachidonic acid [36]. After hydrolysis, arachidonic acid is further metabolized into prostaglandins, thromboxanes, leukotrienes, and lysophospholipids are converted to platelet-activating factors. These compounds play a major role in oxidative stress and neuroinflammation. Since neurological disorders are characterized by oxidative stress, inflammatory reactions, as well as increased activities of brain phospholipase A2, inhibitors of PLA2 should also be given as choices for treating neurological disorders. Inhibitors of PLA2 for the treatment of neurological disorders are shown in Fig. (7).

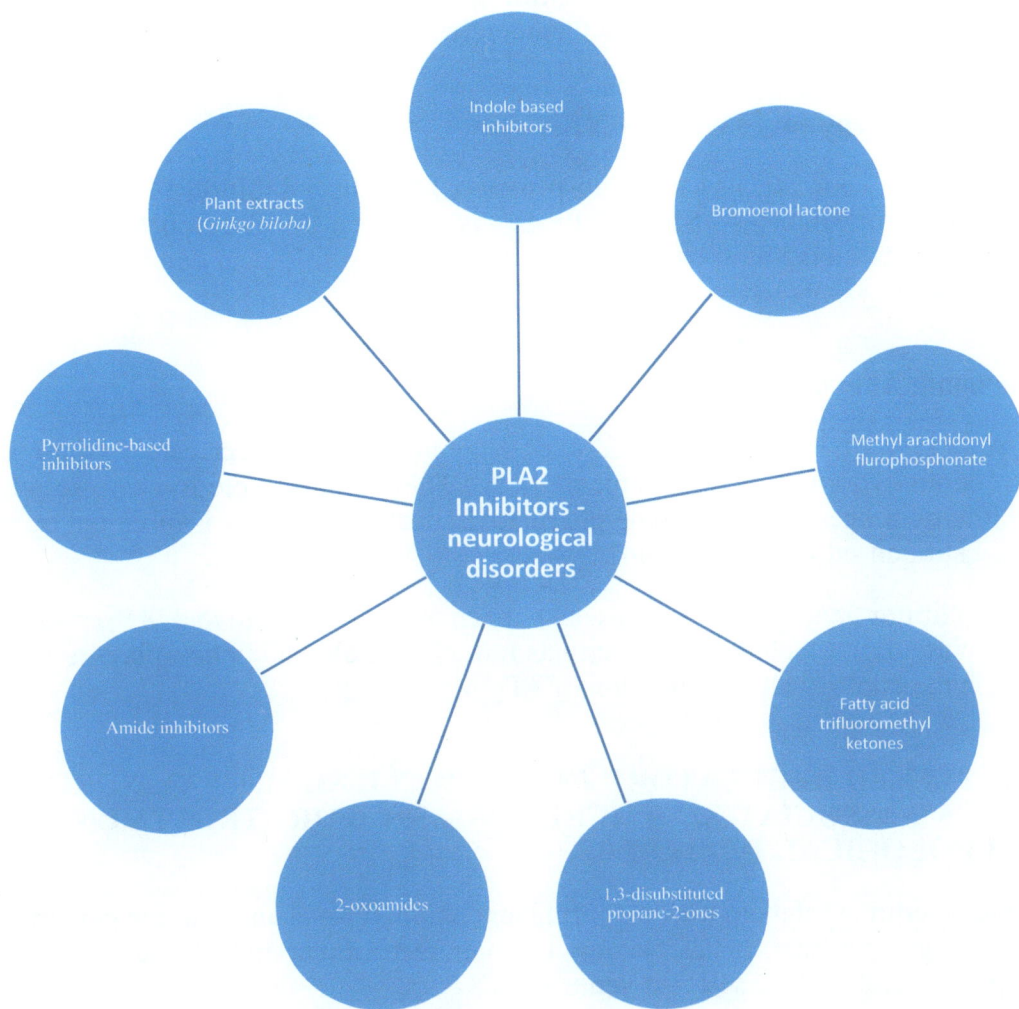

Fig. (7). Old and new synthetic inhibitors of PLA2 for treatment of neurological disorders in cell culture and animal models.

Glutathione S Transferases (GST)

The main function of GST [EC 2.5.1.18] is to protect cells against oxidative stress and toxicity, and are involved in the synthesis and modification of leukotrienes and prostaglandins [41]. GSTs can conjugate glutathione to carcinogens, therapeutic drugs, and products of oxidative metabolism. It makes them less toxic and ready to be discharged from the cell.

Dysregulated GST may lead to the overproduction of oxidants that may cause oxidative stress. Since neurons are highly sensitive to oxidative stress and contribute to the development of neurological disorders, substances that inhibit GST should be investigated for its therapeutic role in alleviating disorders.

Dipeptidyl Peptidase IV [DPP-IV]

DPP-IV [EC 3.4.14.5] is a serine protease [42]. Its inhibition is an effective treatment for diabetes. Since DPP-IV is doubted to develop neurological disorders, its therapeutic use for the treatment of neurological conditions is still under study.

Carbonic Anhydrase [CA]

CAs (EC 4.2.1.1) is wide-spread zinc enzymes and efficient catalysts for the reversible hydration of CO_2 to bicarbonate. They are also able to catalyse other hydrolytic processes. To develop pharmacological agents, CAs is inhibited by the unsubstituted sulphonamides [44].

CA inhibitor methazolamide has also been reported to prevent Aβ (plaques) neurovascular mitochondrial toxicity as mitochondrial dysfunction plays a causal role in the etiology and progression of AD [43].

CLUSTERED REGULATORY INTERSPACED SHORT PALINDROMIC REPEAT-ASSOCIATED 9 (CRISPR/CAS9) FOR TREATMENT OF NEUROLOGICAL DISORDERS

Genome editing refers to making permanent and precise changes in the genome of the living organisms. It has become a repair mechanism and nowadays used to knock out or introduce selected genes [45].

In neurodegenerative disorders like AD and PD, gene-editing technology allows neuroscientists to access and edit the genome effectively and hence paves the way to study the brain-behavior relationships [46]. Among the various gene-editing tools, CRISPR/CAS9 is an advanced technique for accurate, specific and precise genetic manipulation of DNA and applies to the eradication of single and double-stranded DNA viruses that causes neurological disorders and infections. Hence, this technology can be applied to eradicate DNA viruses that cause neurological infections. This facile gene manipulation technology consists of guide RNA [gRNA] to target a specific sequence that is adaptable and flexible to different target nucleotide sequences. CRISPR/Cas nucleases act on the adaptive immune system in nearly all Archaea and most of the bacteria to fight against viruses. The gRNA in CRISPR-CAS9 is used to direct and degrade target DNA in a sequence-specific fashion [47]. With CRISPR technology, mutations can be induced in the gene responsible for neurological disorders and hence could be used to study disease-linked epigenetic changes in cell models [48].

CRISPR/Cas9 system can be tailored to either activate transcription (gain-of-function) or achieve gene silencing (Loss-of-function) [49]. CRISPR/Cas9 can perform both knocking out of genes (inhibition of enzymes) and knocking-in mutations (over-expression of enzymes). This technique can be used to specifically inhibit gene which expresses enzymes whose overactivity is responsible for causing neurological disorders. In addition to that, by creating additional sgRNA sequences directed against different gene targets, it is possible to target multiple genes at one time using CRISPR/Cas9 system. Because of this technology, studying the effects of deleting proteins that are encoded by multiple genes on behavior has become reality [46].

Using CRISPR/CAS9, one can build animal models of neurological disorders and this can create a breakthrough in the neuro research and drug development for neurodegenerative disorders [50]. This kind of gene editing system is seeking attraction for clinical applications (in neurology) as repairing/incorporating genes might be of therapeutic benefit [45].

DNA viruses that can cause neurological disorders and infections can be subjected to CRISPR/Cas9 treatment as they are more amicable when compared with RNA viruses unless a cellular DNA target is mandatory for their life cycle of growth of viruses. For example, varicella-zoster virus or CMV belongs to RNA strand retroviruses have a DNA intermediate in their life cycle. Another example is human T-lymphotropic virus-1 (HTLV-1) which causes myelopathy/tropical spastic paraparesis, a neurological pathology disorder. Thus, the CRISPR/Cas9 technique is a potential approach for treating neurological infections caused by HTLV-1 [47].

CRISPR/Cas9 has the potential to be used as prophylactic and therapeutic treatment strategies. In neuro-AIDS, CRISPR/Cas9 system besides removing provirus present in cells (*i.e.* inactivating gene expression and replication of retrovirus HIV-1 in a variety of promonocytic, microglial and T cells of infected cells), it also prevents the infection by adding Cas9 and HIV-1-specific gRNAs to uninfected cells [47]. Fig. (**8**) shows the CRISPR-CAS9 gene editing/silencing mechanism of HIV.

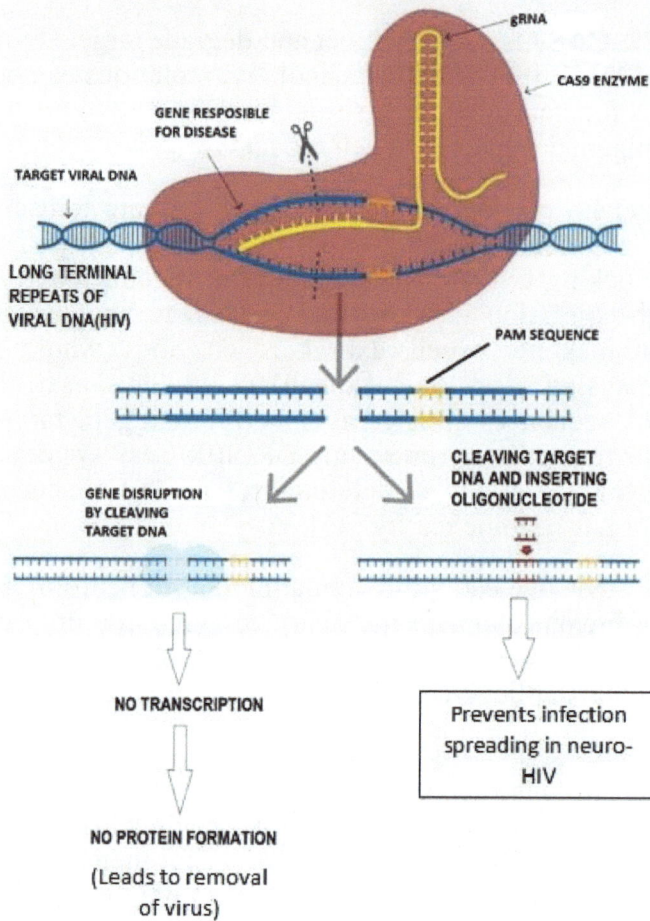

Fig. (8). CRISPR-CAS9 – Mechanism of Gene silencing and Mutation in HIV.

Progressive multifocal leukoencephalopathy (PML) is a severe neurological disorder that causes the immune system to malfunction. It is characterized by an opportunistic infection of the CNS by JCV. CRISPR/Cas9 can be used as a therapeutic agent for PML as it can eradicate JCV infection from the human CNS [47].

HD is another neurodegenerative disorder characterized by dementia, behavior disturbances, and choreatic movements. CRISPR/Cas9 based gene editing approach is proved to inactivate HD-associated mutant Huntington allele and in a mouse model affected with Schizophrenia, brain-specific genes were deleted using CRISPR/Cas9 [51].

It could be a new therapeutic model for AD also as this technology manipulates the Alzheimer's amyloid pathway. While up-regulating neuroprotective α-cleavage, it edits the endogenous APP which attenuates β cleavage, by inhibiting the interaction between BACE-1and APP [52].

Amyotrophic Lateral Sclerosis (ALS) is a rare neurological disease that causes the death of neurons controlling voluntary muscle movements. It is characterized by abnormal protein accumulation which is believed to be the cause of neuron toxicity. Recently, by using genome-wide CRISPR sgRNA libraries, scientists were able to identify about 200 genes that can either protect the cells from or sensitize them to the toxic proteins. It shows that some of these genes are potent protectors of the ALS proteins and might be potential drug targets for ALS treatment. Applying genome-wide CRISPR screens in research is a new exciting approach for understanding the disease causes that we were not able to find before [53].

However, delivering Cas9 nuclease and gRNA efficiently to the brain is limiting the usage of this technology [51].

CONCLUSION

Our understanding of the development of many neurological disorders is still very limited and no effective treatment is available for patients yet. Only inhibitors or antagonists don't need to be used to prevent the overactive enzyme from doing its function in CNS. It can be either silenced or knocked out using the gene-editing technologies.

CONSENT FOR PUBLICATION

Not applicable.

CONFLICT OF INTEREST

The author(s) confirms that there is no conflict of interest.

ACKNOWLEDGEMENT

I acknowledge Mr. Shrihari S, a native English language expert from the Department of Humanities and Social Sciences, Sri Venkateswara College of Engineering, Sriperumbudur for thoroughly extensively editing the book chapter for grammar, punctuation, spelling and overall style.

REFERENCES

[1]   Sharma R. Enzyme inhibition: mechanisms and scope In enzyme Inhibition and Bioapplications.

InTech 2012.
[http://dx.doi.org/10.5772/1963]

[2]   Mohan C, Long KD, Mutneja M. An introduction to inhibitors and their biological applications. EMD Millipore Corp. 2013; pp. 3-13.

[3]   Copeland RA, Harpel MR, Tummino PJ. Targeting enzyme inhibitors in drug discovery. Expert Opin Ther Targets 2007; 11(7): 967-78.
[http://dx.doi.org/10.1517/14728222.11.7.967] [PMID: 17614764]

[4]   Kanwar JR, Sriramoju B, Kanwar RK. Neurological disorders and therapeutics targeted to surmount the blood-brain barrier. Int J Nanomedicine 2012; 7: 3259-78.
[http://dx.doi.org/10.2147/IJN.S30919] [PMID: 22848160]

[5]   Lleó A, Greenberg SM, Growdon JH. Current pharmacotherapy for Alzheimer's disease. Annu Rev Med 2006; 57: 513-33.
[http://dx.doi.org/10.1146/annurev.med.57.121304.131442] [PMID: 16409164]

[6]   Gonçalves S, Romano A. Inhibitory Properties of Phenolic Compounds Against Enzymes Linked with Human Diseases InPhenolic Compounds-Biological Activity. InTech 2017.

[7]   Finberg JP. Pharmacology of rasagiline, a new MAO-B inhibitor drug for the treatment of Parkinson's disease with neuroprotective potential. Rambam Maimonides Med J 2010; 1(1)e0003
[http://dx.doi.org/10.5041/RMMJ.10003] [PMID: 23908775]

[8]   Perez-Lloret S, Rascol O. Safety of rasagiline for the treatment of Parkinson's disease. Expert Opin Drug Saf 2011; 10(4): 633-43.
[http://dx.doi.org/10.1517/14740338.2011.573784] [PMID: 21453201]

[9]   Finberg JP, Rabey JM. Inhibitors of MAO-A and MAO-B in psychiatry and neurology. Front Pharmacol 2016; 7: 340.
[http://dx.doi.org/10.3389/fphar.2016.00340] [PMID: 27803666]

[10]   [homepage on the Internet] Sciencedirect 2018. (cited: 7th Dec, 2018). Available from: https://www.sciencedirect.com/topics/pharmacology-toxicology-and-pharmaceutical-science/selegiline

[11]   Ansara AJ, Nisly SA, Arif SA, Koehler JM, Nordmeyer ST. Aspirin dosing for the prevention and treatment of ischemic stroke: an indication-specific review of the literature. Ann Pharmacother 2010; 44(5): 851-62.
[http://dx.doi.org/10.1345/aph.1M346] [PMID: 20388864]

[12]   [homepage on the Internet] Newyork: Healthline 2018. (cited: 7th Dec 2018). Available from: https://www.healthline.com/health/brain-tumor

[13]   [homepage on the Internet] New Drug Approvals 2018. (cited: 7th Dec, 2018). Available from: https://newdrugapprovals.org/2014/03/25/prinomastat/

[14]   Zhao W, Qiu Y, Kong D. Class I phosphatidylinositol 3-kinase inhibitors for cancer therapy. Acta Pharm Sin B 2017; 7(1): 27-37.
[http://dx.doi.org/10.1016/j.apsb.2016.07.006] [PMID: 28119806]

[15]   Tremont-Lukats IW, Gilbert MR. Advances in molecular therapies in patients with brain tumors. Cancer Contr 2003; 10(2): 125-37.
[http://dx.doi.org/10.1177/107327480301000204] [PMID: 12712007]

[16]   [homepage on the Internet] Chicago: American Brain Tumor Association 2018. (cited: 7th Dec 2018). Available from: https://www.abta.org/tumor_types/glioblastoma-gbm/

[17]   2018.https://www.aans.org/Patients/Neurosurgical-Conditions-and-Treatments/Glioblast-ma-Multiforme

[18]   Lau D, Magill ST, Aghi MK. Molecularly targeted therapies for recurrent glioblastoma: current and future targets. Neurosurg Focus 2014; 37(6)E15
[http://dx.doi.org/10.3171/2014.9.FOCUS14519] [PMID: 25434384]

[19]   Faraco G, Cavone L, Chiarugi A. The therapeutic potential of HDAC inhibitors in the treatment of multiple sclerosis. Mol Med 2011; 17(5-6): 442-7.
[http://dx.doi.org/10.2119/molmed.2011.00077] [PMID: 21373721]

[20]   Dietz KC, Casaccia P. HDAC inhibitors and neurodegeneration: at the edge between protection and damage. Pharmacol Res 2010; 62(1): 11-7.
[http://dx.doi.org/10.1016/j.phrs.2010.01.011] [PMID: 20123018]

[21]   Rajput MS. Natural monoamine oxidase inhibitors: a review. J Pharm Res 2010; 3: 482-5.

[22]   Babitha KV, Shanmuga Sundaram R, Annapandian VM, *et al.* Natural products and its derived drugs for the treatment of neurodegenerative disorders: Alzheimer's disease–A review. Br Biomed Bull 2014; 2: 359-70.

[23]   Suganthy N. KARUTHA P, PANDIMA D. Cholinesterase Inhibitors from Plants: Possible Treatment Strategy for Neurological Disorders-A Review. International Journal of Biomedicine and Pharmaceutical Sciences 2009; 3: 87-103.

[24]   Khatri DK, Juvekar AR. Kinetics of inhibition of monoamine oxidase using curcumin and ellagic acid. Pharmacogn Mag 2016; 12 (Suppl. 2): S116-20.
[http://dx.doi.org/10.4103/0973-1296.182168] [PMID: 27279695]

[25]   Nabavi SF, Daglia M, D'Antona G, Sobarzo-Sánchez E, Talas ZS, Nabavi SM. Natural compounds used as therapies targeting to amyotrophic lateral sclerosis. Curr Pharm Biotechnol 2015; 16(3): 211-8.
[http://dx.doi.org/10.2174/1389201016666150118132224] [PMID: 25601606]

[26]   Kim JY, Park J, Lee JE, Yenari MA. NOX inhibitors-a promising avenue for ischemic stroke. Exp Neurobiol 2017; 26(4): 195-205.
[http://dx.doi.org/10.5607/en.2017.26.4.195] [PMID: 28912642]

[27]   Mojaverrostami S, Bojnordi MN, Ghasemi-Kasman M, Ali M, Ebrahimzadeh HG. A Review of Herbal Therapy in Multiple Sclerosis
[http://dx.doi.org/10.15171/apb.2018.066]

[28]   Kumar GP, Anilakumar KR, Naveen S. Phytochemicals Having Neuroprotective Properties from Dietary Sources and Medicinal Herbs. Pharmacogn J 2015; 7(1)
[http://dx.doi.org/10.5530/pj.2015.1.1]

[29]   Zarmouh NO, Messeha SS, Elshami FM, Soliman KF. Natural product screening for the identification of selective monoamine oxidase-B inhibitors. European J Med Plants 2016; 15(1): 14802.
[http://dx.doi.org/10.9734/EJMP/2016/26453] [PMID: 27341283]

[30]   Singhal AK, Naithani V, Bangar OP. Medicinal plants with the potential to treat Alzheimer and associated symptoms. International Journal of Nutrition, Pharmacology. Neurological Diseases 2012; 2(2): 84.
[http://dx.doi.org/10.4103/2231-0738.95927]

[31]   Murray AP, Faraoni MB, Castro MJ, Alza NP, Cavallaro V. Natural AChE inhibitors from plants and their contribution to Alzheimer's disease therapy. Curr Neuropharmacol 2013; 11(4): 388-413.
[http://dx.doi.org/10.2174/1570159X11311040004] [PMID: 24381530]

[32]   Mazzio E, Deiab S, Park K, Soliman KF. High throughput screening to identify natural human monoamine oxidase B inhibitors. Phytother Res 2013; 27(6): 818-28.
[http://dx.doi.org/10.1002/ptr.4795] [PMID: 22887993]

[33]   Orhan IE. Potential of natural products of herbal origin as monoamine oxidase inhibitors. Curr Pharm Des 2016; 22(3): 268-76.
[http://dx.doi.org/10.2174/1381612822666151112150612] [PMID: 26561069]

[34]   Viña D, Serra S, Lamela M, Delogu G. Herbal natural products as a source of monoamine oxidase inhibitors: a review. Curr Top Med Chem 2012; 12(20): 2131-44.
[http://dx.doi.org/10.2174/156802612805219996] [PMID: 23231392]

[35]   Passos CD, Simoes-Pires C, Henriques A, Cuendet M, *et al.* Alkaloids as inhibitors of monoamine oxidases and their role in the central nervous system.Studies in Natural Products Chemistry. Elsevier 2014; Vol. 43: pp. 123-44.

[36]   Ong WY, Farooqui T, Kokotos G, Farooqui AA. Synthetic and natural inhibitors of phospholipases A2: their importance for understanding and treatment of neurological disorders. ACS Chem Neurosci 2015; 6(6): 814-31.
[http://dx.doi.org/10.1021/acschemneuro.5b00073] [PMID: 25891385]

[37]   Neumann U, Ufer M, Jacobson LH, *et al.* The BACE-1 inhibitor CNP520 for prevention trials in Alzheimer's disease. EMBO Mol Med 2018; 10(11)e9316
[http://dx.doi.org/10.15252/emmm.201809316] [PMID: 30224383]

[38]   Barten DM, Meredith JE Jr, Zaczek R, Houston JG, Albright CF. γ-secretase inhibitors for Alzheimer's disease: balancing efficacy and toxicity. Drugs R D 2006; 7(2): 87-97.
[http://dx.doi.org/10.2165/00126839-200607020-00003] [PMID: 16542055]

[39]   Chopra V, Quinti L, Kim J, *et al.* The sirtuin 2 inhibitor AK-7 is neuroprotective in Huntington's disease mouse models. Cell Rep 2012; 2(6): 1492-7.
[http://dx.doi.org/10.1016/j.celrep.2012.11.001] [PMID: 23200855]

[40]   Quinti L, Casale M, Moniot S, *et al.* SIRT2-and NRF2-targeting thiazole-containing compound with therapeutic activity in Huntington's disease models. Cell Chem Biol 2016; 23(7): 849-61.
[http://dx.doi.org/10.1016/j.chembiol.2016.05.015] [PMID: 27427231]

[41]   Allocati N, Masulli M, Di Ilio C, Federici L. Glutathione transferases: substrates, inhibitors and pro-drugs in cancer and neurodegenerative diseases. Oncogenesis 2018; 7(1): 8.
[http://dx.doi.org/10.1038/s41389-017-0025-3] [PMID: 29362397]

[42]   Al-Badri G, Leggio GM, Musumeci G, Marzagalli R, Drago F, Castorina A. Tackling dipeptidyl peptidase IV in neurological disorders. Neural Regen Res 2018; 13(1): 26-34.
[http://dx.doi.org/10.4103/1673-5374.224365] [PMID: 29451201]

[43]   Solesio ME, Peixoto PM, Debure L, *et al.* Carbonic anhydrase inhibition selectively prevents amyloid β neurovascular mitochondrial toxicity. Aging Cell 2018; 17(4)e12787
[http://dx.doi.org/10.1111/acel.12787] [PMID: 29873184]

[44]   Supuran CT, Scozzafava A. Carbonic anhydrase inhibitors and their therapeutic potential. Expert Opin Ther Pat 2000; 10(5): 575-600.
[http://dx.doi.org/10.1517/13543776.10.5.575] [PMID: 30217119]

[45]   Madigan NN, Staff NP, Windebank AJ, Benarroch EE. Genome editing technologies and their potential to treat neurologic disease. Neurology 2017; 89(16): 1739-48.
[http://dx.doi.org/10.1212/WNL.0000000000004558] [PMID: 28931646]

[46]   Walters BJ, Azam AB, Gillon CJ, Josselyn SA, Zovkic IB. Advanced *in vivo* use of CRISPR/Cas9 and anti-sense DNA inhibition for gene manipulation in the brain. Front Genet 2016; 6: 362.
[http://dx.doi.org/10.3389/fgene.2015.00362] [PMID: 26793235]

[47]   White MK, Kaminski R, Wollebo H, Hu W, Malcolm T, Khalili K. Gene editing for treatment of neurological infections. Neurotherapeutics 2016; 13(3): 547-54.
[http://dx.doi.org/10.1007/s13311-016-0439-1] [PMID: 27150390]

[48]   Ostick J. CRISPR for gene editing in neuroscience and neurological disease. ACNR 2017; 9-10.

[49]   Raikwar SP, Thangavel R, Dubova I, Ejaz Ahmed M, *et al.* Neuro-Immuno-Gene-and Genome-Editing-Therapy for Alzheimer's Disease: Are We There Yet?. Journal of Alzheimer's Disease 2018; 1-24. Preprint

[50]   Vesikansa A. Unraveling of Central Nervous System Disease Mechanisms Using CRISPR Genome Manipulation. J Cent Nerv Syst Dis 2018; 101179573518787469
[http://dx.doi.org/10.1177/1179573518787469] [PMID: 30013417]

[51]   Khan S, Mahmood MS, Rahman SU, *et al.* CRISPR/Cas9: the Jedi against the dark empire of diseases. J Biomed Sci 2018; 25(1): 29.
[http://dx.doi.org/10.1186/s12929-018-0425-5] [PMID: 29592810]

[52]   Sun J, Carlson-Stevermer J, Das U, Shen M, *et al.* A CRISPR/Cas9 based strategy to manipulate the Alzheimer's amyloid pathway. bioRxiv 2018.: 310193.

[53]   Kramer NJ, Haney MS, Morgens DW, *et al.* CRISPR-Cas9 screens in human cells and primary neurons identify modifiers of C9ORF72 dipeptide-repeat-protein toxicity. Nat Genet 2018; 50(4): 603-12.
[http://dx.doi.org/10.1038/s41588-018-0070-7] [PMID: 29507424]

SUBJECT INDEX

A

Acetic acid 64, 123, 127, 155, 156, 157, 158, 167
   toxicity 156
Acetylation 13, 274, 275
 histone 13, 274, 275
Acetylcholine 32, 64, 67, 101, 182, 265
   biosynthesis 265
Acetylcholinesterase 2, 32, 64, 98, 101, 102, 104, 105, 106, 108, 109, 264
  and choline oxidase 106
  immobilized 102
  inhibitors 264
Acids 49, 64, 82, 83, 98, 107, 108, 109, 111, 123, 124, 130, 132, 151, 154, 155, 156, 158, 162, 179, 189, 250, 252, 257, 275, 276, 277
  aliphatic 155, 275
  apetalic 250, 252
  benzoic 98, 109, 156
  betulinic 82
  butyric 64
  cinnamic 156
  concentrated phosphoric 128
  fatty 82
  ferulic 156
  formic 154, 155, 158
  furoic 158
  galacturonic 124
  gallic 156
  glucaric 158
  glutamic 111
  hydroxamic 82, 83
  inorganic 156
  isopetalic 252
  levulinic 156
  mercapturic 189
  oleanolic 277
  organic 132, 155, 162, 179
  ortho-toluic 156
  release uronic 155
  sialic 257
  suberoylanilide hydroxamic 275, 276

  sulfenic 49
  sulfuric 129, 155, 158
  sulphuric 130
  syringic 154, 156
  uric 107, 108
  ursolic 277
  valproic 275, 276
Activated protein kinase (AMPK) 233, 234
Activation 41, 73, 74, 75, 77, 80, 81, 84, 105, 186, 211, 272
  imbalance 73
  -induced cell death (AICD) 186
  intracellular 75
  matriptase-dependent 74
  thiol ester bonds 81
AIDS 248, 251, 252, 254
  developing 254
  inhibitors 252
Alkaline phosphatase 2, 6, 65, 67, 98, 102, 103, 104, 105
  based biosensor 65
  fabricating 65
  inhibition of 103, 104
Alkaloids 239, 233, 242, 255
  coptidis 242
  isoquinoline 233
Allergic reaction 6, 87
Ammonia fiber expansion (AFEX) 157
Amperometric 100, 103
  measurement of inhibition of catalase 103
  transducer measures change 100
Amperometric biosensors 21, 24, 28, 65, 66
  inhibition-based 65
  screen-printed based 65
  tyrosinase-based 66
Amylase 130
Amyloid precursor protein 264, 278
Amyotrophic lateral sclerosis 277, 283
Analgesic drug acetaminophen 107
Antisense 36, 37, 39, 43
  effect 39
  genes 36
  mRNA 37
Antisense RNA 36, 37, 38, 39, 41, 42, 44, 50

C

Cancer 5, 6, 48, 73, 74, 76, 83, 84, 85, 86, 87,
    204, 205, 206, 207, 208, 272
  advanced solid 86
  colorectal 87, 204
  esophageal 74
  gastric 76
  ovarian 76, 87
  prostate 48
Cancer cells 80, 83, 84, 206, 207, 273, 274
  cervical 80
Cancer 84, 216, 217, 272, 273
  drugs 273
    genome atlas 217
    metabolism 272
    recurrence 216
    therapy 84, 272
Cancer treatment 3, 5, 6, 12, 82, 84, 85, 206
  progression 206
Carboxylic acids 83, 154, 157, 158
Carcinogenesis 2, 80, 179, 272, 273
Carcinomas 76, 180, 206
Cardiovascular diseases 73, 74, 83, 85, 87
Catalyses 10, 28, 155, 156, 158, 165, 214,
    216, 280
  ammonia-based 155
  chemical 155
  sulfite 158
Cell morphogenesis 217
Cellobiose 166, 167
Cell proliferation 77, 184, 187, 221, 244
  eukaryotic 244
  progenitor 187
Cellulose 122, 123, 124, 125, 127, 129, 130,
    131, 152, 153, 154, 155, 156, 157, 158,
    159, 160, 163
  break 160
  digestibility 125, 159
  digestion 155
  fibrils 153
  structures 154
Central nervous system (CNS) 87, 181, 182,
    282, 283
Chikungunya virus (CHIKV) 231, 232, 233,
    235, 236, 237
Chronic obstructive pulmonary disease
    (COPD) 2, 5, 76
Collagen 74, 75, 271
  cleaving triple helices 74

Collagenases 73, 77, 78
  interstitial 73, 77
  neutrophil 78
Collagen catabolic process 77, 78
Collagenolytic activity 73
Competitive reversible inhibition 9
Compound annual growth rate (CAGR) 152
Consumption 133, 159
  microbial biomass 159
  nutrient 133
Cyclodextrin 104
Cyclohexenyl chalcone derivatives 240
Cyclooxygenase 210, 211, 270
Cytokines 77, 83, 84, 186, 210, 248
  inflammatory 83, 210
  proinflammatory 77, 248
Cytosolic 11, 194
  aryl hydrocarbon 194
  chemotherapy medicate 11

D

Dendritic cells 248
Diarrhea 266
Diseases 46, 47, 48, 50, 73, 74, 75, 81, 82, 83,
    85, 86, 87, 179, 204, 236, 254, 263, 264,
    275, 276, 277
  arthritic 82
  atherosclerotic 85
  autoimmune 73, 74, 81, 83
  deadliest 204
  gastrointestinal 75
  immune 275
  infectious 236, 254
  neurodegenerative 264
Disorders 2, 5, 85, 178, 196, 264, 275, 277,
    280
  autoimmune 2, 275
  bleeding 5
  cardiovascular 85
  gastrointestinal 2
  human health-related 196
  reproductive 178
DNA 22, 37, 39, 40, 41, 46, 61, 84, 165, 217,
    230, 231, 244, 251, 257, 273, 275, 280,
    281
  activities, dependent 251
  and histone proteins 275
  and RNA virus replication 244
  binding elements 165

www.ingramcontent.com/pod-product-compliance
Lightning Source LLC
Chambersburg PA
CBHW050809220326
41598CB00006B/163